D0846689

HOBBY FARM ANIMALS

i-5
PRESS

Hobby Farm Animals

Project Team
Editor: Amy Deputato
Copy Editor: Joann Woy
Design: Mary Ann Kahn
Index: Elizabeth Walker

i-5 PUBLISHING, LLC™
Chief Executive Officer: Mark Harris
Chief Financial Officer: Nicole Fabian
Chief Content Officer: June Kikuchi
Chief Digital Officer: Jennifer Black-Glover
Chief Marketing Officer: Beth Freeman Reynolds
General Manager, i-5 Press: Christopher Reggio
Art Director, i-5 Press: Mary Ann Kahn
Senior Editor, i-5 Press: Amy Deputato
Production Director: Laurie Panaggio
Production Manager: Jessica Jaensch

Copyright © 2015 by i-5 Publishing, LLC™

All rights reserved. No part of this book may be reproduced, stored in a retrieval system, or transmitted in any form or by any means, electronic, mechanical, photocopying, recording, or otherwise, without the prior written permission of i-5 Press™, except for the inclusion of brief quotations in an acknowledged review.

Library of Congress Cataloging-in-Publication Data
Weaver, Sue, author.
  Hobby farm animals : a comprehensive guide to raising beef cattle, chickens, ducks, goats, pigs, rabbits, and sheep / Sue Weaver, Ann Larkin Hansen, Cherie Langlois, Arie B. McFarlen, PhD, and Chris McLaughlin.
    pages cm
  Includes index.
  ISBN 978-1-62008-152-5
  1. Livestock. 2. Domestic animals. 3. Farms, Small. I. Title. II. Title: Comprehensive guide to raising beef cattle, chickens, ducks, goats, pigs, rabbits, and sheep.
  SF61.W23 2015
  636--dc23
                        2015014330

This book has been published with the intent to provide accurate and authoritative information in regard to the subject matter within. While every precaution has been taken in the preparation of this book, the author and publisher expressly disclaim any responsibility for any errors, omissions, or adverse effects arising from the use or application of the information contained herein. The techniques and suggestions are used at the reader's discretion and are not to be considered a substitute for veterinary care. If you suspect a medical problem, consult your veterinarian.

i-5 Publishing, LLC™
www.facebook.com/i5press
www.i5publishing.com

Printed and bound in China
18 17 16 15   1 3 5 7 9 8 6 4 2

CONTENTS

WITHDRAWN

PART ONE

BEEF CATTLE

BY ANN LARKIN HANSEN

WHY BEEF CATTLE?

Beef cattle are as much at home on the hobby farm as they are on the range.
Adaptable to almost any climate and easy to manage and market, they are well
suited to any farmer with the pasture room and a hankering for a cowboy hat.
Although beef cattle require a higher initial investment than any other traditional
farm animal except dairy cows, they require the least amount of daily maintenance.

SELECTING AND BRINGING HOME

BEEF CATTLE

Attention to the basics of raising beef cattle will reap rewards in the form of a freezer full of homegrown beef as well as extra cash from meat and calf sales. Where cattle are common, so are the auction barns, truckers, and processing plants that make it fairly simple to buy, sell, and process cattle. Americans love beef, so there is a ready market for beef cattle.

There's another benefit to owning beef cattle: they can improve your land. This may come as a surprise, given the reputation cattle have acquired in certain quarters for overgrazing and destroying sensitive lands. But beef cattle are a tool, not a cause. The result depends on how the tool is wielded—just as a hammer can be used to fix a building or destroy it. Research and on-the-ground experience have demonstrated that, when properly managed, beef cattle can be a highly effective tool for restoring health to damaged grasslands and watersheds. On a hobby farm, well-managed cattle can continually increase the richness of your soils, the biodiversity and lushness of your pastures, and the water quality of your ponds and streams.

Beef cattle will also enhance the view from your kitchen window. Every time I look out the window to see our cattle grazing the green slopes of our farm, hear bobolinks singing in our pasture, or prepare homegrown steaks for dinner, I'm glad we have beef!

For hundreds of years, people have bred cattle to develop characteristics that were best adapted to a particular climate and purpose. Eventually, this resulted in distinctive breeds of cattle, each with a distinctive palette of physical traits. Today, a cattle buyer can choose from a wonderful array of color, build, size, growth rate, and potential meat and milk production to fit cattle to the farm, the climate, and the purposes of the owner.

Although not all cattle are created physically equal, they do share general behavior characteristics. Cattle sense the world differently than we do. They eat different foods and digest them differently. Understanding how cattle operate is key to knowing what to expect from them, what they will like and won't like, and how to get them to do what you want them to do. Understanding and working with cattle's natural behaviors will result in calmer, healthier animals.

Beef Breeds

Until the middle of the eighteenth century, cattle were tough, multipurpose animals that were not selectively bred for any specialized purpose. Differences in size, color, and build were simply results of groups' being isolated from one another in remote settlements. Then, in 1760, Robert Bakewell, an Englishman, began the first known systematic breeding program to improve the uniformity and appearance of his cattle. The results were published in 1822 in George Coates's *Herd Book of the Shorthorn Breed*, the first formal recognition of a cattle breed. Other breed herd books soon followed, and, as the concept of breeding for a specific purpose spread, cattle were divided into two main categories: those bred primarily for milk production and those bred primarily for beef production. Even the original dual-purpose Shorthorn breed has been split into Shorthorns for beef and Shorthorns for milking.

More than five hundred breeds of cattle exist in the world today, although only a few are common in the United States. Milking Shorthorn; the ubiquitous black-and-white Holstein; and the rarer Guernsey, Brown Swiss, Ayrshire, and Jersey make up the six primary dairy breeds in the United States. All dairy breeds produce excess bull calves that are raised for beef, and plenty of beef operations are built on dairy calves.

Highland cattle have a distinctive look that appeals to many.

Hereford cattle, with their familiar white faces, red bodies, and white markings, have been the backbone of the American beef industry since a few decades after their arrival in 1847. The Black Angus, first brought to the United States in 1873, is now almost as numerous as the Hereford, while the breed's offshoot, Red Angus, established its own breed registry in the mid-1900s. These three breeds, along with the Shorthorn, Scottish Highland, Dexter, Devon, and Galloway breeds, are the major British breeds of beef cattle, so designated because they all originated in England, Scotland, Ireland, or Wales. In general, the British beef breeds are smaller, fatten faster, and are more tolerant of harsh conditions than the continental breeds.

The continental breeds from Europe are generally larger and slower to mature but offer a bigger package of beef to the producer. The most common are the Charolais, the Limousin, and the Saler from France; the Simmental from Switzerland; and the Gelbviehs from Austria and West Germany.

Historically, the term *cattle* was used to refer to all varieties of four-legged livestock, including horses, goats, and sheep. When referring specifically to bovines, ranchers used the term neat cattle. Only in the past 100–150 years has the meaning of the word cattle changed to refer only to domesticated bovines of the *Bos* genus.

In the United States, new breeds have been developed that tolerate southern heat better than do the European imports. The famous Texas Longhorn, which developed mostly on its own from Spanish cattle brought over by colonists, provided the starting foundation for American ranching. The American Brahman was developed from Indicus-type imports and then was crossed with different European breeds to create the Santa Gertrudis, the Brangus, the Beefmaster, and several other uniquely American cattle breeds.

When well cared for, any breed of cattle will produce good beef. Look for a breed suited to your climate and pasture type and, if the income is important, to the prospective buyers of your beef. Auction-barn buyers and finishers will have definite preferences. Most important, get something you like. You may fall in love with the Oreo-cookie markings of the Belted Galloway, the shaggy look and big horns of the Scottish Highland, or the gentle disposition of the Hereford.

Endangered Breeds

Of the hundreds of cattle breeds adapted to an enormous range of climates and conditions throughout the world, many are now endangered. The Livestock Conservancy lists fifteen breeds in the United States that need help to survive. On the "Critical" list, defined as breeds that have fewer than 200 US registrations each year, are the Canadienne, Dutch Belted, Florida Cracker, Kerry, Lincoln Red, Milking Devon, Milking Shorthorn (native), Randall, and Texas Longhorn. On the "Threatened" list, with fewer than 1,000 registrations each year, are the Ancient Red Park, Pineywoods, and Red Poll. The "Watch" list, with fewer than 2,500 registrations, includes the once-popular Ayrshire and Guernsey, along with the Galloway.

Hooves and Hide

Good feet and legs are important in all cattle. Cows have cloven (two-part) hooves that average around $3\frac{1}{2} \times 4$ inches. If a cow weighs 1,000 pounds, that's a lot of weight coming down on those little hooves every time she takes a step, and she takes a lot of steps in a day to get food and water. Cattle don't like mud or slippery surfaces because

DID YOU KNOW?

Domestic cattle, which belong to the genus and species *Bos taurus*, have no wild siblings. The last known wild aurochs, or *Bos primigenius*, a cow, died in Poland in 1627. Other members of the bovine group are bison and yaks. Sheep, goats, and pigs are more distantly related, belonging to the same family, Bovidae, as cattle.

How Cattle Sense the World

Cattle have excellent eyesight, but it works a little differently from human sight. They can see color to some extent, and they see exceptionally well in the dark. Because their eyes are spaced so far apart, their horizontal vision (side to side) spans an amazing 300 degrees at a time, with their only blind spot directly behind them. However, their vertical (up and down) range of vision is limited to 60 degrees, which means that they have to look down to see where to put their feet when the footing is unfamiliar. What's more, their eyes work somewhat independently of one another, rather than in concert as our eyes do, which gives them poor depth perception. When herding cattle, it is important not to work directly behind them; they can't see you there and will either turn to look at you or spook and run away. When moving cattle into a new area, give them plenty of time to see where to put their feet.

Cattle's sense of hearing is acute, and they can swivel their ears around to hear even better. They dislike loud, sharp noises such as yells from handlers, but they are soothed by soft talking or singing.

Cattle use quite a bit of verbal communication. They know one another's voices, and they'll learn yours. They'll bellow for feed, bawl for their calves, and moo back when you call them. A cow has a special low moo for when her calf is fed and settled and all's right with the world.

Cattle have a superb sense of smell, which they can use to follow the trail of their calves and to tell different plants apart. They also have a strong sense of taste and, as a result, have strong preferences for some plants over others. Research by Utah State University professor of rangeland science Fred Provenza has demonstrated that calves learn their plant preferences from their mothers and remember them all their lives. Year after year, they'll seek out their favorite grazing spots. We have one small patch of bluegrass that always gets grazed to the ground before anything else is touched, although it looks no different from any other bluegrass!

Choosing, Buying, and Bringing Home Cattle

You'll need your fences, feed, facilities, shelter, water tank, and salt and mineral feeder in place (see Chapter 2) before you're ready to go shopping for cattle. By this point, you will have invested more time and money than you would have for any other type of farm animal, except dairy cows, but it will all pay off in cattle that stay home, eat well, and handle easily.

You have a few options for where to buy cattle and several choices in what kind of cattle you buy. Fit your purchase to your

they can fall and hurt themselves; they will walk around bad footing if they can. Cattle hooves grow continuously, and long hooves cause lame cattle. If the herd is getting enough exercise, however, their hooves should not get too long. A few rocks in the pasture will help keep hooves worn down.

Cows also use their hooves to scratch themselves and to kick. They can kick both sideways and backward, and they're quick as lightning. If they have horns, they'll also use these to defend themselves, and a horn can do even more damage than a hoof. For this reason, many cattle owners prefer polled, or naturally hornless, cattle. You can also dehorn calves when they are quite young so that their horns never grow.

For pests that are too small to kick, such as biting flies, cattle have long tails for flicking them off, and their thick hides protect them from some insect species. But several kinds of flies can bite through cowhide, and some will even bore holes in hide and lay eggs there. The irritation and discomfort caused by flies can slow weight gain in calves and keep cows miserable on hot days. To stay warm in cold climates, cattle will grow longer winter coats. Unlike most dairy cows, beef cows will have hairy udders.

As herd animals, cattle thrive in each other's company.

budget, the size of your pasture, how much time you'll have each day for chores, and whether you're interested in beef for your freezer or in building a herd.

Whatever age or sex you buy, the minimum number of cattle you should purchase is two. Cattle are herd animals and hate being alone. They will adopt a goat or a donkey or anything else handy as a companion, but they thrive best when in the company of their own kind. If you're buying a steer primarily for your own consumption, keep in mind that most families will take a year or two to eat a single steer. Plan on selling the extra steer at the auction barn or the extra half or quarter of beef to friends or relatives.

Following are some general guidelines for choosing and purchasing cattle.

WATCH THEM MOVE

When buying cattle, watch how they move and hold themselves. You should observe no lameness or hunched backs when they walk. Cattle that won't relax, that keep their heads high and bodies braced, may be wild and hard to handle. Cattle that won't let you anywhere near them to look them over could be a problem, too, but don't expect to walk up and pet them, either. Unless they're show cattle, most beef cattle aren't accustomed to being approached too closely by strangers.

What to Buy

Steer calves purchased in the fall will be ready for butchering when they are between sixteen and thirty months of age, depending on the breed and your feeding program. This means that a male calf bought after fall weaning could be ready for the processor as soon as the following fall. Heifer calves bought after fall weaning will be ready to breed the following summer, provided they grow well through the winter and spring.

Any calves you buy should have been weaned for at least three weeks. They should also have received their nine-way vaccines, followed by boosters two to four weeks later. Heifers should be wearing small metal ear tags to show that they've had brucellosis vaccinations. Bend down and look behind steer calves to make sure that the castration got both testicles, or you could have a bull on your hands by mistake.

If you want cattle around just for the summer, you can buy steers in the spring and sell them in the fall. If you're buying for your own freezer or to sell as finished (ready-to-slaughter) cattle, they should be started on grain right away. They'll quickly learn to come running in from the pasture for their daily grain rations.

If you're buying breeding stock—whether cows, heifers, or a bull—finding high-quality cattle is more important than if you're raising cattle for processing. Start the search early and take the time to find out about different breeders. If you're planning on showing cattle or enrolling your kids in the 4-H beef program, look for operations with good show records. If your objective is to get a decent cow-calf herd started, it's more important to find sellers with calm, clean, reasonably good-looking cattle.

Three breeds of cattle possess the attractive and very distinctive "Oreo cookie" coloring—black except for a broad white band around the middle: the Dutch Belted, the Belted Galloway, and the Buelingo. The breeds differ from each other, however, in other characteristics and uses. The Dutch Belted was a prized dairy breed in the United States until about 1940, whereas the Belted Galloway, probably descended in part from the Dutch Belted, is primarily a beef breed. The Buelingo is a uniquely American beef breed developed during the 1970s and 1980s by North Dakota rancher Russ Bueling, primarily from Shorthorn and Chianina genetics.

Another option is to buy dairy bull calves. They will produce fine beef. However, it takes a lot more grain to fatten them up, the cuts aren't as nicely shaped, and there's a smaller proportion of meat to bone and by-products. Since only cows give milk, bull calves are not viable in dairy operations and are generally sold sometime between three days and a few weeks old. Consequently, they aren't weaned, so you'll have to bottle-feed them milk until their digestive systems are mature enough to handle grain and forage. Although they're a lot of extra work, these calves are available year-round, are very inexpensive compared with all other cattle, and can be transported in the back of a van.

Occasionally, you may be able to buy an orphaned beef calf or one that's been rejected by its mother. Some dairy farmers breed their heifers to a beef bull for easy calving with their first calf, and these half-beef, half-dairy calves are a bargain.

What to Look for in Cattle

The only way to acquire an eye for good cattle is to look at a lot of cattle. It takes a few years to develop that eye, but there are some things that beginners can spot. Cattle get a little nervous when a stranger shows up in their pasture or pen, so give them time to settle down again. Lean on the fence or stand quietly in the pasture and take a good, long look at their shape and how they act.

Shape

Whatever the breed, beef cattle should ideally look thick and square, like big, hairy rectangular boxes on legs. While a dairy cow will look like a wedge, with the narrow part at the front, a beef cow should be blocky. The back should be straight and the line of the belly nearly so, not tapering too sharply up to the hind legs, with a rib cage that is rounded, not flat. The hindquarters should look broad and meaty, especially in steers and bulls, and the legs should be straight. Hooves should be even and short, definitely not so long that they curl upward, and they should point straight ahead.

Steers should have thick necks and fleshy forequarters. Cows and heifers should be slimmer through their shoulders and necks and have more feminine heads. Cows should have high, well-shaped udders. Both sexes should have wide muzzles, indicating that they can take big bites of grass.

Health

Look closely for any signs of illness or discomfort. Eyes should be clear, not mattered (crusted) or inflamed.

In summer, the coat should be smooth and shiny (except in long-coated breeds). A dull, flat-looking coat generally means internal parasites or poor nutrition. In late fall, winter, and spring, the coat should be uniform and thick. Bare spots could mean ringworm; although rarely a huge problem, it is a chore to treat.

Pay special attention to hooves and legs, checking for growths or swelling, especially at the tops of the hooves and at the joints. There should be no swelling of the jaw, neck, shoulders, or brisket.

An occasional cough in adult cattle usually isn't anything to worry about (cows do get colds and runny noses), but constant coughing signals problems. I'd hesitate before bringing any coughing animal home to my herd. In young calves, constant coughing or labored breathing may indicate pneumonia, a dangerous condition.

Check the hind ends of calves to make sure they aren't matted with manure. Manure that is more water than feces is typical of scours, another common and dangerous affliction of young calves.

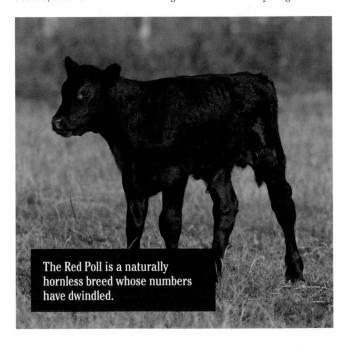

The Red Poll is a naturally hornless breed whose numbers have dwindled.

Auction-barn purchases are best left to experienced cattle people.

Polled cattle are naturally hornless, but if you're buying a horned breed that has been dehorned, make sure that the spot is well healed and not sprouting any more horn.

How Cattle Are Priced and Sold

The pricing and availability of cattle generally follow a yearly cycle, which varies by the age and sex of the animals. Calves and steers are usually sold by weight, whereas heifers and cows for breeding are sold either by weight or by what the market will bear.

Bulls are normally priced according to their quality or what the owner thinks he or she can get. Bulls are most expensive in the spring and early summer, when they're in high demand for the breeding season, and cheaper in the fall.

Feeder calves—those that have been weaned and are ready to go on pasture or on a finishing ration—are usually least

expensive in late fall, when the market is flooded with calves born the previous spring that are being sold before winter. *Stockers* or *backgrounders* (feeder calves headed for a few months on pasture before going on a finishing ration) are expensive in the spring, when landowners are buying cattle to keep their pastures grazed during the growing season.

Open, or unbred, heifers will be most expensive in spring, just before breeding season; the price tapers off through fall, when they're cheap because it's not economical to winter an open heifer. In addition, a heifer that didn't "settle," or get bred, during the summer may be infertile and good only for being finished and slaughtered. Bred cows will be most expensive in the spring, just before calving, and cheaper in the fall, when you will have to feed them through the winter. Before buying a bred cow, always have a veterinarian check to see whether she is really pregnant.

Cattle prices are listed, usually weekly, in local and state farm papers. If you have a farm radio station in your area, you can typically find prices being announced daily or weekly, and some state extension services list current cattle prices on their websites. Cattle prices are listed either as dollar and cents per pound or as dollars and cents per hundredweight. If, for example, I were looking for feeder calves, I would look down the column until I came to that category under the listing for the auction barn closest to my farm, and then I'd start with the midrange, 400–600-pound, category. If steer feeder calves were listed at $1.00, then the

CALVES IN A VAN

Our first beef cattle were dairy calves. To bring them home, we put a tarp down with some straw on top in the back of the van, and I had the kids sit with the calves and keep them lying down. They won't make a mess if they're lying down!

BUYER BEWARE

I like them so they look straight all the way back on their backbone. There's almost an edge to them. They look upright, they look strong; it's almost an eager look. I always check the feet—some cattle, it seems like a hereditary thing, have really long hooves. Temperament is a big thing for me, too, with little kids around. I've got one cow with a little bit of attitude, but she's never offered to come after you. I've got another one that's so meek and mild she always gets pushed back.

—Randy Janke

There's so much going on now with numbers—expected progeny differences, relative this and that—and it's so far extended that the average young person needs to just focus on one or two things when choosing breeding stock. The ones I strongly recommend are calving ease and temperament. You need to stick with cattle that won't explode every time you come around.

—Rudy Erickson

I've never bought from a sale barn. We bought private treaty or from a neighbor's auction where we knew the herd.

—Dave Nesja

price for a 500-pound feeder steer would be $500. Keeping track of prices gives you a good idea of what you should be paying when you buy, but keep in mind that it may be worthwhile to pay a little extra for cattle you know are healthy, have been vaccinated, and come from good parents.

Where to Buy Cattle

Finding cattle for sale is a matter of checking ads in local newspapers or regional farm papers; looking at bulletin boards at the feed store, the farm supply store, and rural gas stations; and just asking around. If there's a beef producers' association in your area, join it. If you don't know whether there is one near you, give your extension agent a call and ask. An association is a great place to network and get some background information on area beef cattle operations and auction barns. County and state fairs are other good places to find beef producers with cattle for sale. Go to the cattle shows, walk through the barns, and visit with the exhibitors. Two additional sources for leads on cattle for sale are your local artificial-insemination service and veterinarian.

Auction barns move a lot of cattle, but they're no place for beginners to buy. If you go, take a friend who is a good judge of cattle and can help you avoid the ones that are sick, are wild, or have bad hooves and legs. You may want to make a few dry runs to the barn, going early to visit the pens and then watching the auction without buying, to give you a feel for how the bidding process works and how cattle are moved in and out of trailers, pens, and the auction ring.

A better idea is to buy cattle directly from a *seed stock producer* or a commercial producer. Seed stock producers raise purebred cattle for sale as breeding stock and are good sources of quality animals. Commercial producers generally have mixed herds of several breeds or crossbred cattle being raised for beef production instead

of breeding stock. These won't be registered purebreds, but often they're of good quality and reasonably priced; sometimes they aren't. Most commercial cow-calf operators sell their calves after weaning in the fall, and this can be an excellent opportunity to purchase.

Dairy bull calves are common at auction barns, but it's better to buy directly from the farmer and save the calf the stress of being hauled twice to strange places and exposing him to who-knows-what at the auction barn. When you buy directly from the farmer, you can make sure the calf is at least three days old and has received colostrum, his mother's immunity-boosting first milk. This is critical to a calf's health.

Bringing Cattle Home

Once you've bought your cattle, you have to get them home. If you're buying from a breeder, he or she may be willing to deliver the cattle for an extra fee. Otherwise, you'll either have to hire a cattle hauler or do it yourself. There are usually haulers for hire in any area where there's cattle, and you can track one down by asking neighbors or calling a local auction barn. If your only option, or the option you prefer, is to transport your new livestock yourself, you'll have to buy, borrow, or rent a trailer (unless you're buying small calves that you can fit into a pickup or small truck).

When the cattle arrive at your farm, ideally you'll turn them into a solidly fenced small pen or barnyard, with water available and some nice hay scattered around. Don't rush them out of the trailer; give them time to look around and step down carefully. Of course, they may decide to all come out in a rush, but let that be their decision. Once they've had a few hours to get a drink, find the salt, and get a bellyful of hay, open the pasture gate. By then, they should be calm enough to walk, not run, out. They might start grazing immediately or go on a tour to figure out where the fences are.

A female cow is kept in the herd as long as she produces good calves.

If you're using electric fencing, and the cattle you've bought are familiar with it, you can turn them out with no worries. If they don't know what an electric fence is, you'll need to train them as discussed in the fence section of Chapter 2. Don't try to train them as soon as they get off the truck, however. That's a lot to ask of already-stressed animals and may send them over the fence and back toward their previous home. You can have the wire ready in the pen, but don't turn it on until they've settled down.

You can also turn new cattle directly into the pasture. If you do this, plan on spending some time watching them to make sure they don't charge and break fences or decide to hop over and head back where they came from. Make sure they find the water, salt, and mineral within twenty-four hours.

Watch your new cattle particularly closely for the first two or three weeks. Are they grazing contentedly, bunched tightly, or spending a lot of time walking instead of chewing? If they're walking all the time,

you've got pasture that's too poor, and you should give them some supplemental hay. If they're bunched, you probably have a fly problem and should provide a shady area for them to get away from the worst of the flies. You may also need to put up some sort of rub—a rope or padded post impregnated with fly repellent—that will put the repellent on the cattle when they scratch themselves. Are the cattle

BODY CONDITION

Body condition scoring (BCS) is a simple method for assessing how fat or thin cattle are, which will help when you're looking at cattle to buy or assessing the cattle you already have. BCS is a simple system, and you can start using it right away, but getting exact results takes experience.

The optimum BCS for a cow is 5–7, or "pleasantly plump." A cow's body condition should be scored three to four months before calving, allowing time to adjust the ration and to let her gain or lose enough weight to be in optimum condition for calving and rebreeding. The numbers break down as follows:

1. Emaciated: No fat visible anywhere; ribs, hip bones, and backbone clearly visible through hide
2. Poor: Spine doesn't stick out quite as much
3. Thin: Ribs visible; spine rounded rather than sharp
4. Borderline: Individual ribs not obvious; some fat over ribs and hip bones
5. Moderate: Generally good overall appearance; fat over the ribs and on either side of tailhead
6. High moderate: Obvious fat over ribs and around tailhead
7. Good: Quite plump; some fat bulges (pones) possibly visible
8. Unquestionably fat: Large fat deposits over ribs and around tailhead; pones obvious
9. Extremely fat: Tailhead and hips buried in fat; pones protruding

Plenty of drinking water is just as important as quality feed.

maturing breed or is on a less intensive feeding program, he may be kept until age two or three.

Female, or heifer, calves intended for breeding are kept separate from their mothers and the bull after weaning, usually until they're fifteen months old or a little older. At that point, they're bred either by using a live bull or by artificial insemination, usually in midsummer of the year after they're born. Nine and a half months later, if all goes as planned, they deliver their first calves and officially become cows, instead of heifers.

As long as a cow raises a good calf each year and doesn't exhibit any major personality problems, she's kept in the herd. When she becomes too old or infirm to get pregnant, the owner culls her. Sending a gentle old cow away to slaughter is difficult. I try to console myself with the knowledge that she had a full and happy life on our place and by arranging for her to go somewhere close and quick.

Bulls are usually kept in groups until they're sold as breeding stock. The bull's first calves will be delivered the next year, just ahead of breeding season, so it's safe to use him for a second year. If you keep a bull for a third year, his daughters will be old enough for him to breed, so you will need to either make other arrangements for the heifers or get a new bull to prevent that from happening. Bulls can be sold to other cattle owners to give them a couple more happy years in the pasture or sent to slaughter.

This cattle production cycle creates a lot of opportunities to tailor your beef operation to your personal preferences and calendar. A cow-calf operator is on the job year-round, but it doesn't take intensive management to keep cows and calves happy and productive. A backgrounder can buy feeder calves in the fall or the spring and keep them on forages until they're ready for the feedlot. It's quite easy with this system to have cattle only for the summer so that you don't have to make or buy hay, and you can take the winter off. Feedlot operators can work on small amounts of land because they don't need pasture; they do, however, need excellent management skills and a lot of knowledge about cattle nutrition.

Finally, seed stock producers, who raise bulls and heifers for cow-calf operations, must have plenty of experience with breeding high-quality cattle as well as good marketing skills.

spending plenty of time lying down and chewing their cud, or are they always standing up and acting nervous? If they aren't lying down, something is bothering them, and you'll need to figure out what it is and fix it. Once, we had a bear stroll through the back pasture, and the whole herd went through the fence! We moved the cattle to a paddock close to the house, where the dogs could keep the bears at a distance and the cattle could chew their cuds in peace.

Make sure, too, that your cattle are drinking enough water. In temperate, reasonably dry weather, they'll come for a drink at least once, and usually twice, a day. If they're not drinking and it's not raining, then there's something wrong with your water setup. I remember one cold winter day when our cattle wouldn't drink, and I found out why when I touched the water. It was carrying an electric charge from a shorted-out heater!

The Cattle Production Cycle

The life of a beef cow or steer follows a pretty standard pattern in most of the United States. After a calf is born, usually in the spring, the infant stays with the mother until it is weaned (between four and ten months of age). By weaning time, a male calf has been castrated and is ready to go on pasture or hay as a stocker calf, or *backgrounder*, for several months. An older calf may skip this stage and go directly on feed (presumably this is why all weaned calves are called feeder calves). *On feed* means putting young cattle in a pen instead of a pasture and feeding them a high-protein diet to accelerate growth and fattening. If the steer is an early-maturing breed and has been well fed, he may go to slaughter as young as sixteen months of age. If he is a slower

You *could* stick your new heifers in the garage until you get a fence up around the pasture, but think of the mess! It's much better to have the "three Fs"— fences, feed, and facilities—in place before you bring home the cattle.

FENCES AND FEED FOR
BEEF CATTLE

Fences

Good fencing is essential to a cattle operation, more important even than shelter. Poor fencing makes for bad neighbors and sleepless nights. If your cattle are constantly in the neighbor's cornfield or causing a traffic hazard on your road, your neighbor and the local sheriff are going to be upset. So you'll need to make sure that your fences are heifer high and bull tight.

Well-fed and calm cattle are about the easiest farm animals to fence. Hungry and scared cattle—and those in heat—will jump, break, or trample a weak fence. The fence around your property boundary should hold your cattle no matter what mood they're in. You also need a fence that will discourage them from reaching over or under for a taste of some fragrant plant on the far side and from using the fence as a scratching post. Those sorts of activities break wires and let the herd out for an unscheduled field trip.

Whether building a new fence or rebuilding an existing one, you'll need to pay close attention to the wire gauges, post spacing, and bracing. General guidelines and options for cattle fencing follow. For more detailed fence-building instructions, find a do-it-yourself book or a neighbor who can show you how to build it right. Fencing projects are best scheduled for early spring while the ground is still soft and the air is cool.

Out with the Old

Most small cattle operations are started on old farms, and old farms generally come with old fences. If the old fence is still in somewhat good condition, you may be able to get a few more years out of it by running a single electric wire along the inside of the fence to keep the cattle from scratching and leaning on it. If the old fence is half-buried in weeds and strung on rotted or rusted posts, the sooner you can take it down and replace it with something new and tight, the better. Otherwise, you'll be lying awake all night wondering if this is the night that the cows will make a break for it.

Taking down old fencing is a slow job best done in cool weather, when it's comfortable to wear the heavier clothing you'll need to protect yourself against those sharp wire ends and barbs. First, clear away as much brush and as many weeds as necessary to uncover the fence, and then you can start on the fence itself. Take along a bucket, fencing pliers, and heavy leather gloves. You'll also need a post puller, a handy device that looks like a tall jack, which you can borrow from a neighbor or pick up at a farm-supply store.

To disassemble a basic barbed-wire fence, start at a corner. With the fencing pliers, remove the metal clip holding the bottom wire to the post and throw it in the bucket. Continue removing all of the clips along a couple hundred feet of fence. Go back to the corner and spool up the wire in a big doughnut shape. When the doughnut gets heavy, cut the wire with the fencing pliers and lean the doughnut against a fence post for later pickup. Then go back to the corner and do the same with the other wires, working from the bottom up.

After you've removed all the wire, use the post puller to yank out the metal posts. (If they are U-posts instead of T-posts, you may have to rig a wire loop on the puller to make it work.) To pull wooden posts, you can pound a long nail into each, leaving a couple of inches sticking out, then rig a rope or wire loop under the nail and around the post to pull it out with the post puller. Even better, if you can get a four-wheeler, tractor, or vehicle with a trailer hitch close to the post, you can put a loop of chain around the post and then run the chain over the top of a tall board stood on end (next to the post) and down to the trailer hitch. Drive away slowly, and the board will tip over and pull the post up and out.

Have a fenced area ready for your cattle before they arrive.

Immediately fill any holes left by pulled posts to prevent animals and humans from stepping in them and twisting an ankle or breaking a leg. Whichever pulling technique you use, be aware that rotted wooden posts will often break off at ground level, which will save you the trouble of filling the hole, although you may have to do so later when the remaining wood rots. For expediency, I have even sawed off posts at ground level instead of pulling them out.

You can burn old wooden posts, but save any usable metal posts for the new fence. Load up the bent and rusted metal posts and the old wire, and haul them to a junk dealer or a recycler.

In with the New

The two most practical options for a new cattle perimeter fence are high tension and barbed wire. A high-tension fence is the Cadillac of fences, long lasting and presenting a significant physical barrier even to a half-ton cow. It is also more costly than barbed wire. However, even though barbed wire is the cheapest type of fence to build, it will still cost some real money for posts, wire, clips, braces, and a few tools. Your agricultural extension office or fence dealers should be able to give you information on costs so you can budget for your fencing project.

High-tension fence works best where you have long, straight stretches to fence and a decent budget. Because the wire is heavy and stretched very tightly, it requires excellent corner braces and some expertise to install. The wire, made extra strong for these fences, is stretched so tightly and anchored so well that tree branches and even cows bounce off after hitting the fence.

If you have lots of curves and corners to fence, old-fashioned barbed wire works fine for cattle, although it's not usually recommended for any other type of livestock. For a perimeter fence, use a minimum of four wires.

High-tension fences are often electrified, adding a psychological barrier to the physical barrier of the wire. An electric fence, on the other hand, creates a purely psychological, rather than a physical, barrier for cattle. An electric fence power unit pulses a static charge through the fence wire. When a cow touches the fence, the charge flows through her to the ground and back to the ground rods attached to the power unit, completing the circuit and giving her

a healthy jolt. When done correctly and well maintained, electric fencing is extremely effective for subdividing pastures and keeping groups of cattle separate.

Always use smooth wire for electric fences; it is illegal in many areas to electrify barbed wire. To augment your barbed wire with an electric barrier, you can use offset insulators to mount a smooth electric wire along a barbed wire fence. This is a good combination, especially if you're keeping a bull separate from heifers.

The standard low-tension, soft-wire permanent electric fence uses permanent posts and works well with three wires. The top and bottom wires carry the charge, and the middle wire is a grounded wire. This is necessary for those times when the ground is covered with snow or is very dry and acts as an insulator rather than the receiving end of the circuit. A cow sticking her head through the fence will connect a hot wire with a ground wire and get a jolt.

Portable electric fencing, built with lightweight step-in posts and usually a single strand of plastic wire, is used during the grazing season to temporarily subdivide pastures into paddocks. This fence technology is slowly revolutionizing grazing in this country. For relatively little time and money, livestock owners can subdivide pastures into paddocks to manage their grazing, which improves pasture growth. You can set up and take down portable electric fencing almost as fast as you can walk because all it requires is stepping a line of posts into the ground, then unreeling the wire and popping it into the clips on the posts; some even come with the wire already attached.

Cattle are quick to figure out when an electric fence is not working and will walk through it if the grazing looks better on the

 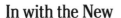

A minimum of four wires should be used when fencing in cattle with barbed wire.

Any fences enclosing a confined area where cattle might be crowded or stressed from handling—such as corrals and holding pens—should be made of heavy-duty wood or metal. These fences should be high enough that cattle won't even think about jumping them: at least 5½ feet for small, calm cattle and 6 feet or higher for large cattle or cattle unaccustomed to people or to being handled. Build these fences low to the ground, too, so your cattle won't try to scramble underneath. It's amazing how small an opening a cow will try to get through when she's frantic.

All pasture fences should be at least 4 feet high and have a wire close enough to the ground to keep calves from scrambling underneath but not so low that you can't trim the grass under the wire. (I put the bottom wire a foot off the ground.) When building a new fence, make sure to leave enough room to use a brush mower or weed trimmer on both sides, and avoid installing the fence on steep banks and close to rock piles and big trees. Keeping fences clear of vegetation at least doubles their lifespans and makes the inevitable repairs much easier to do.

Gates

All fences need gates for moving cattle, people, and equipment in and out of pastures and pens. In your corral and handling facility area, gates should be solid metal or wood and bolted on so a steer can't stick his nose under the bottom rail and flip it off the hinges. Metal and wooden gates are wonderful in perimeter fences, too, but if your budget doesn't allow for as many nice gates as you would like, you can build a "poor man's gate" by extending the fence wires across the gate opening. Instead of attaching them to the post at the far side, attach them to a 4-foot stick. Put a wire loop at the top and bottom of the gatepost. To close the gate, insert the ends of the stick into the wire loops.

Feeding Beef Cattle
Cattle Biology

Cattle are ruminants, members of a class of grazing animals with four-chambered stomachs adapted to digesting coarse forages that other animals cannot utilize. Consequently, cattle—as well as sheep and goats—can make use of land too rough, rocky, dry, or wet to grow crops for humans.

other side, so check electric fencing often. It's also critical to exactly follow instructions for sizing and installing the power unit and for grounding it properly with a series of copper rods.

If your cattle are unfamiliar with electric fencing, you'll need to train them to recognize and respect it. To introduce them to the concept, run a temporary electric wire along the inside of a wooden corral or a holding pen fence. Out of curiosity, the cattle will see the wire and sniff it, giving themselves a jolt on the most sensitive part of their anatomy—the nose. Because the solid fence is in front of them, they'll back away rather than jump forward at the shock. Once they've figured out what the wire means, take it down. You don't want it there when you're working cattle because if one should touch it accidentally, you'd have some upset cattle on your hands in uncomfortably close quarters. If you have cows that know about electric fences with new calves that don't, you needn't worry. The calves will learn without any special training.

DID YOU KNOW?

Lay out your fences and gates for ease in moving your cattle. It's much easier to move cattle through a gate in a corner than through one in the middle of a long, straight stretch of fence. Field gates are usually best located in the corner nearest the watering or grain-feeding area because your cattle will be moving back and forth between water and pasture regularly.

Cattle are ruminants that spend a lot of time grazing and chewing.

Cattle pick their meals by smell and taste, and then they graze until the first chamber of the stomach, the *rumen*, is full. Because they have no front upper teeth, just a hard pad, they tear the grass instead of biting it. (This is also why cows don't normally bite people.) Watch a cow grazing, and you'll see it grip a bite of grass between the pad and the lower front incisors and then swing its head a little to rip it off.

The long muscular tongue, as rough as sandpaper, is useful in quickly conveying grass back to the throat. The tongue is also used for grabbing grass, for licking up those last bits of grain, and for a little personal grooming (although cows aren't flexible enough to reach around too far). Copious amounts of saliva—up to fifteen gallons a day for a mature cow—moisten the grass so it slides easily down the throat.

Once the rumen is full of pasture grass or hay, the cow will lie down in a comfortable spot and, mouthful by mouthful, burp it all back up again. Because it initially swallowed without chewing, the cow now brings those huge rear molars into play and takes the time to grind up the grass into a slimy pulp before swallowing it again, this time into the second stomach chamber, the reticulum. *Chewing cud*, as this process is called, takes eight to ten hours each day and involves up to forty thousand jaw movements.

From the reticulum, the cud moves into the omasum and next to the abomasum, the true stomach, then down the intestines. What's not absorbed comes out the back end. Because a cow's diet is high in fiber and fairly low in nutrients, an awful lot comes out the back end, ten or twelve times a day, for a grand total of up to 50 pounds of manure every twenty-four hours.

Along with all undigested organic matter and dead gut bacteria, cow manure often carries the eggs of internal parasites, or "worms," as most people call them. Cows won't graze near their own manure, an evolutionary response to the parasite problem. But cattle show no discretion as to where they poop, so pastures need to be large enough or rotated often enough that the cattle don't foul the grazing areas to the point that nothing is edible.

In addition to the manure deposits, cattle urinate eight to eleven times a day. Both manure and urine are superb fertilizer for pastures. Although the cattle won't graze those areas right away, they will after the deposits decompose.

Because cattle need to spend so much time resting and ruminating, they'll graze for only about eight hours a day. (When it's hot, they do much of their grazing at night.) The higher the quality of the pasture or hay, the easier it is for cattle to get enough to eat in those eight hours and to gain weight and bear healthy calves. Young, lush pasture is their favorite food, high in muscle-building and milk-making protein. If it's too young and too lush, however, pasture can cause problems. Cattle digestive systems are set up for lots of fiber, which young pasture and legumes, such as clover and alfalfa, lack. Too much of this type of feed can pack the rumen so tightly that digestive gases can't escape, and the cow begins to bloat. If the bloat isn't treated quickly, it will put such pressure on the cow's lungs that she won't be able to breathe, and she'll die. For this reason, you should never put cattle on wet or frosted legume pastures and should always provide some dry hay in the spring when pastures are just greening up. Bloat is a fairly common killer of cattle, although it's more common among dairy cattle than beef, due to the much richer diets fed to dairy cattle—which brings up an important point: when talking to others about and asking for advice on feeding, be sure to mention that you have beef cattle because their dietary needs are very different from those of dairy cattle.

Older pasture and hay composed of mixed grass and legumes are lower in protein but higher in carbohydrates. This keeps cattle's digestive systems in better order, helps fatten the cattle, and keeps them warm in the winter. It is also healthier for pregnant and nursing cows. Mother cows can get too fat on rich pasture, which is hard on their feet and legs and may contribute to difficult calving.

For the small-scale beef producer, feeding doesn't have to be complicated. There are three basic components: pasture, hay,

Dividing your pasture into paddocks lets your cows graze one area at a time while the other areas rest and regrow.

and grain. One important rule to remember is that whenever you change your cattle's diet—whether moving the cow herd from hay to pasture each spring or moving steers on to a finishing ration—do it slowly. The naturally occurring bacteria in their digestive systems, which transform food into nutrients, need time to gear up for a new ration.

Pasture

Your pasture is the centerpiece of your beef operation. It normally makes up the bulk of a herd's diet, and cattle that feed on good pasture are healthy and happy. Providing good pasture also means not having to provide as much hay, and the less time and effort you invest in hay, the more likely it is that you will end up in the black at the end of the year. Call your extension agent to find out how many acres of pasture it takes to support a steer or cow in your area, which can be anywhere from one and a half in the humid Southeast to forty in a semidesert area in the West. You can then estimate how many head of cattle you can theoretically sustain on your land. Keep in mind, though, that this is just an estimate. The actual number will vary considerably, depending on the fertility of your soil, whether it's a dry or wet year, and whether you have uplands, lowlands, or something in between. Keep your capacity on the conservative side, at least until you have a few years of experience under your belt. It's cheaper and less hassle to be long on feed and short on cattle than the other way around.

Pasture Quality

Once you have a rough idea of how many cattle your land may be able to support, take a walk in your pasture. The most critical

ingredient in the recipe for developing and maintaining a high-quality pasture is, as the old saying goes, "the footsteps of the owner." What's growing there? Grasses and legumes that cattle thrive on, or weeds? A weed, in this context, is not necessarily a bad plant; it's just something that cattle won't eat. Quack grass, for example, may be a weed in the yard but is good eating for cattle. After you've evaluated your pasture, you may want to adjust your carrying capacity accordingly.

On a side note, when patrolling your pasture, look for old bits of wire, stray nails, and other metal garbage and get rid of it before the cattle arrive. They will eat this stuff, which could perforate their stomachs and make them ill. This is called "hardware disease," and it's far better to prevent it than treat it.

So, how do you go about improving pasture quality? To help you figure out what to plant, get your soil tested. Some extension services offer soil testing, or you can check with your seed dealer for contact information of soil-testing labs in your area. The test results will indicate what soil amendments you need, and you can proceed accordingly. Be sure to specify that you're testing for pasture because soil amendments and fertilizer recommendations are calculated differently for row crops.

You ideally want your pasture to consist mainly of palatable grasses with a healthy component of legumes. Achieving this happy state may take a few years of managed grazing, mowing, and fertilization. You may want to add plant species by *overseeding*—that is, scattering seed in an established pasture. Much of the fertilization and all of the grazing will come from your cattle. Your job is to manage the cattle so that they do a good job of fertilizing and grazing.

The grasses and clovers that cattle like to eat grow differently from trees, shrubs, and some weeds. If you understand this difference, you'll understand why mowing and grazing are the keys to good pastures. Grasses and clovers have a "growing point" at or near the ground. When a cow bites off a blade of grass or a clover stem, the plant quickly regrows from this growing point. Trees, shrubs, and weeds, however, grow from the tips of their branches and leaves. That's why, when you prune a shrub, it stays pruned for months. By contrast, you have to mow the lawn every week—the cutting actually stimulates it to grow faster by removing the older leaves that are

getting in the way of the growing point at the base of the plant. Grazing has the same effect, so grazing, when correctly managed, results in lush pastures.

Unmanaged grazing, however, can devastate a pasture. This is because when a mower or a cow shears off the leafy part of the plant, it temporarily depletes the food supply to the roots, and some of those roots die. Dead roots put a lot of organic matter into the soil, which is great for holding water and keeping the soil moist, but a great many live, healthy roots are necessary for a thick, lush pasture. You want a balance between dead roots and live roots. If you cut your grass every day or let your cows graze the same plants every day, you kill too much of the root, and the grass will become stunted or even die. If the process goes on too long, the soil loses much of its plant cover and becomes vulnerable to wind and water erosion.

Weeds are especially abundant when the cattle feed in the same pasture for an entire growing season. Because the cows keep the grass and clover so short, the weeds have no real competition for sun or water and thus can grow with little restraint. In the spring, when all of the plants in an extensive pasture get off to an even start and are growing like gangbusters, this type of pasture looks great. By late summer, when the rain has slacked off, the spring growth spurt is over, and the cattle have kept their favorite plants short, a lot of these pastures are full of big weeds, tiny grass plants, and skinny cattle.

By contrast, grass that isn't grazed while it's still fairly young and tender gets stiff from hard-to-digest cellulose as it matures. The tall grass blades shade the growing point near the soil, and growth slows or stops. Some older plants in a pasture are OK to supply some fiber. However, the older the plant is, the slower it grows, and the less palatable it is to cows. Keep in mind, too, that a certain amount of old growth left over the winter can protect roots and growing points from freeze-thaw cycles that heave the soil and break roots. Too much, though, and the ground will be shaded and slow to warm in the spring, and new growth will have a tough time struggling through the old stuff to reach sunlight.

In summary, a thick pasture full of grasses and legumes that cattle like and lacking the weeds they dislike—with grass that isn't too old or too short—is ideal for the health and growth of cattle. This type of pasture provides the added advantages of growing longer into dry spells, greening up sooner in the spring, and staying green longer in the fall, which means money in your pocket that you won't have to spend on extra hay.

Rotational versus Extensive Grazing

It would seem that the best way to graze cattle is to let them graze an area thoroughly for a short period and then put them somewhere else while that area rests and regrows. This is called *rotational* or *management-intensive grazing*. Figured out in the 1960s and 1970s by Allan Savory, founder of Holistic Management International, and a host of other researchers, farmers, and ranchers around the world—and since portable fencing became readily available in the 1980s—rotational grazing has been quietly revolutionizing pasture and range management.

Nonetheless, *extensive grazing* is still by far the most common pasture system in the United States. If you have a lot of land and just a few head of cattle, it may be the most economical choice. The cattle can be turned out for the grazing season and left largely on their own; you only need to make sure that they have water, salt, and a mineral mix on hand at all times. If the pasture is big enough, no one will starve, although if it gets dry in the late summer and grass growth stalls, the cattle will need hay.

The biggest long-term problem with extensive grazing is that it can wreak havoc with the soil, water, and vegetation. Cattle that return to the same areas day after day to graze or rest will kill the plants and compact the soil. If there's a stream or pond in the pasture, they'll trample the banks into mud. With no lush vegetation to shade the water and no roots to hold the soil, the water temperature will rise and the banks erode, clouding the water and silting up the bottom. This is devastating for many aquatic species, especially prized ones

A large pasture with a small herd should be fine with extensive grazing.

Especially in dry climates, where even without cattle it's difficult for vegetation to prosper, extensive grazing can be devastating.

Setting up rotational grazing for beef cattle is fairly simple. Using whatever type of fence you prefer, split your pasture into several paddocks. Step-in posts and plastic electric wire are the cheapest, quickest, and most common choices for paddock fences. If, however, you don't care for the constant maintenance required by electric fencing, you can put up permanent paddock divisions.

such as rainbow trout. When cattle rest in the shade of the same trees day after day, the trampling can destroy the delicate feeder root systems and kill the trees. When palatable grass and clover are constantly grazed into the ground, noxious weeds can flourish.

Paddocks should be sized to provide enough pasture to feed the herd for at least three days but usually no longer than a week. A shorter period tends to make beef cows too fat, and a longer stretch allows them to regraze plants that are

FEED ADDITIVES AND GROWTH SUPPLEMENTS

More than 90 percent of cattle being fattened for slaughter are implanted with hormones and given antibiotics in their feed to make them grow more quickly and to convert feed to muscle more efficiently. When done correctly, the economics of these practices are persuasive. Properly used hormone implants will add 40–50 pounds of weight to a finished steer for a couple bucks' investment, while antibiotics and ionophores (a particular class of antibiotics with a different mode of chemical action in the body) increase the efficiency of digestion by 10–20 percent, which results in quicker weight gain. Antibiotics in the feed at low levels also help prevent illness and disease in what is (in a feedlot) a very crowded and dirty environment for cattle, which are being fed an unnatural diet.

Hormone implants are placed in the middle third of the ear and are either a synthetic estrogen that will increase muscle gain, a synthetic androgen that decreases protein breakdown and thus increases muscle mass, or a combination of the two. However, implants inserted incorrectly or the wrong implant type for the animal can diminish the benefit. An estimated 25 percent of implants have abscesses around the implant site, reducing their effectiveness, and implants in larger breeds can result in overly large animals. Implants can also increase toughness and delay marbling—the fat deposited inside the muscle—lengthening the time the animal must be on feed to reach a higher quality grade.

Concern over the effects of these practices on the environment and humans has grown markedly. An article by Michael Pollan in the March 31, 2002, edition of the *New York Times Sunday Magazine*, titled "This Steer's Life," probably did more than any other single event to galvanize public interest in what had been fed to the beef they were eating. Mr. Pollan revealed that most antibiotics sold in the United States end up in animal feed, not in people, and thereby add considerable impetus to the ominous and accelerating development of antibiotic-resistant human infectious agents. Manure that gets washed into nearby streams and lakes, in turn, leaves measurable levels of hormones in the water, where fish with abnormal sexual characteristics have been found. Some scientists believe, Mr. Pollan wrote, that this build-up of hormonal compounds in the environment may be connected to falling sperm counts in human males and premature maturity in human females.

When the snow melts, the ground will be fertilized by a mix of wasted hay and manure.

just beginning to regrow. If you're grazing steers with the goal of putting on as much weight as quickly as possible, you can shorten the rotation time to twenty-four, or even twelve, hours, although it's not necessary if the pasture is in good condition.

Paddocks need to be rested anywhere from a couple of weeks during the spring flush of growth in high rainfall areas to several months in hot, dry regions. Getting it all right takes some experimentation, talking to other rotational grazers in your area, and practice. Fortunately, rotational grazing is a forgiving process, and the cattle will probably do fine while you tweak your system.

Portable fencing makes it possible to change paddock sizes and configurations at the drop of a hat. In those years when I have more cows, I subdivide the land into more paddocks and graze them for shorter periods. When I have fewer cows, I cut back on paddock numbers and don't graze as tightly. Most years, I also graze all or part of our hayfields, either early in the spring or late in the fall or during a dry stretch when the pastures have given out. Anytime your cattle can harvest forage for themselves will save you time and money.

Overall, rotationally grazing pastures produces significantly more grass—as well as more palatable grass—than extensive grazing does. In practical terms, that means faster-growing animals and fewer out-of-pocket costs for feed. Organic matter in the soil increases, which helps hold moisture in. Consequently, grass growth continues longer into a dry spell. The pasture gets thicker and lusher. Trees are hardier. The tall grass along streams shades the water, and the roots hold the soil of the stream banks tightly, keeping the water clear and the stream narrow and deep.

Grazing-Management Basics

Some grazing-management practices are dependent upon climate. In arid areas, according to Allan Savory, a high density of grazing cattle is necessary to break up the soil crust and to work seed and fertilizing manure and urine into the ground. In the Deep South, where high summer temperatures prohibit grass growth, some cattle owners plant warm-season annual forages for grazing when pastures aren't producing. In our area, the Midwest, the major concern is weed control. I mow paddocks once or twice a season, just after they've been grazed. In general, it takes grasses and

clovers about three days to begin regrowing after they've been grazed. You want to mow within that time window so you're only mowing plants that the cattle didn't graze, not cutting regrowth. Where the ground is too rocky or steep to mow, I hand-cut weeds, preferably before they go to seed.

Every two or three years, it's a good idea to test your pasture soil. This involves taking a small shovel and a bucket and gathering samples from the top few inches of soil at several locations in the pasture. Mix up the samples, put some in a plastic bag, and send the combined sample to the soil testing laboratory. In a few weeks, you'll get back a report showing the pH level of the soil and the nutrient levels for nitrogen, potassium, and phosphorus. When you submit your samples, ask the lab to test for trace minerals, too, especially calcium. If your pasture is deficient in any nutrients or minerals, you'll need to amend the soil accordingly.

To find a soil lab in your area, get precise directions on how to take a sample, or locate lime and fertilizer dealers, start by asking your local agricultural extension agent or feed and seed dealer. Ask what time of year is best for taking samples and spreading amendments in your area, too.

Finally, pastures should not be monocultures (limited to one plant type). You don't want to eat the same thing for every meal, and neither do your cattle. A mix of several types of grasses, a few different legumes, and an eclectic selection of other plants, such as dandelions and plantains, will furnish a nicely balanced diet for cattle. If you're short on one of those plant categories, you can buy the seed and work it into the pasture. Ask your seed dealer, extension agent, and other specialists for advice, and then use what seems to fit your situation.

Hay is a main component of cattle's diet.

of the nest before the hay mower comes through.

Our local dairy farmers like to take three or four cuttings off their hayfields each year, but I take two—I'd rather have the cows harvest it by grazing a couple times than to haul out the tractor and Haybine again. This extends the grazing season and cuts my out-of-pocket feed costs. As an added benefit, the cattle fertilize as they graze.

Winter Hay Feeding

How you winter-feed your cattle depends on your setup and preferences. If you have just a few head of cattle and are using small square bales, it's easy to construct a wooden hay feeder or buy a metal feed bunker. If you're feeding round bales, you'll need round bale feeders sized for the bales you have. These are widely available at farm stores, and the pieces can be hauled or even delivered in a small trailer for assembly at home.

If you've got a shed for the cattle, you can feed hay inside. This is nice in foul weather, but it greatly increases the volume of manure you'll have to clean up next spring because cattle like to stick close to the hay in winter. If you feed outside, you can either feed in the same spot every day or keep moving the hay feeder to a new location.

If you're feeding in the same location every day, the happiest situation is having the hay feeders on a cement pad. This eliminates the mud and simplifies cleanup in the spring. If you don't have a cement pad, your dirt feeding area will become a "sacrifice area," so churned up by the cattle that it's unlikely you'll have anything growing there for a long time. In either case, plan on scraping the feeding area or the shed clean in the spring and spreading the moist mix of manure and old hay on your pasture or hayfields with a manure spreader. This typically requires a Bobcat with a bucket, a tractor, and a manure spreader. If you don't clean up the area, chances are you'll have a terrific infestation of stable flies because the manure-hay mixture is optimal for their breeding.

The other option for winter hay feeding is "outwintering," or feeding hay on pasture away from the barn. You move the feeders each time they're emptied; because the manure and wasted hay is spread as the cattle are feeding, there's not much spring cleanup. Be aware, however, that in areas where winters are wet and the ground doesn't freeze, this system will quickly make a huge muddy mess in

Hay

When pasture is dormant due to drought or cold, your cattle will need hay, and beef cattle don't need top-quality hay. Ideally, the hay should be a mix of grass and legumes cut before they go to seed and baled after they've dried thoroughly. As long as the hay isn't moldy or nothing but stems with no leaves, it'll do for beef cows. Pure, high-quality alfalfa or clover hay, which is high in milk-producing protein and low in body-heat-producing carbohydrates, can be harder on beef cows than old, coarse hay, but you might want to use it if you're fattening steers during the winter. Knowing when to buy or harvest hay saves you money, time, and labor.

When to Buy or Harvest

If you're buying hay for cows and calves, you don't need the expensive stuff. During the growing season, keep a close eye on the weather in your region. Dry seasons quickly create hay shortages, driving prices sky-high. Buying early and in quantity during good years and storing the excess is generally your best bet. In covered storage, hay will last for years, and even round bales stored outside will last two or three years if the bales are tight and kept on dry ground. Don't pay for any hay until you've dug into a few bales and checked for mold, weeds, and stem content.

If you're having hay made for you on your land or making it yourself, don't be in too much of a hurry. Waiting a little longer into the season to make your first cutting of hay gives you several advantages: The hay will be taller, giving you more volume. It will be higher in carbohydrates and lower in protein, which is good for keeping cattle warm in the winter. The weather in most areas will be more settled, with less likelihood of a surprise rainstorm ruining the cutting. In addition, grassland nesting birds will have a better chance of getting their babies fledged and out

FEEDING CATTLE

We have no barn. The round bales are stored outside. The cattle are wintered in a half-wooded horseshoe "coulee" of about 40 acres, with a year-round spring. We feed the round bales right on the ground, and the cows clean it up pretty good. What little they do not eat is their bedding. No sense in having the cows eat all the hay and then haul in straw to a barn just to have a damp place where the sun doesn't shine and disease builds up.

—Mike Hanley

I probably have them on a finishing ration for four to seven months because I don't feed a lot of grain. I put them in a little early, and I don't feed them really heavily, maybe 15–20 pounds of a mix of corn and barley. The rest is silage and hay. I think you get a better meat and fewer health problems.

—Donna Foster

For finishing at home, doing it by the pail method—buying your corn and oats and protein supplement and mixing it yourself—saves the extra expense of having the mill mix it. A lot [of supplements] contain antibiotics and stuff, and we don't use those.

—Rudy Erickson

Consider putting up oat hay, sweet clover, sorghum-sudan hybrid mixes, forages mixed with grains, and soybean hay. All these are things I've seen put up in my lifetime, and with the new equipment today it can be done a lot easier than in the old days. Any of these fed properly to livestock is very good feed.

—Dave Nesja

pastures. Where the ground freezes, it's a terrific way to renovate poor spots in pastures: because it's frozen, the sod won't be badly cut up by the cattle, and, in the spring, the ground will be covered with a layer of manure mixed with hay, the best fertilizer there is.

With outwintering, you can either haul hay out to the feeders a few times a week, or you can set all of your bales out during the fall and not have to start a tractor all winter. In October, I calculate how many bales I think I'll need for the number of cattle I'm carrying through the winter and for the length of time I think the ground will be frozen. With a tractor and hayfork, I set out the round bales in three or four rows, spaced 15–20 feet apart on all sides. When the bales are in place, I cut and pull off all of the twine (or netting wrap) by hand. I build a three-strand electric fence around three sides of the rows, leaving one side open. I roll the round-bale feeders out of storage and pop them over the bales at the head of the rows on the side that I left open. Then I stick plastic step-in posts into the next row of bales and run two strands of electric wire across them to keep the cows from the rest of the hay.

Constructing this "hay corral" takes me 20–30 hours, but it's pleasant work in moderate weather, and it means I won't have to start the tractor on cold mornings or fight with twine or round bales frozen to the ground. Instead, once or twice a week, I unplug the electric fence, walk out to the hay corral, move the portable electric wire back one row, tip the feeders on their sides, roll them over to the new row of bales, then walk back and plug in the fence again. This takes all of fifteen minutes. What's left of the old bales is bedding for the cows, keeping them clean and out of the snow. In the spring, what was a poor area of pasture will be fertilized, and, by late summer, the pasture will be deep green and growing taller and lusher than anywhere else.

Grain

Cattle love grain. It brings them running when we want them, and it can produce tender, tasty beef from even mediocre animals. Although the most common grain fed to cattle is field corn, they will eat a wide variety of other offerings. For the small operator, corn is usually the cheapest, most available, and easiest grain to buy in small quantities (under a ton). Corn should be ground or rolled so the cattle can digest it better.

Talk to your feed store about other feed options in your area or additional additives, especially if you are fattening a steer for slaughter. Corn alone isn't high enough in protein to satisfactorily fatten an animal during a short period. Corn and good pasture will do the job, but if you don't have lush pasture or high-quality legume hay during the finishing period, you should talk to your feed dealer about formulating a finishing ration. In addition, in some regions, there are cheaper alternatives to corn, making it worth your while to inquire.

Beef cows whose purpose is to produce calves, not meat, don't need grain if they're on good pasture in summer and adequate hay in the winter. But I give them a little anyway, as do most beef producers I know. It's called "training grain." A small amount—a pound or less for each cow—brings them running every morning when I call, and getting them in when I need to work with them is never a problem. As a bonus, the cows teach their new calves about grain each year, so when it's time to start the calves on grain, they know exactly what to do.

You should start calves on grain no later than when you wean them. If you want to start them before weaning, set up a "creep feeder" that will keep the cows out of the calves' grain. A creep feeder is a pen or shed with an opening too narrow for cows but wide enough for calves. Inside is a bunker feeder for the calves. I generally put a board over the opening as well to make it too low for a cow to

Your cattle will come running when it's time for their grain ration.

Calves will try to suck on each other, which isn't a great idea, so distract them with calf feed. This is a sweetened grain mix that should be fed free-choice (available at all times) from the time they're a few days old. Take some in your hand and stick it in their mouths after each bottle feeding until they figure out how to eat it themselves. Feed it in a bucket or box attached to the side of their pen, placed high enough that they won't poop in it (too often). In case you can't get them outside in a small area with some green grass to nibble once they're a few days old, keep some high-quality hay available for them. They need to get used to hay while they're still young and open-minded about trying new things.

One sack of milk replacer and one to two sacks of calf feed will raise a dairy calf until weaning at eight weeks of age or older. Grain feeding should continue according to the directions on the sack of calf feed, with a gradual transition to an adult ration as the calf's digestive system matures. Get a calf outside and on pasture as early as possible. You can buy calves in winter, but they're more susceptible then to pneumonia and scours (diarrhea), so provide them with a draft-free, deeply bedded pen, and keep it clean. Give them a good grooming with a cattle brush every day. This mimics the cow's licking and is stimulating and comforting for the calf. A happy calf is more likely to be a healthy calf.

squeeze under. If you have an old shed not being used for anything else, it might work well for a creep feeder.

Until you put weaned calves on a finishing ration, grain isn't essential to their diets, but it will help them grow a little faster. Many cattle owners "rough" calves through the winter on hay alone and don't start them on grain until four months or so before slaughter. But feeding weaned calves a pound or two of grain per head per day will quickly teach them to come when called, help them grow, and keep them tame.

Feeding Dairy Calves

Dairy bull calves need a lot of extra care and special feeding for the first few months. While a beef calf gets as much of his mother's milk and affection as he wants for the first six months or more of life, a dairy calf loses his mother and his mother's milk within three days of birth. So be kind to these babies, even though they're often incredibly stubborn. They'll be a little lost and stressed and vulnerable to sickness.

If you buy dairy calves, pick up a sack of milk replacer for each calf from the feed store and a two-quart calf bottle and nipple per calf from the farm supply store. Follow the directions on the sack for how to mix the replacer, how often to feed, and at what temperature.

At first, it may take a little persuading to get the calf to drink from the bottle. If he won't take the nipple, try backing him into a corner and then straddling him with your legs. Pull his head up and hold it with one hand, stick the nipple in his mouth with the other hand, and squeeze a little milk onto his tongue. Calves usually catch on pretty quickly. To prevent choking, don't hold the bottle any higher than the height of the calf's shoulder. Calves drink amazingly fast, and the milk will be gone long before their sucking instinct is satisfied.

Finishing Rations

There are two approaches to fattening a steer for your freezer. The first is to use time and low-cost inputs, and the second is to speed up the process with a formulated ration fed at a high rate. The first approach usually makes the most sense for small farms because it doesn't require an expensive ration or a separate pen. If you don't have any land for grazing, it's possible to put a weaned calf directly onto a finishing ration, provided it's the right breed and a fast-growing animal. However, the animal will still need plenty of forage-based fiber in its diet. Most calves need some time to mature before they will fatten and are better off on pasture or hay until they're at least a year old.

You also have some choice as to when you send an animal to the processing plant. You can have "baby beef" from a steer as young as a year, although steers are normally kept until they're more mature

Walking back and forth to the barn for a drink encourages cattle to get moving.

and have put on some exterior fat. A steer from one of the English breeds can be ready for slaughter as young as sixteen months, while a steer from a continental breed may not finish until it's two or more years old, depending to a large extent on how much grain you feed. On a small farm, it's practical to keep the steers on pasture and feed them grain once or twice a day. If the pasture is excellent, 4 or 5 pounds of corn (usually with a protein supplement) each day will have most steers ready for slaughter in three to four months. Generally, it's a good idea to finish a steer before the age of twenty months to ensure a tender carcass. Please remember, however, that these are just rules of thumb; finishing cattle is not an exact science.

If the steer is on good pasture, it's helpful if you can time the finishing so that it coincides with the end of the grazing season. Cattle gain weight faster and more cheaply on good pasture than on hay. If the steer is out with a cow herd, he can be trained to come to a separate pen for his ration. To do this, watch where the steer normally is when the herd comes in for the morning drink or grain ration. If he's at the front of the line, close the gate behind him and move him forward into another pen. If he's at the back of the line, close the gate behind the cows and feed the steer with a low bucket in the pasture. If he's in the middle, you'll have to finesse getting the cows ahead and keeping the steer behind for a few days until he figures out to stay behind for his ration.

FINISHED OR FAT?

Most of us can't just walk up to a steer, jam a thumb into his back fat, and know that he's ready. Two beef-raising friends, Barry and Libby Quinn, told me that you just have to develop an eye for finish. Former extension agent, current friend, and lifelong beef producer Dan Riley told us that a steer is ready for slaughter when you can see the fat around its cod, over its pinbones, and on the rear flank. If a steer has fat around the tailhead, it's close to grading prime; if it has a fat brisket, it's too fat.

Grass Finishing

Cattle can be finished on grass, but it takes expertise to turn out high-quality beef without grain. If you are interested in grass-finishing, you first will need to buy the right cattle. Short-legged animals from the English breeds are probably your best bet. Second, you will need superb pasture, lush and high enough in protein that it will enable a steer to gain no fewer than 1.7 pounds per day for the last ninety days before slaughter. In most areas, this takes a combination of pastures and planted annual forages plus experienced management. However, any cattle can be raised to maturity on grass and slaughtered for edible beef.

Water

Clean water should be available at all times for your cattle. If the water isn't fresh, they may not drink as much as they should. Tip the water tank a couple of times a season and scrub out the algae. A float valve, available at farm supply stores, will keep the tank full when you're not around. In below-freezing weather, install a tank heater and plan on filling the tank daily because a float valve freezes up in cold weather.

Unless you run a hose out to the paddocks, your cattle will need to come into the barnyard for a drink. When building your paddocks, create lanes that give your cattle easy access to the barnyard. These can be built with the same portable fencing used for the paddocks and should be about 10 feet wide. Gates can be made by tying a gate handle onto the wire and adding a loop at the far gatepost to hook the handle through.

It's not necessary to have water available in the pasture. Cattle will walk a long way to get water, even when there's snow on the ground, and that's usually good for them and their hooves. Cattle

Feed salt and minerals in a separate feeder.

Salt and Minerals

Salt is essential to cattle, and the best way to make sure that they get enough is to provide free-choice loose salt in a feeder protected from rain and snow. Buy or build a two-compartment feeder and put salt in one side and a mineral mix geared for your area in the other side.

Mineral deficiency used to be a common cause of disease in cattle. The diseases varied from region to region, depending on what was deficient in the soils. That's why it's important to get a mix formulated for your area of the country. As with salt, it's easier for cattle to get enough minerals when the mix is loose rather than in a block.

For steers being finished for slaughter, getting enough minerals and vitamins into their feed is especially important. These can be mixed in the finishing ration according to your feed dealer's directions or fed free-choice in a separate feeder as you would normally do with the cow herd. If fed free-choice, the vitamin and mineral mix should be freshened at least once a week.

need regular exercise to stay healthy, and they can be a little lazy about it. Some producers still rely on snow to water their cattle in the winter, but it's difficult for them to get enough to stay fully hydrated. Eating snow also chills them, and they'll lose weight burning calories to stay warm.

POISONOUS PLANTS

No matter where you live, chances are that some plants in your pasture could poison your cattle. Fortunately, though, most (though not all) poisonous plants taste icky. If your cattle have enough to eat, they probably won't touch anything that's bad for them. But if your pastures are stressed by drought or have been heavily treated with nitrogen fertilizer, or if it's very early in the spring and the only plant that's green is also poisonous, you should be alert for problems.

Six different classes of poisons have been identified in various plants, the most important being the alkaloid and glycoside groups. Alkaloids affect the nervous system, causing loss of motor control, bizarre behavior, and death. Jimson weed, a common species of the western United States, is probably the best-known example of a plant that kills with an alkaloid poison. Glycosides basically cause death by suffocating cells. The animal is breathing, but the oxygen in the bloodstream is blocked from being transported into the individual cells. The buttercup, which brightens low pastures in early spring, is a familiar glycoside-containing plant.

Other familiar plants that are dangerous to cattle include black locust, black nightshade, bracken fern, castor bean, curly dock, death camas, dogbane, horsetail, locoweed, lupine, milkweed (several species, but not all), oleander, pigweed, and tobacco. White snakeroot, common throughout the Midwest, causes the "trembles" in cattle and can kill humans who drink milk from cows grazing it. Thousands of settlers in the Midwest died of milk sickness in the early 1800s, including Nancy Hanks Lincoln, Abraham Lincoln's mother.

For more information, contact your local agricultural extension agent.

Although they are not as bright as dogs and cats, cattle are intelligent in their own way. Usually good at taking care of themselves, they'll find a windbreak when it's cold and shade when it's hot, and they'll never forget where their calves are. While cattle learn more slowly than horses, they do quickly learn to come when called if you reward them with grain or a new pasture when they get there. And when they learn something, they never forget it—especially bad experiences, which is an important point to remember when you're working cattle. When introduced to new experiences slowly and patiently, they're also quite curious. If I walk into the pasture and sit down, I'm soon closely surrounded by a ring of sniffing cows, wondering what in the world I'm doing.

BEEF CATTLE
BEHAVIOR AND HANDLING

Cattle are not democratic. In every herd, there is a hierarchy, from the bossy top cow to the shy one at the bottom of the pecking order that always gets shoved away from the best feed. Dominance is determined by strength and aggressiveness, but it's all kind of low key. Cows don't fight so much as they test to see who can push whom around. You'll sometimes see a pair of cows head-to-head, hooves planted, just pushing. The winner is the one that pushes hardest and longest. Cows will push around a young bull, too. (Don't worry, he'll get over it.)

Cows have different personalities. One of the delights of having a small herd of cattle is getting to know each cow on an individual level. Some are inherently nervous and will never let you get close. Some are docile and like their heads scratched, while others like to spend most of their time shoving everyone else around. Some are overprotective mothers, while others are lackadaisical about their calves. Most will nurse only their own calves, but a few will let any calves help themselves—kind of like those human mothers we fondly remember from our childhoods who would feed any of the neighborhood kids who happened to be around at dinnertime.

Some behaviors are common to all cattle. Because they are, and always have been, a prey species, cattle are hardwired to run at any sign of danger. Danger, to nervous cows, could be anything from a strange human in the pasture to a funny smell. They don't stop to think about whether there is really a threat; they just take off. They'll run right over you if they're in a panic, and they panic fairly easily. For animals that are normally fairly slow moving and clumsy, they can move surprisingly fast when frightened or agitated, and they'll jump gates and fences—or trample them.

Cattle would rather run than fight, but they will also fight if they have to. A cow will defend its calf; a bull that decides for some reason that a person is competition or a threat will charge and can kill that person. Although when kept with the cows, beef bulls will usually behave themselves and run away with the cows, they are naturally more aggressive than cows, and they are unpredictable. You can't trust a bull not to charge you instead of running away. Never, *ever*, turn your back on a bull.

As a prey species, cattle learned long ago that there is safety in numbers. They graze in a group because it's easier for a group to spot and defend itself against predators than it is for a single animal. The herding instinct is not completely dominant, however. Groups of two or three will wander off, although usually not too far, from the main herd in pursuit of some promising grazing.

If the pasture is large enough, cows prefer to go off by themselves to calve and may remain away for a day or two before returning with their new babies. Cows also use babysitters. Often, you'll see a single cow watching over a group of calves while all of the other mothers are off grazing and, I presume, gossiping.

Shelter

Cattle tolerate an amazing range of weather conditions and do fine outside year-round in nearly all climates. There are a few times, however, when they should have some sort of shelter. In the "ice belt" (those states between the snowbelt and the no-snowbelt), where there can be months-long stretches of damp, almost-freezing weather, cattle are healthier and happier if they can get out of the mud and the wet. It doesn't have to be anything fancy; an old shed open on one side or an old barn bedded with old hay, straw, or sawdust will work. However, you'll need to borrow or rent a skid steer with a bucket to clean the place out in the spring.

Most types of existing outbuildings can be converted into indoor cattle-handling facilities.

Cold winds can be hard on cattle, too. If you can arrange a place where they can get out of the wind, they'll do much better. This can be anything from a belt of trees to the side of a building. Extremely hot weather is also tough on most types of cattle, so having shade available, again either under trees or near a building, makes a big difference to them.

Handling Facilities

Cows are bigger, stronger, and faster than we are, so when it's time for vaccinations or shipping, we need a way to make them hold still (other than trying to rope them or grab a tail). You can't hold a half-ton cow if it doesn't want to be held, and generally it doesn't. That's why cattle owners should—for their own safety, if nothing else—have a cattle-handling facility.

Cattle facilities have four parts: a corral or holding pen, a crowding tub, an alley, and a chute with a headgate. The small operator really needs only these four basic components. Even if you have just a couple of steers for the summer, you'll still need some sort of pen and a chute to load them onto the truck at the end of the year. Make sure to locate your facilities where they will allow easy access for a truck and trailer.

The pen or corral holds the cattle until you're ready to work them. A few at a time are herded into the crowding tub, which is a small, circular pen with a swing gate that pushes them into the alley. The alley is narrow, so cattle don't have room to turn and must go single file. At the end of the alley, a grate is raised to let a single animal at a time into the chute, where the cow or steer's neck is caught in the headgate. This holds the cattle in one place so you can safely administer shots, put on ear tags, or do whatever else may need to be done. Once the vaccinating, breeding, or castrating is done, the headgate is released, and the animal moves back out to pasture. The headgate is then reset for the next patient.

When planning your handling facilities, follow these few basic principles. First, unless it never rains in your area, put your facility inside a building or on cement, or both. It's no fun working in the mud. Old dairy barns can be easily converted for working beef cattle, as can nearly any sort of shed. Our facility is on a cement pad outside the old barn.

Second, plan for curves, good footing, and good lighting. A slippery surface will make cattle nervous, and they will balk if they see a barrier ahead. They will, however, follow a curve around to its end. They also will move more readily toward a well-lit area than a dark one.

Finally, if you're building with raw materials rather than buying manufactured components, follow the recommended dimensions exactly. The recommended alley width for your breed of cattle may seem incredibly narrow, but anything wider and you'll have cattle trying to turn around, which can result in injuries. Never force a heavily pregnant cow into the alley or chute—it might get stuck.

With a little creativity, it's possible to build a fine facility at a reasonable cost in a fairly small area. Ours is built on a 75 × 50-foot concrete pad next to the old dairy barn. We also use the area for moving tractors and equipment in and out of a shed. We hung a lot of gates that we usually leave open for mechanical traffic but close when handling cattle in order to subdivide the area into three separate pens that wrap around to feed into each other and then into the crowding tub. I bought a manufactured tub, alleyway, and headgate, all of which we bolted to heavy-duty wooden posts that we cemented in so the cows couldn't push things out of alignment. The result may not be aesthetically beautiful, but for working cattle, it functions wonderfully. The water tank and grain feeders are also situated on the concrete inside the holding pens so the cattle go in there to eat and drink and are very familiar with the setup, which makes for calm and easy handling.

Handling Beef Cattle

Large cattle operations have always known that handling cattle effectively on a regular basis requires a properly equipped facility. Even a small operation with a handful of cattle can build a scaled-down version of a good facility for a reasonable cost, and the right handling methods can make a bare-bones facility work smoothly.

The premise of moving cattle is based on the animal's flight zone.

pressure point—it will move ahead. If you're directly behind it, where it can't see you, it'll turn around to look at you. If you take a very visible stick, such as a white step-in fencepost or something with a small flag on the end, and you hold it out to the side, you'll look three times as wide to the cattle and they'll turn even more easily.

This is critical information to understand and use when you're trying to move cattle. After you've spent some time in the pasture with these concepts, you can get good enough to sort calves from cows for weaning in a holding pen without having to run them through a chute. You may be able to cut out a pair of cows from the herd when you need to get one in the chute. It's important to note that you should never cut out and isolate a single cow. Cattle are herd animals, and they get panicky if made to go off by themselves; a cow may run over you to get back to her friends. Always work with at least pairs, if not triplets, and you'll have much calmer animals.

There's been significant rethinking of traditional cattle-handling methods, thanks to the work of such pioneers as University of Colorado professor Temple Grandin and cattle-handling and marketing consultant Bud Williams, as well as many others. Their work has focused on understanding the natural instincts and behaviors of cattle and using that knowledge to design systems and handling methods that keep cattle calm and tractable. Temple Grandin's cattle-handling layouts and techniques have been adopted by many major US processing facilities and innumerable small farms and ranches. Bud Williams's herding methods have set the gold standard for moving and holding cattle in the open, particularly in rangeland grazing.

Cattle Handling Fundamentals

The most important concepts to understand when handling cattle are *flight zone* and *pressure points*. *Flight zone* refers to an imaginary circle around the animal that, when you cross into it, causes the animal to move away. By working on the edge of an animal's flight zone, you can gently move it in the direction you want. Because the flight zone varies by the tameness of the herd in general and by each individual animal, and can even be affected by weather conditions, it takes a little finessing to find it. Approach slowly until the cattle begin to move away from you. That's the edge of the flight zone. Go farther, and they'll move faster. Back off, and they'll stop.

Pressure points are specific spots along the edge of the flight zone where, if you stand, you can turn cattle in a different direction. The most important pressure point is at the shoulder. If you're in front of the shoulder on the edge of the flight zone, the cow will turn back. If you're behind it, it will move ahead. If you're to the side and slightly behind the rear of the animal—a second

Establishing a Routine

Cattle love it when the same thing happens at the same time every day. Use this to your advantage by establishing a routine that makes your cattle familiar and comfortable with your handling facility. For example, if you're feeding grain, you could feed it in the holding pen at about seven o'clock every morning. One side of our holding pen is lined with bunker feeders, so our cows like being in the pen. In fact, they're usually waiting there for me at feeding time. When you feed, wear the same cap, carry the same grain bucket, and call with

DID YOU KNOW?

You can avoid getting kicked when working around the back ends of cattle by "tailing" them. Raise the tail straight in the air with one hand, grab it around the base with the other, and keep it pulled toward the head. This keeps the animal's hooves on the ground because kicking with the tail in this position hurts the animal's back. Don't be tentative about tailing; if you don't have a good grip at the base, you can still get kicked.

Cattle are often excited and energetic when they get onto fresh grass.

the same call. For the first week or so of establishing a routine, you may need to walk out in the pasture with a grain bucket to where the cattle can see you and then call to them. With new animals, I've even done a Hansel-and-Gretel routine, leaving a trail of little piles of grain leading back to the pen, where the big feed happens. Cattle catch on to the call and the grain bucket very quickly.

You could put the water tank in the holding pen, preferably on a nice cement pad to eliminate the mud. Leave the gate open so cattle can wander in anytime they're thirsty. Once in a while, after they've had their grain or come up together for a drink, shut the gate behind the herd and make them stand in the pen for an hour or so. The more the cattle are accustomed to being in the pen, the easier it will be to put them there and keep them calmly waiting when you need to work with them. Even if you have only a couple steers for the summer, take the time to train them to be penned.

If you're rotationally grazing and switching paddocks regularly, that's another opportunity to teach the cattle to come when they're called. Each time you change paddocks, do it at about the same time of day. Call the cows and then lead them to the next paddock. They'll quickly learn to follow, and it's wonderful to see them romp when they hit the fresh grass.

Of course, the cattle won't know what you're trying to do the first few times you switch paddocks. Wait until they come into the pen for grain or water and then close the gate behind them. While they're getting a drink, go out and close the old paddock gate and

open the new one. Then go back, open the pen gate, call the cows, and walk slowly out of the lane to the new paddock, calling as you go. Eventually, they'll follow, and, after a few repetitions, they'll follow immediately. They may even try to push past you in the lane. If they do, spread your arms and give them a dirty look and a firm "cut it out" so they'll learn that you're the leader. If one gets really pushy, rap it on the nose with a stick. This is another important point: remember that cattle have a herd hierarchy, and you should be at the top of it.

If you've got one big pasture, aren't feeding grain, and aren't even around on a daily basis, it's still a good idea to teach the cattle that when you show up and they come to your call, good things will happen. Call them, let them see the bucket, and then give them a treat in the holding pen. Use apples, carrots, or dried molasses. At first, just a few cattle will figure it out, but eventually they'll all come. If there's a holdout cow, you can either pen the others and wait until it shows up out of curiosity, or you can pen the others, go out in the pasture, and gently herd it in. It'll usually be quite willing to go where it knows the others went.

After a few months of calling and leading, the herd will usually come when you call and follow when you lead. This minor investment of effort is worth its weight in gold when it comes time to bring the cattle in for handling.

If you want to really get ahead of the game, train your cattle to exit the holding pen through the crowding tub and chute, just as they will on handling days. Close the gate behind the herd when they've entered the pen, and open the chute. I have a board cut to fit over the top of the headgate and hold it open so there's no chance of it slamming shut on a cow. This teaches the cattle that

ALL IN THE TIMING

Unless you're an early riser, I recommend scheduling paddock switches for the late afternoon. I used to do it first thing in the morning, until Gretel the cow took to standing at the barnyard gate around five o'clock, bellowing for me to hurry up. Since the gate is directly across from our bedroom window and I don't like getting up at the crack of dawn, I soon realized that I needed to change the routine.

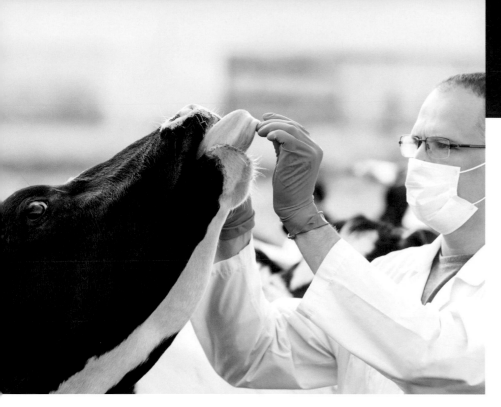

A little advance training will help your cattle be more relaxed on handling days, such as visits from the vet.

Handling Days

Beef cattle are handled just a few times a year. In most areas, calves are castrated, and often vaccinated and ear-tagged, in the spring. If you're using artificial insemination, you'll need to get the cows in the chute whenever they're in heat, and you'll want them to be calm. In the fall, you'll vaccinate the whole herd, and heifers will get their brucellosis vaccinations from the veterinarian. If you haven't ear-tagged before, you'll do this then. Two or three weeks later, the calves will need a booster vaccination. They're also weaned then or shortly thereafter. In late fall, calves and cull cows are loaded into a truck or a trailer for a ride to the auction barn or new owners, and fat cattle for your own freezer are loaded and sent to the processor. You may also occasionally have to pen a cow or calf that is sick or injured so you can treat her.

To accomplish these things on a cattle-handling day, you'll need to sort off the cows, calves, or steers you want to work with and then move them into the chute and headgate. You'll catch them one by one in the headgate, where you'll administer vaccines, put on ear tags, or give some other treatment, and then release them back to pasture. If you're weaning the calves, you'll need to sort them into separate pens or pastures—one for the cows and the other for their calves—as they come out of the headgate. The veterinarian, the artificial insemination technician, or the truck for loading cattle may be around on handling day, too, and because cattle are very aware of strangers, this will add to the stress they're already feeling from the change in routine.

Once you have the cattle quietly penned and have given them half an hour or so to calm down, the quickest way to get the job done is to move slowly. The faster you move, the more agitated the cattle will become, and the tougher it will be to get them into the chute. Don't yell. Research has shown that loud noise, including yelling, is more upsetting to cattle than getting slapped or prodded. Use flight zone and pressure points to move a few cattle at a time from the pen into the crowding tub. Don't fill the tub more than half full because cattle don't like to be tightly packed. Wait until at least two or three are facing into the chute, then slowly move in the crowding gate. The cattle should start filing into the chute. Sometimes, you'll have to back off on the gate to let a cow turn in

the chute is the way out, and it saves you a lot of trouble getting them into the chute on the day you really need them to go. They'll go by themselves, and they'll think it was their idea!

Getting Ready for Handling Days

Cattle are pretty specific about what they do and don't like. As previously mentioned, they like firm, nonslippery footing, and they like moving uphill and toward light. They dislike going into dark, unknown spaces; they detest loud noises; and they are afraid of strange objects flapping in the wind or glinting in the sun. They will follow each other around a curve, but they'll balk at going into a chute when they can see that the end is blocked. Keep these things in mind when you're getting ready for handling days.

Presumably, your facility is already set up with curves and has good lighting. Now, walk around and make sure that there are no soda cans, flapping chains, or fluttering debris anywhere. Make sure that there are no broken boards on the floor of the chute. Oil the gates and levers on the crowding tub, alley, and headgate so they operate silently and smoothly.

If you know ahead of time how you're going to move the cattle through the pens, tub, and chute, it'll be much easier for you to communicate the plan to them. Have a Plan B in case they don't like Plan A. If you need only the calves, for example, or only the open cows, figure out how you can quietly sort off the ones you want and turn out the rest. Decide where you can keep the syringes, needles, and vaccines so they're handy but out of the way of random hooves. Have the ear tags numbered and ready, with extras in case one falls into a fresh cow pie.

the right direction. You can wave a hand or herding stick gently in its face or pat a rump to get it to turn, but don't yell and don't hit. Things won't go any faster, and for sure the cattle will be more difficult to work the next time.

Raise the gate at the end of the alley and let a single animal into the chute. In a perfect world, the cow will walk up to the headgate with just the right momentum to make the gate close on the animal's neck. In the real world, cattle sometimes come in so gently that the gate doesn't close or so hard that you'll worry they've bruised their shoulders. With some of our old cows, I don't bother putting them into the headgate if they're calm enough to stand still while I give an injection.

For the cattle you need in the headgate that aren't cooperating, one helpful trick is to walk quickly from their heads to their rumps, which usually makes them jump ahead. If the animals are reluctant to come down the alley, it may be because they see you standing at the end. Duck out of sight or walk quickly to the rear. Sometimes a pat on the rump or a tap with the stick on their hocks will do the trick. Be patient. In the end, staying calm will take less time than trying to rush them because too much pressure only makes cattle

panicky and balky. A panicky cow may try anything to get away—running over a handler, trying to jump a fence, or trying to crawl under a gate.

In the full swing of things, you'll have a cow in the headgate, a couple behind it in the alley, a couple in the crowding tub, and the rest still in the holding pen, waiting their turns. Take care of the first cow, release the headgate, and make sure that it gets safely out of your way into another pen or back to the pasture before resetting the headgate for the next patient. Look to see whether the next up is a calf or an adult, and adjust the gate accordingly. With practice, patience, calm, focus, and quiet, you should be able to vaccinate a herd of twenty to thirty animals in well under two hours without getting any manure splattered on your pants. But wear old pants just in case.

Loading and Transporting Cattle

Getting cattle on a truck or trailer breaks some of the rules for cattle handling. You're asking them to go into something that's dark and a dead end, and it's an unfamiliar step up to get there. Once again, have some patience. If you are loading into a low trailer, you can load directly from your chute. You can use a pair of loose gates on either side to fill in any gap between the headgate and the trailer gate; if you do this, tie the gates down. If you've got enough cattle to warrant bringing in a semitrailer, you'll need a ramp or some other device to get the cattle up to the level of the truck. Ramps are available commercially, or you can build your own. Semitrailers need a lot of room to maneuver, so put the ramp where the driver will be able to turn and back up easily.

If you're moving just a few cattle, get them into the crowding tub well before the truck pulls in. Because the noise will upset them, it's best to have them as far along in the process as possible before

HOW NOT TO HANDLE CATTLE

In the movies, cattle are handled with horses, lariats, and a big crew of cowboys. Calves are roped, flipped on their sides, and branded. It's strenuous work, even with little calves, and for some reason filmmakers never do show how the cowboys vaccinate the full-grown cattle—not by flipping them on their sides with a rope, I'm sure. In real life, some owners with small herds will pen cattle against a wall or in a corner with a loose gate; some will crowd them tightly in a small pen; and others will somehow get a rope around a cow's neck and tie it up tight to a stout post. What these methods have in common is that they're highly stressful to cattle, fairly unreliable as far as getting every head in a herd treated, and quite dangerous for both cattle and handlers.

Cattle will likely be apprehensive about going into a dark trailer.

they begin to tense up. Once the truck or trailer is in position, begin moving the animals down the chute. The one in front, with the best view of where this is all going, will probably be reluctant. The ones behind may begin pushing the leader along, which is usually OK. Don't rush things. Give the steers time to look at the step.

You will probably need to use a little more persuasion than you would in other handling situations. Have any helpers stand out of sight of the cattle, or try walking quickly from the head of the line to the back. You can slap a rump, but don't yell and don't whip out the electric cattle prods. For one thing, it will cost you money because stressed cattle on their way to auction, the processor, or a buyer have more "shrink," or weight loss, than do unstressed cattle. Stressed animals won't eat or drink, either, so they'll lose more weight in the pens at the auction barn. Stressed animals also release a lot of adrenaline into their bloodstream, which darkens and toughens their muscle. These animals are known as "dark cutters," and their meat is less valuable.

After Handling Days

The day after you've worked your cattle, go back to the same old routine. You want to erase any bad impressions of the day before with all of the good things that will continue to happen if the cattle come when called and follow when led. They won't forget about getting shots or tags, but getting their grain or water and remembering that the quickest exit is through the chute will gradually become uppermost in their minds again.

ADVICE FROM THE FARM

HANDLING CATTLE

At times, it is helpful to impress the cows with size. I either hold my arms out to my sides, like I learned to do when playing basketball as a kid, or carry my trusty 'cow chasing stick,' which extends my reach to the side. This stick is not used to strike an animal; it just gives the illusion of size. When I get in pretty close proximity to the cows, I start making a kind of *sshshsshsshh* noise. It gets their attention. When I want them to move, I change the noise to more of a *shhooo*, soft and low-pitched. I want them to move quietly, not run.

—JoAnn Pipkorn

One of the most important things we have now in the industry that we didn't used to have is the self-catching headlock. Have a swing gate on at least one side. It makes you willing to do some of those chores you're supposed to do.

—Rudy Erickson

One thing I always do, and it's something my Dad did, is whenever we have beef cattle out in the field, we always have a piece of machinery, like a wagon, in the field. I tell the boys if something starts coming after you and you can't get under the fence, then get under the machinery or on the other side of it. I keep the boys out of there when we have the bull. And you don't want kids running around in the pasture when the cattle are just getting outside in the spring and they're feeling goofy. You can never trust a bull, and you have to have an exit strategy.

—Randy Janke

Beef cattle raised outdoors on pasture and hay are naturally hardy animals that tend to have few health problems. When they do become ill or injured, it's important to identify the problem and treat it quickly, before it worsens and causes permanent damage to or kills the animal. As a cattle owner, you have a responsibility to take the necessary steps to prevent problems, to recognize abnormal behavior immediately, to treat minor health problems, and to call the veterinarian for conditions beyond your skill. Your cattle health program, then, should have three parts: prevention, diagnosis, and treatment.

KEEPING

BEEF CATTLE

HEALTHY

Prevention

Preventing your cattle from becoming sick or injured is much easier than dealing with a half-ton patient that has no interest in your nursing or a vet's ministrations. Start your cattle wellness program with these few simple actions.

Balanced Diet, Exercise, and Shelter

Like humans, cattle are healthiest when they get enough exercise and plenty of fresh air, are kept warm and dry in cold weather and cool in hot weather, and eat an adequate diet that includes the necessary vitamins and minerals. As previously discussed, they also need salt and clean water available at all times, although it doesn't hurt them to have to walk a distance in order to get to the water—that can be part of their prescribed exercise program. In addition, maintaining a low-stress environment and a regular routine will contribute a great deal to the overall health of the cattle.

Vaccinations

Along with good nutrition, water, and shelter, the best and cheapest insurance against cattle disease is vaccination. Every cattle owner should work with a veterinarian to develop a vaccination program appropriate for the ages and types of his or her cattle and for the region.

Types of Vaccinations

Every program should include vaccinations for calves and annual booster shots for adult cattle. In the past, combination vaccines were often formulated to address only diseases that were a problem in the region in which the cattle lived. Today, with cattle traveling so often and so far, it's standard to give all cattle a nine-way vaccine that covers bovine rhinotracheitis, viral diarrhea, parainfluenza, leptospirosis, and several other diseases.

A nine-way does not cover the clostridial diseases, which include tetanus, botulism, anthrax, "wooden tongue" (actinobacillosis), and blackleg. There are vaccines for these, but because they're hard on the animal and because those diseases aren't common in cattle in northern Wisconsin, where we live, they aren't usually recommended here. They are recommended, and sometimes required, in many other areas. Check with your

CATTLE DISEASES AND PEOPLE

Tuberculosis can be transmitted to people through raw milk from infected cows as well as through the air in poorly ventilated barns and sheds. Brucellosis, or Bang's disease, in cattle is transmitted to people through raw milk or during the delivery of calves from infected cows. Anthrax can be transmitted when people handle infected meat or hides, as can foot-and-mouth disease. Cattle owners can also acquire mange, ringworm, toxoplasmosis, leptospirosis, and tapeworms from their animals, though, thankfully, none of these afflictions is common in people today. All the same, it's a good idea to keep your cattle's vaccinations up to date and your premises clean and well ventilated.

veterinarian to determine whether any of the clostridial diseases are common in your area. If so, even if there's no legal requirement, it's worthwhile to vaccinate because these illnesses are difficult to treat and can kill animals in as few as twenty-four hours. It's also possible to vaccinate against rabies, but, again, this usually isn't done if rabies is not a problem in the area.

Brucellosis, also called Bang's disease, causes abortions and fever in cattle and *undulant fever* in people. Undulant fever is all but forgotten now in the United States but is still fairly common in other parts of the world. It's transmitted through contaminated meat and unpasteurized dairy products. Brucellosis vaccination is cheap insurance against abortions in your herd. The shot must be administered by a veterinarian when the heifer is between four and eleven months of age. After giving the vaccination, the veterinarian will attach a metal ear tag to the heifer's ear and officially record the shot.

When to Vaccinate

Calves often receive certain vaccinations soon after birth, but many beef cattle owners rely on the immunity that the babies get from their mothers' colostrum, or first milk, for the first few weeks and wait until it's nearly weaning time—at four to eight months of age—to vaccinate. These older calves should be given booster vaccinations two to four weeks after receiving their first vaccinations. Dairy and

orphan calves should be vaccinated by two weeks of age, because they aren't getting ongoing immune protection from their mothers' milk, and again at about six months.

Adult cattle should get annual booster vaccinations. Years ago, cows needed to receive their booster vaccinations before they were bred because some of the vaccines could cause abortions or birth defects. A new generation of vaccines now allows most cows to be vaccinated after they've been bred, but check with your veterinarian to make sure that these boosters are given at the proper time for your cattle and for your region.

Keep vaccines properly refrigerated and replace them once they pass their expiration dates. Buy your vaccines from your veterinarian, who will have been careful to keep them cool. Don't mix vaccines unless it says specifically on the label that it's safe to do so.

Injection Methods

Most vaccines are injected into the muscle. This is called an intramuscular (IM) injection, and it's the simplest type to give. Use a 2-inch, 16-gauge needle for adult animals and a 1½-inch, 18-gauge needle for calves. Don't give IM shots in the rump because this tends to make a permanent lesion in some of the most valuable meat.

Give injections in the neck. Slap the animal's neck a few times, pop the needle in quickly and deeply with a firm stroke, and depress the plunger. The animal will probably jump, so don't have your arm between him and any sort of bar, where you could get caught and end up with a broken limb. Change needles often because tough hides quickly dull the needles, and a dull needle hurts more and is harder to push in than a sharp one.

A subcutaneous (sub-Q) injection is given just under the skin and is used for many types of medications. Sub-Q injections take two hands and a few seconds longer than IM injections do.

DID YOU KNOW?

When a cow has "hardware disease," it means that it has eaten stray bits of metal, which have perforated its stomach, making it sick. As a preventive measure, some farmers use a balling gun to feed cow magnets to their cattle. These are cylindrical magnets about 5 inches long and an inch in diameter. The magnets lodge in the reticulum and attract all metal bits that may be there, keeping them from washing around and doing damage. Not all veterinarians believe in the effectiveness of using magnets, however, so it's better for you to harvest any metal pieces on your land before your cattle do.

CATTLE MEDICAL KIT

You should have a few basic supplies on hand for treating sick or injured animals.

- **Balling gun**—for getting pills down the throat
- **Rectal thermometer**—tie a long string around it when using; many thermometers have disappeared into cows because of unexpected muscle contractions
- **Rope halter and stout rope (a couple lengths)**—use the halter and rope for holding a head still in the headgate and use just the rope for pulling a calf during a tough delivery
- **Stomach tube**—for administering fluids and medications orally and relieving bloat if you're a long way from a veterinarian

- **Suturing needles and thread**—for stitching cuts
- **Syringes and needles**—for administering medications
- **Trocar or sharp knife**—for sticking bloated cows
- **Baking soda**—for easing stomach upsets in calves
- **Epsom salts**—for digestive upsets and soaking sore or infected feet
- **Iodine**—for treating wounds and for dipping navels on newborn calves
- **Topical antibiotic**—for treating pinkeye and other skin infections

You'll also need a loaded ear-tagger with extra tags and studs, a bottle of nine-way vaccine, a syringe gun, and a notebook and pencil for record keeping.

Making sure your arms aren't in a position in which they might get trapped by a plunging cow, grab a pinch of skin between your thumb and forefinger, use your other hand to quickly push the needle lengthwise into the bottom of the fold, and then depress the plunger.

The third type of injection, into a vein (intravenous), is used infrequently and should be done by a veterinarian or someone with experience.

The veterinarian can do all of your vaccinating for you, but most cattle owners learn to do it themselves. I use a handy device that I found at the farm store that's shaped like a pistol. It contains a big syringe that holds up to ten doses of vaccine. The calibrated trigger delivers exactly the right dose with each injection, and I can vaccinate a whole line of cattle without having to reload the syringe.

When vaccinating, change the needle every two or three animals and always use a new separate needle to draw the vaccine out of the bottle into the syringe so you don't contaminate the vaccine. Syringes can be reused if you clean them carefully with soap and hot water and dry them thoroughly. Needles can be cleaned, sharpened, and reused, but they're cheap, and it's generally easier to replace them.

Finally, set up a little table or some other clean and convenient place with the vaccine bottles, syringes, needles, record book, ear tags, and other equipment you'll need. Having everything handy and organized but out of the way of the cattle is easier—and safer—than trying to hold the ear-tagger in your teeth while storing the syringe behind your ear.

Internal Parasites

Many different types of worms like to live inside cattle. The most common parasites are roundworms, lung worms, liver worms, liver flukes, and pinworms. Most cattle owners treat with dewormer medication once or twice a year: either in early spring, before grazing starts; in late fall, after grazing is done for the season; or at both times. For the most effective worm control, discuss with your veterinarian the best time of year to deworm in your area and how to rotate deworming medicines for better results. Worm medicine is widely available at farm-supply stores and from your veterinarian. It is either poured along the back or injected.

Concern has risen about the increased resistance of internal cattle worms to available medications, and some veterinarians are recommending that cattle not be treated unless worms are really causing a problem. Worm infestation levels are calculated by taking a manure sample and examining it under a microscope for worm eggs, something that can be done only by your veterinarian (although you need to collect the sample). It's easier to first keep a close watch for external signs of a worm problem—weight loss and a dull coat in the summer (lacking the smooth, glossy shine).

Another common internal parasite is *cattle grub*, the immature stage of heel flies. They live inside cattle until they become adults and then emerge by drilling holes in the hide of their host. Timing of treatment is crucial to controlling cattle grubs and varies by region. Talk with your veterinarian.

Because internal worms are transmitted by eating grass infected with worm eggs from previous manure deposits, it's possible to reduce worm infestations by managing pasture rotations. Don't return cattle to a paddock until worm eggs deposited in the manure from their last rotation have had time to mature and die. For specific information on the life cycles of different worm species (which vary somewhat by moisture and temperature), consult your veterinarian or university extension service.

Some level of internal parasites is almost inevitable in pastured cattle, but it's usually not a problem unless the infestation is affecting a cow's general health and reducing her resistance to other diseases or slowing weight gain. Gear your treatment to both the nature and the extent of the problem.

Common flies are not true parasites, but they can be big nuisances.

External Pests

Several species of flies delight in tormenting cattle. Cows afflicted with horn flies, face flies, heel flies, or stable flies might run around with their tails in the air, trying to get away, or quit grazing and bunch together tightly, even on hot days, trying to reduce the amount of hide exposed to bites. When this happens, it's time to take action.

If your cattle are tame and you have time, you can spray them daily with a fly repellent made for animals. More economical are medicated ear tags or a repellent-soaked rope or post in the barnyard that the cattle will use for scratching. Inside sheds, hang flypaper or use a light trap or baited trap.

Taking preventive measures to destroy fly-breeding areas will save a lot of money in fly killer. Horn flies lay their eggs in fresh cow pies. If you develop a horn fly problem, try dragging paddocks with a harrow or any sort of homemade drag after grazing to break up the manure pats. Something as simple as an old bedspring behind a car or an ATV (all-terrain vehicle) will work. Stable flies, by contrast, breed best in the mix of old manure and hay that builds up in a ring around round bale feeders. Cleaning up feeding areas in the spring by removing the detritus will nip a lot of stable-fly infestations in the bud. To reduce other flies, in general, keep feeding and watering areas as dry and as free of manure as possible. Any cattle that have considerably more fly problems than the rest of your herd should be considered for culling.

Cattle can also suffer from ticks, lice, and mites. Cattle that spend a lot of time scratching themselves on feeders and fence posts probably have lice and should be treated with a spray or powder. Follow the application directions exactly, and don't mix lice treatments with other insecticides. Tick control comes in the form of sprays, dips, and dusts. Scabies mites cause small sores and scabby areas, and mange mites cause thick, wrinkled skin. Roundworm, common in the winter, causes bare patches on the hide. For these problems, talk to your veterinarian about establishing effective control and treatment.

Record Keeping and Animal Identification

Keeping a notebook to record vaccination, birth, weaning, breeding dates, and other important information for each animal in your herd helps keep your herd's wellness and breeding programs on time and on track. After a few years of record keeping, you'll be able to pick out which animals are healthiest, which breed back the quickest, and which raise the best calves. You'll also be able to cull the cows that are most susceptible to health problems, are poor performers, or have bad dispositions. The result will, I hope, be a herd of uniform, docile, healthy cows that are a pleasure to own and for whose calves you will always have buyers.

Record keeping also plays a crucial role in US animal-health initiatives. Outbreaks of foot-and-mouth disease in Britain, anthrax and tuberculosis in the United States, and bovine spongiform encephalopathy (mad cow disease) in several countries fueled a federal government initiative to establish the National Animal Identification System (NAIS). The hope is that by identifying each animal individually and tracking the animal's movements, any disease outbreak can be quickly traced to its origins and then suppressed. In my state, Wisconsin, all livestock owners are required to register their location and type of livestock with the state.

Many cattle owners already identify individual animals for their own records and breeding programs. The standard identification method for cattle is numbered ear tags. The tags and the tool for attaching them are available at farm supply stores. Get the large tags; the hairy ears of cattle make it hard to read the small ones. You can buy tags that are already numbered, or you can get a special marker and blank tags and do the numbering yourself. Attach the tag in the center of the ear when you've got the animal in the headgate for vaccination. Tags too close to the edge have a tendency to rip out. Make sure that the tag is on the front side of the ear and that the number is facing forward.

TO THEIR HEALTH

How do you spot a sick animal? With my cattle, it's really easy—if I take feed out and one is slow coming up, I know something's up. The best way is to check on them every day. I have it a little easier because they're close (to the house). If I walk out to the barn, they're talking to me. I keep a real close eye on them. The meek one, for example, will walk up to me, and I'll pat it on the forehead and I know it's feeling fine. If it doesn't do that, I know something's up.

—Randy Janke

When action is needed, your response may range from catching the animal for closer examination and treatment to calling the vet. When to call the vet is a matter of your comfort level, knowledge, and experience. This is where a knowledgeable and experienced neighbor can be very helpful. Such a neighbor can demonstrate the way to take a temperature, or inspect feet, or take one look and say, "Call the vet." If there is ever a question in your mind about whether to call or not to call, call! It's better to overreact than to lose an animal.

—Sherri Schulz, DVM

How to Tell When an Animal Is Sick

You can't identify a sick animal until you know how a healthy animal looks and behaves. That's why it's important to spend time on a routine basis just watching your herd. Become familiar with how your cows walk, lie down, get up, chew their cud, lick themselves, stretch, scratch, push each other around, and take care of their calves. This can be a delightful task. When our cows are in the paddock next to the house, there's nothing more pleasant than taking a cup of coffee out onto the deck and watching them. Summer evenings are especially nice, when it cools off and the calves get frisky. They'll race around the cows, butt heads, and buck. And sometimes the cows even come over and watch us.

Once you know what's normal for your cattle, you'll be better able to spot abnormal behavior or appearance. If you get into the habit of regularly taking a good look at each animal, you'll catch problems early. Early diagnosis and treatment is half the battle when it comes to curing a sick or injured animal. A cow that goes off by itself and is not grazing with the others is reason for concern. A cow that is not chewing its cud, spends an abnormal amount of time lying down, has inflamed eyes, shows no interest in its surroundings, or moves only with effort should be examined more closely. A cow that kicks at its belly, stands still with its back arched, has watery diarrhea, or looks like it's been pumped up with

an air compressor needs immediate attention. A calf with labored breathing should also be examined immediately.

What to Do When an Animal Is Sick

A good veterinarian, a manual of cattle ailments, and a basic medical kit (see "Cattle Medical Kit") are your lines of defense against problems. When you have an animal that is injured or not acting or looking normal, and it's not obvious what you should do, call your veterinarian for advice. You may be able to deal with the problem yourself, or the veterinarian may prefer to make a farm call. If the animal is walking, get it into a clean, bedded pen (or into the chute, if that's more appropriate) for observation and treatment.

As you gain experience with cattle, you'll learn to recognize and treat minor problems. With good preventive measures and good luck, the only time you'll need your veterinarian is once a year for brucellosis vaccinations and to fine-tune your cattle wellness program.

A Glance at Beef Cattle Afflictions
Acidosis

Acidosis is a bad stomachache that is caused by a significant and sudden increase in the grain ration for feeder cattle. Prevent it by increasing grain rations slowly and gradually over a period of a couple of weeks. If an animal is kicking at its belly, goes off its feed, or shows other signs of distress, call your veterinarian. Untreated acidosis can progress to fever, diarrhea, and laminitis, a dangerous inflammation of the hooves.

Birth Defects

Calves are occasionally born with any number of malformations, from extra toes to crooked legs. Their causes range from genetic defects to the cow's consumption of poisonous plants during pregnancy.

Occasionally, a calf will be born with hooves that knuckle under. You can wait for a couple of days to see if they straighten out. If they don't, the calf may have to be culled. For all other birth defects, consult your veterinarian about the calf's prognosis.

DID YOU KNOW?

In his 1974 book, *Farm Animal Behavior*, Andrew Fraser notes that no concrete evidence exists that adult cattle actually sleep. If they do, it is only for short periods. Normally, they will rest by standing or lying on their stomachs, although they will lie on their sides for short periods. If you see a cow on its side for longer than an hour, something may be wrong.

Bloat

Bloat is caused when gas builds up in the rumen because a mass of feed is blocking the exit to the esophagus, preventing the animal from belching. Bloat usually occurs when cattle are switched too rapidly to a very rich diet, such as young, lush pasture high in legumes or a lot of grain. Bloat can also occur after a frost or even as the result of heavy dew on rich pasture.

Bloat occurs rapidly and can kill cattle if the gas is not released. In the early stages, release of the gas may be possible by putting a tube down the animal's throat, but in the late stages it may be necessary to punch a hole in the rumen. If you see one of your herd looking like a swelled-up balloon, call your veterinarian immediately.

Colds

Cattle can catch head colds just as humans do, complete with a cough and runny nose. As long as the sick animal is eating well, breathing normally, and not showing any obvious signs of discomfort, you can allow the cold to run its course. But if symptoms worsen, call your veterinarian. Especially in young calves, a cold can turn into pneumonia.

Diarrhea (Scours)

Most often seen in young calves, scours is caused either by a bacterial infection or a virus. In older calves or feed-lot steers, it's often caused by a coccidiosis (bacterial) infection. Knowing the cause is not as important as treating the resulting dehydration, which is the primary killer of calves with scours. If you see a calf with watery diarrhea, call your veterinarian.

Foot Rot

Caused by a bacteria and most common where cattle are kept in muddy or wet areas, foot rot makes cattle lame. The area just above the hoof or between the toes may swell, and often the animal will lose its appetite. Untreated foot rot can affect joints and tendons and may cause permanent lameness. Foot rot is treated with antibiotics.

Hardware Disease

Cattle often consume some weird stuff along with their hay and pasture, including fencing nails, sticks, bits of wire, and other small metal objects. Usually the junk winds up in the animal's reticulum, the second chamber of the stomach, and causes no harm. Occasionally,

a sharp piece of metal will work its way into and even through the wall of the stomach, sometimes as far as the adjacent heart sac and diaphragm. This can cause infection and death.

Many times, a cattle owner will feed an animal a cow magnet (available at farm-supply stores) if he or she suspects hardware disease, hoping that any stray metal will stick to the magnet and stay in the stomach. If you notice that an animal has gone off its feed and is acting abnormally, hardware disease is a possibility. Call your veterinarian.

Laminitis (Founder)

Laminitis is a severe inflammation of the hoof that can cause the hoof wall to separate from the underlying structure of the hoof and lead to permanent lameness and deformity. Any animal displaying sudden severe lameness should be examined for laminitis.

The causes of laminitis range from the aforementioned acidosis to other shocks to the system, such as having to suddenly walk a long way over rough ground or taking a too-big drink of cold water. Fortunately, laminitis is uncommon in beef cattle, but if a case occurs among your cattle, call your veterinarian.

Lumpy Jaw

The fungus that causes lumpy jaw, actinomycosis, is naturally occurring in the environment and usually does no harm unless it gains entry to an animal's mouth tissues through a cut or sore. An animal with lumpy jaw will have a painful swelling on either the upper

Ear tags are the typical
way to identify cattle.

swelling, mattering (discharge), and watering of the eyes. Pinkeye is carried by face flies.

Most cattle recover from pinkeye without treatment within about three weeks, but some cases can cause permanent eye damage or even blindness. Serious cases can be treated with antibiotics on the advice of your veterinarian.

Pneumonia

Pneumonia is a respiratory disease that often follows a period of high stress or another illness that has lowered an animal's resistance. Symptoms are labored breathing, coughing, fever, and listlessness.

Pneumonia is a primary killer of young calves, and any calf showing symptoms should be immediately moved to somewhere warm and dry and given plenty of water to drink. Pneumonia affects all ages, and any sign of it should prompt a call to your veterinarian.

or lower jaw. As the swelling increases, chewing becomes painful, and the animal loses its appetite, which then results in weight loss. Treatment of lumpy jaw should be handled by a veterinarian.

Mastitis

A bacterial infection of the udder, mastitis causes abnormal-looking milk as well as a painful udder and sometimes fever. A nursing cow with an udder that is abnormally swollen on one side or one quarter should be examined for mastitis. Untreated mastitis can turn into a dangerous systemic infection. Prevent mastitis by keeping cows on clean bedding or pasture.

Navel Infection

When a newborn calf's navel is contaminated with feces or mud, a bacterial infection can result. Symptoms are swelling around the navel, followed by loss of appetite, diarrhea, and fever over the course of a few days or weeks. The infection may eventually affect the calf's joints or form abscesses in other parts of the body.

Treatment of navel infection is difficult, so prevention is the best policy. See that calves are born on clean bedding or pasture. If calving conditions are less than ideal, navels of newborn calves should be dipped in iodine as soon after birth as possible.

Pinkeye (Infectious Bovine Keratoconjunctivitis)

Pinkeye is a contagious and painful eye infection that causes

Poisoning (Toxic Plants)

There is a long list of plants found in pastures that are poisonous to cattle. Fortunately, cattle usually won't eat these toxic plants, but if you recently moved your cattle to a new pasture or started feeding them hay from a different source and some of them suddenly take sick, suspect plant poisoning.

Symptoms of poisoning vary widely, depending on the plant that caused it, and can run the gamut from constipation to diarrhea or from extreme nervousness to total collapse. Call your veterinarian if you suspect poisoning. Obtain a list of problem plants that are common in your region from your local agricultural extension agent and be sure to get those plants out of your pastures.

Tuberculosis

A respiratory disease, tuberculosis is contagious to people, and infected cattle must be destroyed.

Cattle infected with tuberculosis often show no symptoms at all, but watch for a chronic cough and labored breathing. Stay informed of any tuberculosis cases in your area through your veterinarian or your extension agent.

Warts

Cattle warts are most common on the head, neck, and udder but can appear anywhere on the body. Treatment generally isn't necessary or recommended, and most warts will eventually disappear on their own.

Raising your own calves is interesting and rewarding. You'll get to pick the parents according to color, build, temperament, and growth rate. You'll have the pleasure of seeing calves born, watching them grow up, and assessing how well you've done in achieving your goals. Then you get to do it all over again the next year!

BREEDING
BEEF CATTLE

Choosing Cows and Heifers

A beef cow's primary job is to raise a big, healthy calf every year. So, the first thing you need to find out about a cow is whether she calves easily and is a good mother.

Easy calving depends as much on the genetics of the bull as it does on those of the cow, so look for this trait when you are evaluating either parent. A heifer's mother should have a record of easy calvings, and reputable bull producers will have records going back several generations that can be correlated with performance records in breed directories. (Ideally, the previous owner will have kept good records.)

If you're looking at a heifer to buy, find out what kind of mother her mother is. A good mother is protective of her calves and diligent about cleaning them up and nursing them quickly after birth. A good mother always knows where her calf is located and produces plenty of milk. By weaning time, the calf is chubby, frisky, and a good grazer.

A cow also needs to be easy for you to deal with. A good mother is no good for you if she goes berserk in the chute or is in the habit of jumping fences. Look for cows or heifers that are calm, can be herded quietly, and don't take off for the next county when a stranger enters their pasture.

Get a cow or heifer that can physically handle the job of carrying and birthing calves. Look for wide-set pinbones as well as an udder that is well attached front and back and won't eventually sag so low that the calf will have to kneel to suckle. The cow should have reached puberty early for its breed, a sign of fertility, and be cycling regularly.

Finally, a cow or heifer should look nice, with that beefy boxcar body, a clean-cut head and neck, a wide muzzle, and calm eyes, and she should not be so fat that she waddles or so thin that her ribs show. Strong, straight legs and tidy hooves are signs that the cows will hold up for years of grazing and wandering around after calves.

Prioritize your wants. My primary concern is to get a heifer that's going to be easy to live with and won't have calving or milk problems. I'd also rather compromise on body type than on disposition. You'll have to decide for yourself what is most important to you.

Breeding Cows and Heifers

Breed your cows nine to nine and a half months before you want calves. A cow should start cycling, or going into heat, within three weeks to sixty days of having a calf and will cycle about every three weeks after that until being bred again. Many cows will breed back (conceive) on their first cycle, while some will take two cycles; those that take three or more or don't get bred at all should be considered for culling.

A heifer is usually bred for the first time in the summer after her first birthday, at about fifteen months of age. If a heifer is small for her age, you should give her a little longer or cull her. Slow growth is not a desirable trait in beef-cattle production.

Artificial Insemination

Artificial insemination (AI) allows you to pick from the very best bulls available nationwide, based on detailed statistical information provided by the AI company. In addition to the physical characteristics and growth records of bulls, the company's catalog will list the expected progeny differences (EPD) numbers. This is the statistical chance that a bull's progeny will be above or below the breed average for that characteristic.

If you don't dehorn as soon as the horn buds appear, you will have adult cattle with horns (unless you've chosen a polled breed.

run a high risk of incomplete castration, a rapidly increasing number of auction-barn buyers and feedlots pay a lower price for cattle castrated using either of these methods. A bull calf that still has a working testicle is still half a bull and is called a *stag*. That's what you get when the rubber band slips or you miss a spermatic cord with the crusher.

Dehorning

If you're going to dehorn, do it as soon as the horn buds appear. This is best done with an electric dehorner, which is held on the horn bud until it's burned to a crisp. The process is highly painful for the calf, much worse than castration. I strongly recommend having your veterinarian do this or getting your vet to teach you how to do this properly because if the growth ring isn't completely destroyed, your calf will grow misshapen horns. If you have polled, or naturally hornless, cattle, you won't have to deal with this chore at all.

ADVICE FROM THE FARM

BIRDS, BEES, AND AFTER

We run the bull year-round with the cows. We start with a new bull and turn it out with the cows in late August or early September. Each year, the calving date is moved up because the bull is with the cows and heifers year-round. Then, in the fall of the third year, we sell the bull and don't buy a new one until the following fall. The calving date is back to late spring and early summer, and we start over again, moving the calving date back.

—Mike Hanley

Everyone knows that bulls are dangerous and that a mother cow will protect its calf, but no one told me about heifers in heat. I was in the barnyard one day making a minor fence repair when some of the heifers wandered over to see what I was doing. This is pretty normal, so I said hello and went on with my work. As I turned away, a big heifer that must have been going into heat tried to mount me. If I had been just a little closer to the fence, I would've been squashed against it. As it was, I was knocked down but was able to scramble away. So don't turn your back on a bull or a heifer!

—Ann Hansen

We calve late now, in May. You have to be either ahead of the mud or behind it—if the cows drop the calves in 6 inches of mud, they're wet and cold.

—Rudy Erickson

The rare times a cow needs help calving, we put the cow in the corral and also put in a round bale and dig out the end of the bale for the calf. The cow and calf have shelter with the bale, plus feed from the bale, easy access to the headgate, and the water tank in the corral.

—Mike Hanley

A new mother is very unpredictable. The cow will chase anything, including you and the dog, so never turn your back.

—JoAnn Pipkorn

When a cow has a calf, she can be very protective. It's not the time to be sending a ten-year-old child to check what kind of calf it is. We had a cow that was just nuts when it calved every year; after three calves, we shipped it because we couldn't handle it.

—Rudy Erickson

Weaning Calves

Weaning involves separating the cows from the calves so the calves can't nurse anymore and the cows' udders dry up, giving them a rest and a chance to regain a little weight before calving again. Weaning is the most stressful thing that ever happens to a calf, and it's hard on the cows, too. Make it as gentle as possible.

The first rule is *not* to wean by loading the calves onto a truck and sending them off to the auction barn or feedlot. Although this method is still common, it inevitably results in a lot of sick, and a few dead, calves. Consequently, buyers have learned to pay less for "truck-weaned" calves.

With a small herd on a small farm, it's generally impractical to separate cows and calves by enough distance that they won't be able to hear each other bawling during weaning. And they will bawl. It may take as long as three days and nights of steady noise before they either get tired of it or get too hoarse to go on. There's nothing much you can do about it.

The most practical method for the small operator, which also tends to minimize the noise, is called "fence line" weaning. This involves splitting the cows from the calves and then putting the calves back into their familiar pasture and the cows in an adjacent pasture. This at least gives them the comfort of seeing each other, even if the calves can't nurse. If you do this, be aware that a single-wire electric fence will not keep the two groups apart. They need to be on either side of a sturdy, perimeter-type fence, and it doesn't hurt to have an electric wire as reinforcement. You'll also need to set up a separate water tank and a separate salt and mineral feeder for the calves. There will still be an amazing amount of noise, but the noise doesn't seem to last as long with this method.

Alternatively, you can move the calves into a pen with plenty of hay and water and put the cows back on pasture. The calves may take a little longer to settle down than if they were in their familiar pasture, but they eventually will. If you use this approach, remember that calves should not go directly from pasture to hay. Instead, gradually introduce hay into their diet well before weaning to allow their digestive systems time to adjust. Weaning and a sudden change in diet at the same time are a recipe for illness.

Buyers prefer that calves be weaned for a minimum of three weeks before shipping, so plan accordingly. Heifers or steers that you'll be keeping for your own herd need to be kept separate from their mothers for about three months, and heifers should be kept separate from the bull until they're ready to be bred.

Rebreeding Cows

New calves are left with the cows throughout the breeding season (they don't seem to notice their parents' amorous activities). Many producers leave the bull with the cows year-round because it results in a much happier bull, and cows generally begin cycling again about the time you'd turn a bull out anyway. Next year's calves may be a little earlier, but you can correct this when you change bulls by not having the new bull delivered until it's a little later in breeding season. If you're using AI, then the timing of the breeding is at your discretion, not the bull's.

As mentioned earlier, a bull that has bred your cows for two years should be sold to prevent him from breeding with its daughters. If you really like a bull, you may be able to board and breed the heifers at someone else's farm, perhaps in exchange for doing the same for their heifers, in order to keep the bull for one more year.

When it's time to sell the bull, you can look for a private buyer, ship him to an auction, or eat him. Bull meat is prized by some, although others think it's a little too flavorful.

WARM-WEATHER WEANING

I try to wean during the first cold snap in fall, when everyone's windows are closed, but it isn't always possible. If you wean in warm weather, when everyone's windows are open, you may not get a lot of sleep that first night, and you risk complaints from nearby neighbors. At any rate, the noise tapers off after a few days, and the cows and calves settle into the new routine within a couple of weeks.

It's time to say good-bye. The steers are fat, the calves are weaned, the cow has gotten too old to calve, or the bull needs to move on. You have four choices: sell at auction, sell to a grade and yield processor, have the animal processed locally for your own freezer or for sale directly from your farm, or sell to a neighbor who wants stocker calves. With feeder calves, you can also sell directly to a feedlot.

Selling at Auction

Sending animals to an auction barn for sale is probably the most common way for small producers to market their beef cattle. An estimated 85 percent of beef cattle in the United States are sold through a local auction barn at some point in their lives. These barns hold regular auctions for a variety of different ages and weights of animals, from special fall feeder-calf auctions to weekly auctions of fat cattle and cull cows.

To find the auction barns in your region, check your area farm paper, ask the neighbors, or call your extension agent. If you're fortunate enough to have a choice of two or three different barns, find out which type of cattle each barn specializes in, and keep an eye on their sale prices. You don't want to send a load of beef feeder calves to a barn specializing in dairy cattle because the buyers obviously won't be too interested in beef calves. If they were, they'd be at the auction barn having the feeder-calf sale. Somebody at a dairy barn will probably buy your feeder calves, but they won't have to bid nearly as high to get them, which means that you, the owner, lose money.

Once you've decided on an auction barn, call to get more information and instructions for selling your cattle there. Many auction barns have field representatives who will talk to you over the phone or even visit your farm to discuss the best time to sell cattle and how to prepare and transport them to obtain the best price. Ask if you should have the cattle there the day of the auction or the night before. Find out what the availability and charges are for penning, feeding, and watering your cattle before the sale.

Most types of beef cattle don't require any special preparation before an auction, except to make sure they are well watered and fed and are loaded calmly onto the truck. Feeder calves do require some special measures to obtain a top price. Buyers are especially interested in uniform groups of calves that will grow and finish at the same rate. Usually, buyers will pay a bit more for calves that are already vaccinated and weaned, have been started on a grain ration (also called *bunker broke*), and have been knife-castrated.

This is a very important point, so I'm going to repeat it: calves that are to be sold as feeders in the fall, whether directly to a feedlot or at auction, should have been weaned for a minimum of three weeks, should know how to eat grain from a bunker feeder, and should have been knife-castrated. They also should have been vaccinated in the neck and boostered two to four weeks later, again in the neck. Heifers that are to be sold as breeding stock should have their brucellosis vaccinations.

Some auction barns have programs in which the cattle owner or the owner's veterinarian signs a document stating that these procedures have been done. This will be mentioned during the auction and usually affects the sale price. Yet it's not all about money. When you've preconditioned your calves, you know that their sale and move to a new home will be much easier on them. Calves that are preconditioned have lower rates of illness and death compared with calves that are not. The former are less stressed and are more likely to settle into their new homes quickly and to gain weight rapidly. Those calves build your reputation among buyers.

Grade and Yield

Grade and yield buyers are just what the name implies: they pay according to the carcass grade (prime, choice, select, standard, commercial, utility) and the amount of meat it yields. In other words, they pay so much per grade and per pound. USDA quality grades are a subjective measure of the meat-palatability traits of flavor, juiciness, and tenderness. A grade and yield operation can be a good place to send cull cows, which often

To sell beef from your farm, you must have it butchered by a qualified processor.

bring a little more sold by grade and yield than they would if sold at a regular auction barn. Just as with auction barns, asking around is the best way to find grade and yield operators.

Getting Your Beef Processed

Decades ago, every neighborhood where beef cattle were raised had a small-scale processing plant. Today, if you're lucky, there will still be one in your area. If you're not, you may have to drive a distance. Small-scale plants are completely different from the huge meat-packing plants with their assembly-line cattle processing. At a small plant, each animal is handled individually, and the plant owners and workers are often your neighbors.

Find a reputable plant by asking other beef producers their opinions. A reputation for making good sausage is a plus. You should then pay a visit. Don't go in the morning, when most of the heavy work is being done and the staff may not have time to talk. Instead, head over in mid-afternoon, when it's less hectic. The plant should look and smell clean. Talk to the manager, who is often also the owner. Most managers are willing to deal with beginners and will take the time to explain the entire process, from loading your steer onto a truck to picking up the meat. Ask how long it takes from the time the animal arrives at the plant to when the animal gets slaughtered; if it's going to be a few hours, ask whether they make water available to the steer.

Ask about the processing charges. Most plants charge a per-pound processing fee, with extra charges for fancy butchering or sausage-making. If the plant picks up the steer from the farm, then there will also be a trucking charge.

Once you find a processor and make arrangements for your steer, you'll be asked to decide how you want the meat butchered and packaged. Typically, you'll need to tell them how thick to cut steaks, how many to put in a package, how big you want the roasts, and whether you want all of the round steaks and ribs or you want these ground for hamburger. You can specify that the heart, liver, and tongue be saved or discarded and that the tail be saved, if you're fond of oxtail soup.

There will be a lot of meat. A typical 1,000-pound steer with good beef genetics will generally dress out to around 60 percent or better; dairy steers yield somewhat less. Sixty percent translates to 350-plus pounds of meat, an amount that will cram a large chest freezer right up to the top. If you don't have room in your own freezer, most processing plants will rent you locker storage space for a very reasonable fee.

If there is no processing plant within a reasonable distance from your farm, you could check into on-farm butchering services. These operators will come to your farm, dispatch your steer, and do the initial skinning and dressing. You may have to do the packaging yourself, or you may be able to pay a little more and have them do it.

Selling Beef from the Farm

Selling processed beef directly from your farm is known as *direct marketing*, and it used to be how most rural residents bought their beef for the year. Traditionally, beef is sold in wholes, halves, and quarters, referring to the proportion of the animal sold to an individual buyer. For example, when a customer buys a half of beef, they get all the meat that came from that half of the steer, from burger to steak. Processors are familiar with this method and will divide up the meat for you if you ask ahead of time. If you're selling quarters, you can sell by front quarter (less valuable), back quarter (more valuable, with lots of steaks), or mixed quarters. With mixed quarters, you take half a carcass and evenly divide the types of meat into two portions. That way, each buyer gets the same amount of steaks, roasts, and burger. There aren't a lot of good steaks on a front quarter.

GRADING BEEF

In addition to being graded by quality, beef can be graded by yield. The *yield grade* simply refers to the percentage of the total carcass that is usable meat, and it runs from a 1, the highest yield of edible meat, to a 5, the lowest yield. In general, a good-quality beef steer will yield about 40–50 percent of its total body weight in usable meat.

Another term you may run across is *carcass weight*, the weight after the animal has been slaughtered and the head, hide, feet, organs, and digestive tract are removed. Carcass weight is also called *dressed weight* or *hanging weight*.

Removing the bones and excess fat from the carcass gives you the *cutting yield*, which is expressed as a percentage of the carcass weight, not of the live weight. For example, let's say a 1,200-pound steer yields a hanging, or carcass, weight of 725 pounds. If the cutting yield is in the 60–70 percent range, as it should be for a good beef animal, the amount of meat you'll take home in little white packages will be 440–520 pounds. So a 60-percent cutting yield is the same as 40 percent of the live weight for this animal.

Some direct marketers sell by the piece or by marketing each cut individually. This method means you'll probably sell all the steaks and hamburger quickly and have a lot of roasts and round steaks left. It does attract buyers who don't have the freezer space for an entire quarter as well as buyers who would like to try the meat before committing to a larger purchase. If you have a lot of buyers interested in hamburger, you can have cull cows made into terrific burger instead of sending them to a grade and yield operator or to auction.

Direct meat sales are regulated by the federal government. It's illegal to sell meat across state lines unless it's been processed in a federally inspected plant. The majority of meat-processing plants are state inspected, not federally inspected, so this is a major difficulty for direct marketers living near a state line with a good potential market on the other side. Some have solved this by having their customers drive to the farm to pay for and pick up their meat, while others truck cattle long distances to a federally inspected plant.

Finding buyers for home-grown, grain-fed beef is usually no problem. Relatives, friends, neighbors, and coworkers are all potential customers. But if you sell tough or gamey-tasting beef, you won't get a lot of repeat business. Begin by raising and eating your beef yourself. When you're sure of your beef's quality, then it's time to contact people and ask whether they're interested in buying your beef.

Direct-marketed beef can be priced by live weight, hanging weight, or dressed weight (see "Grading Beef"). This can be very confusing to customers. The easiest method is to charge by the weight of the meat that the customer actually takes home to the freezer. With wholes, halves, and quarters, generally one per-pound price is charged, regardless of the type of meat. This means that on a per-pound basis, the burger is expensive and the steaks cheap; but averaged across the entire purchase, the price is fair. When selling by the piece, the per-pound price varies by whether it's steak, roast, or burger, just as in a grocery store. Many direct marketers add a per-pound processing charge to the price.

Some processors will let your customers pick up their meat directly from the processing plant, while others prefer that you take the meat home, where you can either have your customers pick it up at your place or deliver to them. Specify the date and time of pickup or delivery to avoid having big heaps of meat thawing in your garage, waiting to be picked up or delivered to customers. Advance scheduling enables you to get the meat from the plant to your customers immediately.

Transport and Shrink

Unless you own your own cattle trailer and pickup truck, you'll need to hire someone to move your cattle from the farm to the auction barn or processing plant. The auction barn or processor will be able to put you in contact with truckers who work in your area, or a neighbor may have a trailer and be willing to do the hauling for you. If you're a member of a beef producers' association and participating in a group sale, trucking arrangements are often included as part of the sale. Some small processors may pick up cattle as part of their services.

Set a date and time for the trucker to show up, anticipating some slack in the schedule. Truckers often make stops at several farms to fill a trailer, and, once in a while, somebody has forgotten to pen the cattle or a steer hops a fence. Have your animals in the pen, or even in the crowding tub, before the trucker arrives.

As previously discussed, traveling is stressful for cattle, and they show it by losing weight. Some of the weight is lost in manure or urine, but most is lost by dehydration. This loss is called *shrink*, and the farther cattle are hauled, the more shrink they will experience. Because most cattle are sold by the pound, shrink reduces their price. To reduce the amount of shrink, see that your cattle get a good drink before being loaded and that they are loaded calmly. Hiring a calm trucker helps a lot. A trailer with good footing, so they don't slip, also goes a long way toward reducing stress.

Truckers generally charge by miles traveled divided by the number of cattle being hauled. Trucking charges add up quickly, especially when gas prices are high, so it's a good idea to coordinate cattle shipments with neighbors. Often, the trucker will be able to help with this if you call him far enough in advance. A cattle trucker usually keeps a list of people with cattle to ship, and when he's got a full load, he can give you a call and set the date.

PART TWO

CHICKENS

BY SUE WEAVER

WHY CHICKENS?

In the early to mid-twentieth century, throughout the countryside and in cities large and small, backyard chicken coops were the norm. Chickens furnished table meat and eggs, and most everyone kept at least a few hens. Years passed and attitudes shifted; small-scale chicken keeping became gauche. By the end of the twentieth century, while agri-biz egg and meat producers, immigrants, rustics, and hippies were keeping chickens, urban dwellers and

SELECTING YOUR

CHICKENS

The times they are a-changin' once again. As our world becomes increasingly frenetic, violent, and stressful, a burgeoning number of Americans are seeking a quieter existence. "We'll move to the country," some decide. "We'll live on a small farm and commute or work from home!" "We'll garden!" "We'll have chickens!"

Nowadays, from Minneapolis to New Orleans, from Los Angeles to New York City and all points in between, throngs of city and suburban residents raise and praise the chicken. A few miles farther out, more hobby farmers are likely to raise chickens than any other farmyard bird or beast. Hens are the critter du jour.

Why keep chickens? For their eggs, of course, and (for those who eat them) their healthier-than-red-meat flesh, whether strictly for your own table or for profit as well. Chickens are easy to care for, and you needn't break the bank to buy, house, and feed them. You may also wish, like many hobby farmers, to keep livestock for fun and relaxation. Surprisingly, chickens make unique, affectionate pets. They offer a link to gentler times; they're good for the soul. It's relaxing (and fascinating) to hunker down and observe them.

This information is meant to educate and entertain rookie and chicken maven alike. Are you with me? Then let's talk chickens!

Chicken Classifications

Early on, the American Poultry Association (APA) devised a system for classifying chickens by breed, variety, class, and sometimes strain. A *breed* is a group of birds sharing common physical features such as shape, skin color, number of toes, feathered or nonfeathered shanks, and ancestry. A *variety* is a group within a breed that shares minor differences, such as color, comb type, the presence of a feather beard or muff, and so on. A *class* is a collection of breeds that originate from the same geographic region.

The APA currently recognizes twelve classes: American, Asiatic, English, Mediterranean, Continental, All Other Standard Breeds, Single Comb Clean Legged, Rose Comb Clean Legged, All Other Clean Legged, Feather Legged, Modern Game Bantam, and Game Bantam. A *strain*, when present, is a group within a variety that has been developed by a breeder or organization for a specific purpose, such as improved rapid weight gain and prolific egg production. Chickens may also be classified as light or heavy breeds or as layers, meat, dual-purpose, or ornamental fowl.

When the APA published its first Standard of Perfection in 1874, the following chickens were recognized: Barred Plymouth Rocks, light and dark Brahmas, all of the Cochins and Dorkings, a quartet of Single-Comb Leghorns (dark brown, light brown, white, and black), Spanish, Blue Andalusians, all of the Hamburgs, four varieties of Polish (white crested black, nonbearded golden, nonbearded silver, and nonbearded white), Mottled Houdans, Crevecoeurs, La Fleches, all of the modern games, Sultans, Frizzles, and Japanese Bantams. Today, close to 120 breeds and more than 350 combinations of breeds and varieties are described.

Bantams are one-fifth to one-quarter of the size of regular chickens. They come in sized-down versions of most large fowl breeds, although they aren't scale miniatures: their heads, wings, tails, and feather sizes are disproportionally larger than those of their full-size brethren. A few Bantam breeds have no full-size counterparts. In addition to being cute, bantams can be shown, they make charming pets, and their eggs and bodies—small as they are—make mighty fine eating. The APA issues a standard for bantams, as does the American Bantam Association (these standards don't always agree).

The Turken is known as the "Naked Neck" for a reason!

Which Chickens Are Best for You?

There are countless varieties and hundreds of breeds from which to choose, and it's important to pick the ones that will meet your needs. With the passage of time, humans have designed chickens to fill every niche: cold-hardy chickens, heat-resistant chickens, chickens that don't mind being penned up. We haven't designed the perfect chicken—yet! All breeds have certain shortcomings. Furthermore, a breed that would be a bad choice for one chicken keeper (such as hens meant to be confined who can fly out of enclosures) would be perfect for another (as free-ranging chickens, those flying hens would be able to evade dogs).

Before you can settle on the kind of chickens to buy, you need to determine what purpose they'll serve and what environment they'll live in. Do you want them for their eggs? Sunday dinner? Feathery companionship? Will they spend most of their time inside or out? Will they have to contend with sweltering summer days or frigid winter nights? All of these factors make a difference in your choice of breed.

Next, you must decide whether you want day-old chicks or full-grown birds, as well as how many of them to get. What advantages are there to buying a pullet rather than a chick? Is it better to start with a small flock? If you haven't already done so, you should find out what zoning laws may apply to your keeping chickens and how they affect your decision. Do you need birds on the quiet side?

Ask yourself the following types of questions:

- Will your birds be sequestered in a chicken house, or do you favor free-range hens? Certain breeds don't like being confined. A cramped coop of ornery Sumatras is a disaster waiting to happen, and find-your-own-feed Cochins might starve.
- How much room do you have to devote to chickens? A few bantams can thrive in a doghouse. A dozen 10-pound Jersey Giants? They'll need a heap more space.
- Are your neighbors close by? Squawking, freedom-craving, fence-flying breeds likely won't do. This is especially true if you live in the city or suburbs.

A MATTER OF BREEDING

Some of today's purebred fowl (chickens whose parents are of the same breed), such as the gamecock breeds, trace their roots to the distant mists of antiquity. Egypt's elegant Fayoumi dates to before the birth of Christ. Stubby-legged, five-toed Dorkings came to Britain with the Romans. Squirrel-tailed Japanese Chabo Bantams, miniature chickens weighing between 1 and 3 pounds, emerged in the seventh century AD. Dutch Barnevelders were developed in the 1200s, about the time that Venetian merchant Marco Polo wrote of the "fur-covered hens" (Silkies) of Cathay. Another Dutch chicken, the deceptively named Hamburg, has existed since the late 1600s and is likely far older than that. The crested fancy fowl we call the Polish was developed even earlier, and France's V-combed La Fleche dates to 1660 AD. Naked Necks, also called Turkens (possibly the weirdest-looking chicken of them all), originated in Transylvania before the 1700s. The first all-American fowl, the Dominique, is an early nineteenth-century New England utility fowl.

NAME THAT COMB

The fleshy protuberance atop a chicken's skull is called its *comb*. Roosters' combs are larger than those of same-breed hens. The American Poultry Association recognizes eight basic types:

Buttercup		Cup-shaped with evenly spaced points surrounding the rim.
Cushion		Low, compact, and smooth, with no spikes.
Pea		Medium-low with three lengthwise ridges. The center ridge is slightly higher than the ones that flank it.
Rose		Solid, broad, nearly flat on top, low, fleshy, and ending in a spike. The top of the comb is dotted with small protuberances.
Single		Thin and fleshy, with four or five points, extending from the beak to the full length of the head.
Strawberry		Compact and egg-shaped, with the larger part toward the front of the skull and the rear part no farther than its midpoint.
V-shaped		Two hornlike sections connected at their base.
Walnut		Resembles half of a walnut.

The fancy-feathered Polish is a favorite pet breed.

- Are there toddlers in your family? Testy roosters of certain breeds can injure an unwary tot.
- Do winter temperatures plummet below zero where you live? Roosters with huge single combs get frostbite easily, and some breeds simply won't thrive in this type of weather.
- Are you in a region with hot temperatures? Fiery summer heat wilts heavy, soft-feathered breeds such as Cochins, Australorps, and Orpingtons, while other breeds take heat more in stride.
- Can you keep your top-knotted, feather-legged friends confined when the weather turns bad? Mud, slush, and fancy-feathered fowl usually don't mix.
- Would you like to preserve a smidge of living history and raise old-fashioned or endangered breeds?
- Finally, if your chicken is a pet, will you keep it outdoors with the rest of the chickens or as a household pet?

Although we can't tell you exactly which breed to buy, we can offer general advice and suggest birds that will meet certain criteria.

Chickens for Eggs or Meat

Birds with the greatest egg-laying capacity are not the same as those who plump up into the best candidates for the local chicken fry. Still different are those chickens that are the best choices for providing both eggs and meat.

DID YOU KNOW?

- The genus name for chicken-like fowl is *Gallus*, which means "comb."
- Insulating a rooster's comb with a layer of petroleum jelly during extremely cold weather usually prevents freezing.
- The large single combs of the hens of certain breeds flop over in a jaunty manner instead of standing up like those of roosters.
- Chickens recognize some colors and are attracted to red combs. However, Silkies, Sumatras, and several varieties of game fowl have purple combs, and Sebrights' combs are deep reddish-purple.

If you want eggs—and a whole lot of them—Mediterranean-breed chickens are just your thing. Small, squawky, and hyperactive, these birds mature quickly, and then everything they eat goes into laying eggs. Undisputed queens of the nesting box are white Leghorns and hybrid layers based on this breed. Other impressive Mediterranean-class layers are the Minorca, Ancona, Buttercup, Andalusian, and Spanish White Face.

Some chickens from other classes are laying machines, too. The Campine (Belgium), Fayoumi (Egypt), Lakenvelder (Germany), and Hamburg (Continental Europe) are popular examples. Like their Mediterranean sisters, they tend to be flighty, specialist hens.

Meat chickens (called *broilers* or *fryers*)—usually White Cornish and White Plymouth Rock hybrids—have broad, meaty breasts and white feathers, and they mature at lightning speed. Broilers are ready for the freezer in about seven weeks, and *roasters* (which are just larger broilers) are ready in just three more.

Be aware that because they're hybrids, these birds don't breed true—meaning that their chicks won't possess these stellar features. They also require careful handling; because of their abnormally wide breasts and rapid growth patterns, most become crippled as they mature.

Dual-purpose breeds lay fewer eggs than superlayers and mature a lot more slowly than meat hybrids, but they're ideal all-around hobby-farm birds. They're quieter, gentler, and friendlier than the specialists, and they're hardy and self-reliant to boot. They

There are not many things cuter than fluffy yellow chicks.

a year. If a rooster fertilizes their eggs, and you allow it, most dual-purpose biddies will hatch chicks. Some ornamental breeds are friendly and lay well, too. But avoid flighty, sometimes pugnacious hybrid superlayers and breeds from the Mediterranean class. They don't want to be your friend; they just want to lay eggs. Choose something a tad more laid back.

Chicken Little or Big Bird?

Once you've chosen a breed, you'll have to decide: chicks or full-grown birds? In most cases, the correct answer is chicks. In addition to getting the most for your fowl-shopping dollar, you'll know exactly how old they are. Plus, when purchased from reliable sources, chicks are nearly always healthy.

The Little Guys

Order day-old chicks from commercial or specialty hatcheries. The former sell dozens, sometimes hundreds, of breeds and varieties of quality chicks at modest prices. For most of us, this is the logical way to fly. Specialty hatcheries are run by knowledgeable poultry aficionados who specialize in specific sorts of fowl. You'll pay more at a specialty hatchery, but if you want to show chickens or to one day breed show-quality fowl, paying extra for specialty-hatchery chicks is the way to go.

are broody, so hens will set and hatch their own replacements. Nearly all lay brown eggs and are meaty enough to eat, should you wish to do so.

With a few notable exceptions, dual-purpose birds hail from the English and American classes. There are scores of interesting breeds and varieties.

Chickens as Pets

Do chickens make good pets? Absolutely! They're smart and affectionate, and a chicken costs little to maintain. You can teach your chicken to do tricks—it'll sit on your lap, and it may even sing if it likes you a lot. You don't need a lot of space to keep a chicken. It won't bark at the neighbors while you're at work. You can raise it from a peep for just a few dollars. All in all, a chicken makes a mighty fine friend. You can even take it along when you run errands; a chicken in your car turns heads!

If pets are your pleasure, but you don't plan to handle them, almost any sort of fowl will do. If you want pet chickens that are tame, that's another proposition.

Some breeds are rowdy, antisocial, and just not much fun to have around; others are downright cuddly. You want to choose pets from the latter group. Silkies, Cochins, Brahmas, Naked Necks, and Belgian d'Uccles, for example, are easy to tame and make quiet, affectionate, companion chickens. Flighty Leghorns and their ilk can be tamed—but it takes a lot more time and effort.

If you'd like eggs from your pets, that narrows the equation. Not all hens lay scads of eggs. However, most young hens of the generally calm and amiable old-fashioned, dual-purpose breeds crank out one hundred to two hundred (or more) tasty brown cackleberries (eggs)

MAKING A CHICKEN A PET

When you brood your next batch of chicks, pick one to hand tame. Carefully pull her out of the brooder for short periods every day. Cup her between your hands and hold her near your face. Speak gently for a minute or two and then put her back. If you work with her, she'll bond with you. By the time she leaves the brooder, she'll be your chick.

To domesticate an older bird, work quietly and carefully. Hold her securely so she can't flop. Stroke her wattles—chickens like that—and offer her goodies, such as bits of fruit or veggies. It won't be long until she's tame!

WHICH BREED?

Breeds most likely to make great pets: Barnevelder, Belgian d'Uccle, Cochin, Dorking, Jersey Giant, Naked Neck (Turken), Orpington, Polish, Plymouth Rock, Silkie, Sussex

Other easygoing, friendly breeds: Ameraucana, Araucana (usually), Aseel (cocks are aggressive toward one another), Brahma, Dominique, Faverolles, Java, Langshan, Sultan, Welsummer, Wyandotte (usually)

Cold-hardy breeds: Araucana, Ameraucana, Aseel, Australorp, Brahama, Buckeye, Chantecler, Cochin, Dominique, Faverolles, Hamburg, Java, Jersey Giant, Langshan, Old English Game (dubbed), Orpington, Rosecomb, Silkie, Sussex, Welsummer, Wyandotte

Breeds prone to frostbitten combs: Andalusian, Campine, Dorking, Leghorn, New Hampshire Red, New Hampshire White, Rhode Island Red (Roosters are more likely than hens to suffer frostbite; their combs are larger, and they don't tuck their heads under their wings while sleeping as hens do.)

Heat-tolerant breeds: Andalusian, Aseel, Brahma, Buttercup, Cubalaya, Fayoumi, Leghorn, Minorca, Modern Game, New Hampshire Red, Rhode Island Red, Rosecomb, Silkie, Spanish White Faced, Sumatra

Flying breeds: Ancona, Andalusian, Campine, Fayoumi, Hamburg, Lakenvelder, Leghorn, Rosecomb, Sebright, nearly all bantams

Noisy breeds: Andalusian, Cornish, Cubalaya, Leghorn, Modern Game, Old English Game

Flighty breeds: Ancona, Andalusian, Buttercup, Fayoumi, Hamburg, La Fleche, Lakenvelder, Leghorn, Minorca, Sebright, Spanish White Faced

Aggressive breeds: Ancona, Aseel (cocks), Old English Game, Cornish (cocks), Rhode Island Red (cocks), Cubalaya, Modern Game, Rhode Island Red (some strains), Sumatra, Wyandotte (some strains)

Self-reliant breeds (good foragers, ideal free-range chickens): Andalusian, Australorp, Belgian d'Uccle, Buckeye, Buttercup, Campine, Chantecler, Dominique, Fayoumi, Hamburg, Houdan, Java, La Fleche, Lakenvelder, Marans, Minorca, New Hampshire Red, Old English Game, Orpington, Plymouth Rock, Rosecomb, Sebright, Silkie, Sussex, Naked Neck (Turken), Welsummer, Wyandotte (Note: Avoid all-white individuals; they're more easily spotted by predators than colored and patterned varieties of the same breeds.)

Breeds that tolerate confinement reasonably well: Araucana, Ameraucana, Australorp, Barnevelder, Brahma, Buckeye, Cochin, Cornish, Crevecoeur, Dominique, Dorking, Faverolles, Houdan, Java, Jersey Giant, La Fleche, Lakenvelder, Langshan, Leghorn, Naked Neck (Turken), New Hampshire Red, Orpington, Plymouth Rock, Polish, Rhode Island Red, Silkie, Sultan, Sussex, Welsummer, Wyandotte

Breeds that don't tolerate confinement well: Ancona, Andalusian, Buttercup, Cubalaya, Fayoumi, Hamburg, Malay, Minorca, Modern Game, Old English Game, Spanish White Faced, Sumatra

---

A newly hatched chick can live for three days without food and water, subsisting solely on nutrients absorbed from its egg. Therefore, you can purchase chicks from hatcheries on the other side of the country, and—shipped overnight air—they should arrive safely at your nearest post office without a hitch. However, sometimes a chick does die in transit; thus, it's wise to order from the closest responsible source so that your chicks needn't travel farther than necessary. Some hatcheries will replace chicks that are dead on arrival, but others won't. Read the seller's guarantee before ordering chicks. If the service is available, pay to have your chicks vaccinated for Marek's disease. This can only be done when they're newly hatched, meaning it's now or never, and it's better to be safe than sad.

Be aware that you can't mail-order five or six chicks. For the birds to stay warm enough in transit, a certain number of bodies must be in the shipping box, generating heat. It generally takes about twenty-five large-fowl chicks or twenty-five to thirty-five bantams to do the trick. Some hatcheries allow you to order Guinea keets or other similar-size hatchlings to fill the quota. You can also find others interested in buying a few chicks and place a combined order that will be shipped to one address.

If you don't want to deal with roosters, buy sexed pullets. Straight-run chicks (an equal mixture of males and females) are cheaper, but at least half will be cockerels. If you can raise and butcher the excess roosters, fine. Otherwise, buy just two or three sexed "roos" to add to the mix—or buy none at all. Hens don't need roosters to lay eggs.

It's not unusual to find a chicken in a suburban backyard.

The Big Guys

If you don't want to deal with tiny chicks, you might be able to buy sixteen- to twenty-two-week-old, almost-ready-to-lay females called *started pullets*. Initially, they cost more per bird, but you won't have the expense of brooding them and feeding them for months, so they can actually be a great buy.

Breeders will sometimes part with a few hens or a breeding trio (a cock and two hens), and you can often find chickens for sale at country flea markets, poultry swap meets, or via classified and bulletin board ads. However, buying adult chickens can be risky. Not all sellers are honest, and it's easy to buy someone else's problem hens.

Ideally, you should buy fowl only from flocks enrolled in the USDA's National Poultry Improvement Plan (NPIP). These birds are certified free of pullorum (a severe, diarrheal disease) and typhoid and are healthier than your run-of-the-mill chickens. Choose active, alert, clear-eyed chickens with smooth, glossy feathers and bright, fleshy, waxy combs and wattles. Refuse birds that cough, wheeze, or have discharge or diarrhea. Tip the chicken forward and scope out the area around its vent, and also check under its wings. If you spy insects or eggs and you don't want to deal with parasites, you'd best not buy the bird.

If you want eggs or plan on eating the chickens, you must buy young ones. Young adults have smooth shanks; older birds' shanks are dry and scaly, and their skin is thick and tough. Cockerels have wee nubs where their spurs will grow, and some pullets have them, too; long spurs denote an older bird. Press on the chicken's breastbone; a youngster's is flexible, while old chickens have rigid breastbones.

Before your chicks arrive, assemble everything you'll need to feed, water, and brood them (*brood* means to keep them warm inside a heated enclosure). Have the brooder box ready and waiting.

Plan to be home the day your chicks are scheduled to arrive. In most cases, they won't be delivered to your door; someone from the post office will call you to pick them up. When you arrive for the delivery, open the box of chicks in the presence of a postal worker who can verify your claim should any of them be dead. Then rush your new birds straight home to a cozy brooder box, water, and feed. Don't take side trips with your chicks in tow.

When you get them home, remove the chicks from their shipping box one by one and examine them. If a chick has *pasty butt* (an affliction where crusty, dried droppings block a chick's vent, making it impossible for the bird to eliminate), gently wash its little behind with a soft cloth dampened in warm water. This problem is common with mail-order chicks, especially in their first five or six days after arrival.

Check the toes. When caught early, crooked or curled toes can be splinted using wooden match sticks and strips of adhesive bandage snipped to size. Some straighten, some don't, but you won't know unless you try! If a chick looks normal, dip its beak in water so the chick knows where the water is and starts drinking, and then place the bird gently under the heat source.

Feed stores frequently offer day-old chicks for sale. Breed selection maybe limited, and feed-store chicks aren't often sexed. However, you can choose the ones you want, buy just a few, and get them home quickly. Select bright-eyed, active chicks with straight shanks, toes, and beaks as well as clean, unobstructed bottoms.

How Many Chickens?

If you're new at chicken keeping, don't overextend yourself. Start small and learn as you go. The downside to this advice is that adding new birds to an established flock upsets its pecking order and spawns stress. Overall, though, it's better for your birds to hatch out a new hierarchy than for you to bite off more than you can chew.

By the same token, if you're experienced or you're certain about how many you want to keep, you'll save your chickens a lot of stressful infighting—and possibly disease—by buying all the birds you need up front and then maintaining a closed flock (meaning that you don't add

Proceed with caution if acquiring adult hens.

new adult chickens to an established flock) until you start back at square one again.

Unless you can spend a lot of quality time with a pet chicken (as you might with a house chicken), buy at least two. Chickens are sociable birds; a solitary cock or hen will be lonely.

You should also buy at least one layer hen per family member—or more if your family eats a lot of eggs or if you choose a dual-purpose breed. If you plan to maintain a closed flock, you should allow for several years' flock mortality. To do this, purchase 10–20 percent more chicks than you initially think you'll need, but don't buy more birds than you can properly house.

City Chickens

If you live in the suburbs or in a city, you can still probably keep a few hens. True, there are limits and stipulations, and some of them are strict, but chicken fanciers across the land in cities as diverse as New York City, Los Angeles, Minneapolis, St. Louis, Seattle, Des Moines, and Fort Worth keep city chickens. Chances are, your city or town allows them, too.

If you're looking for a way to justify raising chickens, look no further. Here are seven very good reasons to raise city chickens:

1. Bring a sense of country to the city and a touch of the past to our busy, modern lives. Slow down. Watch your chickens scratching in the dirt, loping after bugs, being chickens. Kick back and dream of less hectic times.

2. The eggs! The yummy, fresh-from-the-hen eggs, with yolks so rich that your mouth will water. Eggs to bake with. Eggs to share with friends. And you don't need a huge flock to supply them. Three or four young hens keep the average family supplied with tasty cackleberries, often with some to spare. And there's a bonus: research indicates that chickens allowed to roam freely and nosh on grass and bugs lay eggs that are higher in omega-3 fatty acids and vitamin E. They're lower in cholesterol than commercial eggs, too.

3. What to do with the leftover salad or last night's broccoli with cheese? It's bonus feed for the chickens, of course! Chickens safely and happily devour almost anything except large portions of meat or fat; raw potatoes, potato peelings, and potato vines; tomato vines; avocados, guacamole, or avocado skins and pits; tobacco (pick up those cigarette butts); and spoiled or excessively salty or sugary foods, especially chocolate. Avoid onions and garlic, as their lingering flavor

---

ADVICE FROM THE FARM

THE CHICKEN CARRY

When you go someplace to buy full-grown chickens, go prepared! Unless they've been hand-tamed, they won't sit quietly in your lap on the way home. Airline-style plastic dog crates make good chicken limousines.

Another option is a sturdy, lidded cardboard box punched with holes. If you do put a chicken in your lap, bring a towel to cover it, and wear long sleeves because scared chickens scratch.

Another thing beginners may not know: don't carry chickens by their legs! It can hurt them, it's undignified, and it scares them silly.

You should carry a chicken close to your body with your right arm hugging his body against yours. Then, your right hand can hold his feet while your left hand can support his chest.

—Marci Roberts

A social hierarchy exists among the members of the flock.

taints those yummy eggs, and citrus or citrus peels tend to lower egg production.

4. Chicken manure in your yard and garden? Oh, yes! Chicken droppings are high in nitrogen and make excellent natural fertilizer. Your gardener friends and neighbors will stand in line for your chicken-coop cleanings, and if you let your hens wander around the backyard for a few hours a day, they will help green your lawn.

5. Chickens rid your yard of bugs. Cockroaches, aphids—they're fair game to hungry chickens, and as the hens scratch around, seeking tasty bugs, they automatically aerate your soil. Chickens also eat grass and weeds, which can cut down on their feed bills. If you build a chicken tractor for your urban hens, you can put them to work anywhere in your yard.

6. By keeping your hens, you're saving lives. If you aren't familiar with the conditions under which factory-farmed hens are kept, research the subject. By keeping your own infinitely healthier and happier layers, you're reducing the demand for store-bought eggs. Less demand means fewer cruelly imprisoned commercial hens. Simple.

7. And, finally, chickens are funny, low-maintenance pets. They're surprisingly intelligent and have quirky, endearing personalities. Tame chickens like to be picked up and cuddled. What's not to like about urban chickens?

First Things First: Check Those Statutes

Before you scope out a place for your coop and before you pick out hens, you must carefully—and I mean *carefully*—investigate your municipality's laws and regulations regarding chicken keeping. Just because your neighbors down the block have hens doesn't mean that you can have them, too. Perhaps they are illegally keeping chickens or have applied for a personal zoning variance. The lay of their property

might allow compliance to setback regulations that yours doesn't.

Many cities publish municipal statutes online. If yours doesn't, take a trip to city hall and ask to be directed to the office that oversees municipal laws. Request a printed copy of the appropriate statutes and then take them home and go over them with a fine-tooth comb. While most towns and cities do allow chicken keeping within city limits, a long list of stipulations may apply. You will be limited in the number of hens you can keep (roosters are nearly always verboten) and where and how you can keep them.

Consider some of the statutes in Duluth, Minnesota, where urban chicken keeping was legalized in 2008.

- A license is mandatory and costs $10 per year. Persons convicted of cruelty to animals in Minnesota or in any other state may not obtain a license. A representative of the animal-control authority in Duluth must inspect chicken facilities prior to licensing. If the chickens become a nuisance, as evidenced by three violations of Duluth City Code within twelve consecutive months, the license will then be revoked.

- Chickens may be kept only at single-family dwellings as defined by Duluth City Code. No person may keep chickens within the single-family dwelling.

- The maximum number of chickens allowed is five hens. The keeping of roosters is forbidden.

- Chickens must be provided a secure, fully enclosed, well-ventilated, windproof structure in compliance with current zoning and building codes, allowing 1 square foot of window to 15 square feet of floor space. It must have a heat source to maintain adequate indoor temperatures during extreme cold weather. The floor area or combination of floor area and fenced yard for keeping chickens shall be not less than 10 square feet of space per chicken.

- Fences around yard enclosures must be constructed with mesh-type material and provide overhead netting to keep chickens inside and predators out.

- Chickens must be kept in their roofed structure or attached fenced yard at all times.

- No chicken structure or fenced-yard enclosure shall be located closer than 25 feet to any residential dwelling on adjacent lots.

- All droppings must be collected on a daily basis and placed in a

Chickens are social creatures who enjoy their feathered friends.

fireproof covered container until applied as fertilizer, composted, or transported off the premises.

• No person shall slaughter chickens within the city of Duluth.

These are examples of typical statutes. Many cities allow renters and dwellers in two-family housing units to keep chickens but require them to first obtain permission from their landlords and neighbors. Some allow free-ranging chickens. Some municipalities, such as Duluth, do not allow the keeping of house chickens.

Once you have a copy of your city's chicken laws, keep it handy in case a neighbor complains. Go to great pains, however, to keep your neighbors mollified.

Choosing a Breed

Nervous, squawky chickens that fly are not good prospects for city living. Don't overlook bantams. Bantams take up much less space than full-size chickens. As with full-size birds, some breeds are more suited to become urban layers than others. I suggest bantam Cochins, sometimes also called Pekins, like my birds Dumuzi and Marge. Cochins are gentle, quiet, attractive chickens, and they aren't prone to flying. Bantam Cochins generate less manure (and

SEX-LINKS AS CITY (AND COUNTRY) CHICKENS

There are two basic types of hybrid sex-link chickens, red and black, although each goes by several names.

Black sex-links, usually called Black Stars, Black Rocks, or Rock Reds, are crosses between a Rhode Island Red or New Hampshire rooster and a Barred Rock hen. Both sexes hatch out black, but a cockerel will have a white dot on his head. Pullets feather out black with a hint of red on their necks; cockerels have Barred Rock-type plumage accented with a few red feathers.

Red sex-links are produced by a number of different crosses. White Plymouth Rock hens with the silver factor are crossed with New Hampshire roosters to produce Golden Comets. Silver Laced Wyandotte hens are crossed with New Hampshire roosters to produce Cinnamon Queens. Additional red sex-link combinations are Rhode Island White hens with Rhode Island Red roosters or Delaware hens with Rhode Island Red roosters. Cockerels hatch out white and then feather out pure white or with a hint of black feathering mixed in, depending on the cross. Pullets hatch out buff or red, depending on the cross, and they feather out buff or red with flecks of white throughout.

Bantams can be a good choice for suburban or urban living.

Both sex-link colors are calm, cold hardy, quiet, and friendly birds with an unusually efficient feed-conversion ratio. They work well in confinement or free-range situations. Hens weigh about 5 pounds each, begin laying earlier than most breeds, and lay a lot of brown eggs. Sex-links are outstanding city chickens.

Neighbors won't complain about clean, attractive chicken accommodations.

as far away from property lines as possible.

- Opt for a generously sized coop, based on the number of hens you plan to house. More space means less crowding and happier hens, along with less-concentrated waste and less smell.
- While some urban chicken keepers go for cute- or quaint-looking coops, in many cases, natural camouflage is more in order. A privacy fence or shrubs planted around your chicken facilities make them less obtrusive and also serve to help deaden sounds.

thus less odor) than full-size breeds, and their small, light brown eggs are so tasty.

If you plan to raise your birds from chicks, you must buy sexed chicks, not straight-run packages, to get all pullets. Even then, there's a certain amount of error in sexing day-old chicks. If you get a cockerel in your box of pullets, what will you do? A very workable solution, if you want good layers but don't have your heart set on the standard breeds, is to buy hybrid sex-link chicks. Newly hatched sex-link pullets and cockerels are colored differently, so you can tell them apart. You'll know the sure the sexes of your chicks, and sex-links are very good hens.

A Home for Your City Chicks

We'll discuss coops and runs for your chickens in Chapter 2, but keep the following thoughts in mind.

- Most city statutes spell out how big your coop and outdoor exercise area should be and which amenities you must provide. What they often don't stipulate—and this is vitally important—is that city chicken-keeping facilities must be *attractive*. Your neighbors won't be pleased if you create an eyesore in your backyard.
- Buy or build the best coop and fencing you can afford. Sturdy prefabricated units are especially appealing in urban situations because they're engineered to combine safety, convenience, and beauty. If you are going to build your own coop, collect ideas by surfing the Internet or visiting other chicken keepers, buying plans, or perusing a copy of Judy Pangman's *Chicken Coops: 45 Building Plans for Housing Your Flock.*
- Obey your municipality's setback laws. Measure to be certain. In fact, to preserve your neighbors' good will, build your facilities

- Plan outdoor facilities for your hens. An exercise pen attached to their coop, a well-fenced backyard (providing your hens aren't flyers), or a chicken tractor all work well. Otherwise, plan to stay outdoors with your hens while they free-range on bugs and grass. They need you to protect them from predators (particularly dogs and humans) and prevent them from wandering into neighbors' yards or the road.

Keep It Clean

A single full-size chicken can produce up to 50 pounds of solid waste per year. If you don't keep your facilities ultra-clean, they will smell. Nothing turns off picky neighbors faster than *eau de barnyard* wafting over the property line. While we usually advocate deep litter bedding for chickens, in a city setting, it's better to pick up messes daily and completely strip and re-bed your henhouse once a week.

Store waste in covered trash receptacles and find a place to dispose of it on an ongoing basis. If you garden, compost it. If you don't garden, compost it anyway and present finished compost to gardening neighbors and friends. Otherwise, take waste to a farming friend in the country so that he or she can dispose of it. Don't let it accumulate, uncovered, on your property for very long.

Be prepared to deal with flies and rodents. Earth-friendly fly sprays and fly traps are the way to go. Rodents are a bigger problem, but one you must face, because chicken feed is ambrosia to mice and rats. Store feed in covered metal containers. Trash cans work exceptionally well. Don't use plastic containers; rats chew through them without a twitch of a whisker.

Eradicating existing rodents is tricky. Don't use poisons that your neighbor's cat or a wandering toddler might find. Traps work well,

The Sultan breed was recognized in the first Standard of Perfection but is now an endangered breed.

but, better yet, adopt a friendly cat or a Parson Russell or Rat Terrier that needs a good home; they are natural vermin control at its finest.

Keeping City Chickens Is a Privilege

It's important, both for you and for fellow municipal chicken keepers, to comply with chicken laws to the letter and not let your birds create a disturbance. It's also important to get along with your neighbors: their complaints could bring animal control to your front door. If enough neighbors complain on a citywide basis, residents' rights to keep chickens could be revoked.

Consider sharing eggs with your neighbors. Be considerate; even if noisy roosters are legal, don't keep one. Invite neighborhood children to meet your hens and distribute chicken feed. Happy neighbors = happy you.

Finally, consider joining or creating a local chicken keepers' association. Encourage members to teach community classes in urban chicken keeping. Take programs to schools and local events. Show naysayers that city chickens aren't the smelly, noisy barnyard fowl that they expect. Maybe they'll take up the banner and get some chickens, too.

Heritage Breeds

Fortunately, growing legions of poultry fanciers and small-scale chicken raisers are stepping forth to reclaim our forebears' poultry breeds. This rare-breed renaissance is occurring throughout the world for numerous reasons.

Some conservators long for the mouthwatering fried chicken Grandma used to serve for Sunday dinner or for yummy, orange-yolked eggs with divine flavor. Some yearn to preserve living remnants of our distant past. Others do it in the name of biodiversity—they feel that if disease or genetic malady should strike down America's beleaguered battery hens and broilers, there must be hardy Heritage breeds ready to take up the slack. Some simply prefer breeds created for specific environments and needs, such as Buckeyes and Hollands for free-range eggs, Chanteclers for winter laying in the far North, or heat-tolerant Cubalayas for the steamy South.

Before you join the rare-breed contingent, get to know The Livestock Conservancy (formerly the American Livestock Breeds Conservancy). This is a nonprofit membership organization devoted to the promotion and protection of more than 150 breeds of livestock and poultry. In service since 1977, it's the primary organization in the United States working to conserve rare breeds and genetic diversity in Heritage livestock. In 2009, The Livestock Conservancy launched its Heritage-chicken promotion, and it's eager to provide new Heritage-breed producers, large and small, with materials to help get started and, later, to promote and market eggs and meat from their Heritage chickens.

Endangered Breeds

If you join The Livestock Conservancy, you'll receive the organization's bimonthly print newsletter and an annual directory full of contacts. Livestock Conservancy breeders make up an active network of people who participate in hands-on conservation, marketing, and public education; if you are getting into chicken keeping, particularly if you plan on raising, showing, or breeding endangered breeds, they are definitely people you want to know.

To find out which chicken breeds meet Heritage-breed requirements, visit The Livestock Conservancy's website (www. livestockconservancy.org) and click on "Heritage Breeds," then "Poultry Breeds," and then "List of Chicken Breeds." This will bring you to the Conservation Priority List (CPL) for chickens, where breeds are categorized according to the following criteria as defined by The Livestock Conservancy.

Critical: "Fewer than 500 breeding birds in the United States, with 5 or fewer primary breeding flocks (50 birds or more), and estimated global population less than 1,000." Breeds in the Critical

The distinctive-looking Shamo is in the Livestock Conservancy's "Watch" category.

Watch: "Fewer than 5,000 breeding birds in the United States, with 10 or fewer primary breeding flocks, and estimated global population less than 10,000. Also included are breeds with genetic or numerical concerns or limited geographic distribution." Breeds in the Watch category in 2015 were the Ancona, Aseel, Brahma, Catalana, Cochin, Cornish, Dominique, Hamburg, Houdan, Jersey Giant, La Fleche, Minorca, New Hampshire, Old English Game, Polish, Rhode Island White, Sebright, and Shamo.

category in 2015 were the Campine, Chantecler, Crevecoeur, Holland, Modern Game, Nankin, Redcap, Russian Orloff, Spanish, Sultan, Sumatra, and Yokohama.

Threatened: "Fewer than 1,000 breeding birds in the United States, with 7 or fewer primary breeding flocks, and estimated global population less than 5,000." Breeds in the Threatened category in 2015 were the Andalusian, Buckeye, Buttercup, Cubalaya, Delaware, Dorking, Faverolles, Java, Lakenvelder, Langshan, Malay, and Phoenix.

Recovering: "Breeds that were once listed in another category and have exceeded Watch category numbers but are still in need of monitoring." Breeds in the Recovering category in 2015 were the Australorp, nonindustrial Leghorn, Orpington, Plymouth Rock, nonindustrial Rhode Island Red, Sussex, and Wyandotte.

Study: "Breeds that are of interest but either lack definition or lack genetic or historical documentation." Breeds in the Study category in 2015 were the Araucana, Icelandic, Manx Rumpy (or Persian Rumpless), and Saipan.

WHAT IS A HERITAGE BREED?

In order to support the American Poultry Association (APA) in bringing the Heritage breeds back to popularity, The Livestock Conservancy has created a list of criteria that chickens must meet to be called Heritage. The following is the definition set forth by The Livestock Conservancy. Heritage chicken must adhere to all of the following conditions.

1. **APA Standard Breed.** Heritage chicken must be from parent and grandparent stock of breeds recognized by the APA prior to the mid-twentieth century, whose genetic line can be traced back multiple generations, and with traits that meet the APA Standard of Perfection guidelines for the breed. Heritage chicken must be produced and sired by an APA Standard breed. Heritage eggs must be laid by an APA Standard breed.

2. **Naturally mating.** Heritage chicken must be reproduced and genetically maintained through natural mating. Chickens marketed as Heritage must be the result of naturally mating pairs of both grandparent and parent stock.

3. **Long, productive outdoor lifespan.** Heritage chicken must have the genetic ability to live a long, vigorous life and thrive in the rigors of pasture-based, outdoor production systems. Breeding hens should be productive for five to seven years and roosters for three to five years.

4. **Slow growth rate.** Heritage chicken must have a moderate to slow rate of growth, reaching appropriate market weight for the breed in no less than sixteen weeks. This gives the chicken time to develop strong skeletal structure and healthy organs prior to building muscle mass.

Chickens aren't choosy. Whether you provide a simple shack or luxurious villa, as long as the accommodations meet their basic housing needs, your birds will be tickled pink with them. A coop must shelter its inhabitants from wind, rain, snow, and sun and protect them from predators. It also needs to be reasonably well lit and ventilated and roomy enough for the number of birds it houses. When your chickens go inside, they should find sanitary bedding, roosts, nesting boxes, feeders, and waterers. For your flock's continuing comfort and health, the coop should be clean and easy for both you and the birds to access.

HOUSING AND FEEDING YOUR

CHICKENS

Consider the breed and type of your chickens. For example, when it comes to indoor living space, laying hens demand more space than broiler chickens, which have much shorter life expectancies. Bantams require less indoor space than 10-pound Jersey Giants. Outdoors, a 3-foot uncovered enclosure will keep Jerseys safely contained but will never do for flying bantams. If you don't provide the latter with a tall, covered run, you may find your entire flock going over the fence.

In northern climes, chicken abodes must be insulated to spare your birds frostbitten wattles, combs, and toes. In torrid southern locales, how to afford relief from the heat will be a major concern.

Costs, time, and aesthetics should be factored in as well. For example, a chicken keeper without a lot of spare cash might decide to build a coop rather than hiring a carpenter or buying a prefab unit. Almost anyone can construct a functional basic coop from scratch, recycling materials at very little cost. Other keepers may have the requisite carpentry skills but not the time to create their own chicken villas.

No matter how excited you are to get started, don't pick up that hammer until you've made sure that the site is right. The *where* of coop building is very important. You don't want to have to raze a half-constructed henhouse after you realize that it's too close to the neighbor's fence. Chicken keepers in suburban and urban areas are subject to municipal codes.

Also factor in your own preferences. For example, if watching hens peck in the yard will soothe your soul, it makes little sense to shut them away where you can't see them.

Your Coop: Basic Requirements

Access, lighting, ventilation, insulation, and flooring all need to be carefully considered as you plan your coop. For example, how do you provide sufficient lighting and ventilation without compromising the effectiveness of your insulation? Which flooring materials are both sturdy and easy to clean? You must also think in terms of easy access for you and your flock—but *not* for predators.

Access

Your coop will need at least two doors: one for you and one or more for your birds. If your coop is low and close to the ground (a good design in northern climes, where body heat is wasted in taller structures), your door might simply be a hinged roof. With this kind of simple opening, you can easily feed and water your birds, tidy the coop, and gather eggs. If the coop is a standard, upright model, the roof should swing inward so chickens are less likely to escape when you open it.

Cut a chicken door (or more than one), 14 inches tall by 12 inches wide and 4–8 inches from the ground, in an outer wall. Use the cut-out piece of wood to make a ramp. Affix full-width molding (for traction) every 6 inches along its inside surface and then hinge it at the bottom so that the ramp doubles as a door that swings out and down. Fit it with a secure latch so you can bar the door closed at night. If raccoons are a problem in your area, choose a fairly complex latch—if a toddler can open the lock, then a raccoon can unlock it easily.

Strips of wood or molding give the chickens traction on a ramp.

Lighting and Ventilation

Light is essential to chickens' health and happiness. Natural lighting is better than bulbs and lamps, but if you want your hens to lay year-round, you must wire your coop and install fixtures.

Sliding windows work best for lighting and ventilation because chickens can't roost on them when they're open. Every window must be tightly screened, even if your chickens can't fly. If predators can wriggle their way around or through those screens, they will. You'll need ½- to ¾-inch galvanized mesh to keep wee beasties, such as weasels and mink, at bay.

If you live in frigid winter climes, large south-side windows are a must; they admit lots of winter light and radiant heat. In general, allow at least 1 square foot of window for each 10 square feet of floor space. If you live where temperatures rarely dip below freezing, install even more windows. It's hard to let in too much light.

Extra windows also create cooling, healthful cross-ventilation when summer heat is an issue. Install the extra windows on your coop's north wall and possibly the east one, too. Your coop must be properly ventilated. Chickens exhale up to thirty-five times per minute, releasing vast amounts of heat, moisture, and carbon dioxide into their environment.

Faulty coop ventilation quickly leads to respiratory distress. Where large windows (and lots of them) aren't possible, saw 6-inch circular or 2-by-6-inch rectangular ventilation openings high along one or more nonwindowed walls. Unplug these vents when extra air is needed, and close them tightly when it's frigid outside. Chickens can weather considerable heat or cold when their housing is dry and draft-free, but they don't do well in smelly, damp conditions. If your nose smells ammonia as you enter or open your coop, it is not adequately ventilated. Fix this problem immediately.

Insulation

To get your chickens through winters as unforgiving as those in northern Minnesota, their coop must be well insulated. If money is

FLOOR SPACE REQUIREMENTS

The minimum amount of floor space needed per chicken depends on several factors, including bird type, the presence of indoor roosts, and the size of the outdoor run.

Free-range chickens and chickens with adequate outdoor runs and indoor roosts:

Heavy breeds: 5 square feet per bird (2 square feet if slaughtered before sixteen weeks of age)

Light breeds: 3 square feet per bird

Bantams: 2 square feet per bird

Confined chickens without access to outdoor runs:

Heavy breeds: 10 square feet per bird (6 square feet if slaughtered before sixteen weeks of age)

Light breeds: 8 square feet per bird

Bantams: 5 square feet per bird

year, strip everything back to floor level and start again.

Your Coop: Basic Furnishings

Basic coop furnishings include roosts, nest boxes for laying hens, feeders, and watering stations. Roosts are the elevated poles or boards on which chickens prefer to sleep. Roosting helps them feel safer at night, making these perches a must if you want happy hens. Nesting boxes are a necessity—unless you like going on egg hunts or prefer your eggs precracked. You'll also need more than one set of well-placed waterers and feeders.

Roosts

In the winter, roosted birds fluff their feathers and cover their toes; they tend to stay warmer that way. A roost can be as simple as an old wooden ladder propped against a chicken coop's inner wall and tacked to the wall at the top. Place it at enough of an angle so that birds settled on one rung don't poop on flockmates roosting on rungs below.

If you build traditional stair-stepped roosts for your birds, set the bottom perch about 2 feet from the floor and set higher rungs an additional foot apart. Two-by-two boards with rounded edges make ideal roosts for full-size chickens, and 1-inch rounded boards or 1-inch dowel rods are fine for bantams. Tree branches of the same diameters make fine roost rails, too. Don't use plastic or metal perches; chickens require textured perches that their feet can easily grip. Allow 10 inches of perch space for each heavy-breed chicken in the coop; provide 8 inches and 6 inches for light breeds and bantams, respectively.

Wherever you place your roosts, make certain that sleeping chickens won't be perched in cross-drafts. Check frequently and move the roosts if necessary.

Nesting Boxes

Unless you provide nesting boxes, free-range hens will sneak off to lay in nooks and crannies, and you may never find the eggs! Confined hens will plop eggs out wherever they can, which can result in poop-splotched, cracked eggs. Specially designed nesting boxes with slanted tops and perches in front work best, but any sturdy cubicle

scarce, you can insulate only the coop's north wall and bank outside by using hay or straw bales stacked at least two deep. Another trick: bank snow up against the coop by shoveling or pushing it as far up the sides as you can. If it's still too cold inside the coop, you'll need a heat lamp. But remember: fallen heat lamps can, and often do, spark fires, so install it in a reasonably safe location and use it only when really needed.

Chickens can die in temperatures higher than 95° Fahrenheit. Situate your coop and outdoor enclosures in partial shade—or plant vegetation around your chickens' lodgings to partially shade it. Insulation helps repel daytime heat, and fans generate badly needed airflow. Opt for light-colored or corrugated metal roofing and paint external surfaces a matte white color to reflect the heat. Allow additional space for each of your birds; overcrowding leads to higher indoor temperatures and humidity.

Flooring

Your coop's floor may be constructed of concrete, wood, or plain old dirt. Concrete is rodent-proof and easy to clean, but it's comparatively expensive. Wood must be elevated on piers or blocks; it looks nice, but it can be hard to clean and periodically needs replacing. Well-drained dirt floors work fine. However, if a dirt floor is poorly drained or allowed to become mucky, you'll have a sheer disaster on your hands.

Deep bedding nicely insulates a coop floor, it is simple, and it works! For the best-bet deep-bedding system, blanket your floor of choice with a cushy layer of absorbent material to keep things tidy and fresh. Chopped straw (wheat straw is best) or wood shavings are ideal, and rice or peanut hulls, sawdust, dry leaves, and shredded paper work well, too. Line the floor 8–10 inches deep; remove messes and add more material when necessary. Once or twice a

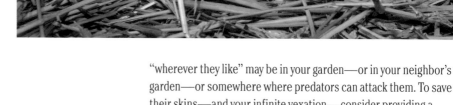
Straw works well as bedding to insulate the coop.

with a top, a bottom, three enclosed sides, and bedding inside will do nicely.

Make sure the box is larger than the chicken. A 14-inch wooden cube with one open side makes an ideal nesting box for a full-size hen; a 12-inch cube accommodates bantams with ease. Leave the top off in steamy summer weather.

If you use a regular box, attach a 3–4-inch lip across the bottom front to keep bedding and eggs from spilling out. Mount the unit 2 feet from the floor in the darkest corner of the coop. Provide one nesting box for every four or five hens.

Feeders and Waterers

Place at least two sets of commercially made feeders and waterers in every coop and locate each set as far from the others as you can to prevent guarding by high-ranking flock members. Instead of setting units on the floor, install them so the bottoms of the waterers and the top lips of the feeders are level with the smallest birds' backs. They'll stay cleaner that way, and your chickens will waste less food. Be sure to provide one standard hanging tube-style feeder per twenty-five chickens. If you prefer trough feeders, allow 4 inches of dining space per bird when deciding what size to purchase.

Outdoor Runs: Sunshine and Fresh Air

Another way to keep your chickens happy and healthy is to get them into sunshine and fresh air almost every day. To do that, you'll probably need a chicken pen, or run, attached to your coop. You can allow your chickens to free-range (wander wherever they like), but

---

THE COMFORTS OF HOME

Be sure to provide a shaded area for your flock in the outdoor run. No trees? Stretch a tarp across one corner and hook it to the fence with bungee cords.

Chickens enjoy lounging under outdoor shelters when it rains. So, if you leave the tarp up when it rains, poke a few holes in it for drainage—just don't poke holes near the center, where your birds gather. While you're at it, add a sand pit for dust bathing as well.

"wherever they like" may be in your garden—or in your neighbor's garden—or somewhere where predators can attack them. To save their skins—and your infinite vexation—consider providing a fenced-in area.

Chicken runs are traditionally crafted of chicken wire (also called poultry netting), which is a flimsy, 1-inch mesh woven into a honeycomb pattern. If a dog or larger varmint wants to get at your chickens, this lightweight mesh is not going to be a deterrent. If you value your birds, don't use chicken wire as a barrier. Workable alternatives include substantial posts with attached medium- to heavy-duty yard fencing or sturdy wire sheep panels (sometimes installed two panels high), or electroplastic poultry netting.

If predators, including dogs, are an ongoing headache, a strong electric charger and two strands of electric wire fencing can provide effective but cheap insurance. String one strand on 10-inch extension insulators 4 inches from the ground, along the outside bottom of the run. Using the same type of insulator, stretch another strand parallel with the top of your existing fence. These wires will prevent hungry critters from scaling or tunneling under your chicken-run fence.

Allow at least 10 square feet of fenced run for each heavy chicken in your flock, 8 square feet per light-breed chicken, and 4 square feet per bantam. In general, you can contain your chickens using a 4–6-foot fence, but most bantams and certain light full-size breeds can neatly sail over 6-foot barriers. Keep them in by installing netting over the enclosure.

Location

Don't create your coop area too far from utilities, especially if you must carry water or run a hose or electrical extension cord from your house, garage, or barn. Place your coop area at a reasonable distance from neighbors' property lines, especially when setback regulations are part of municipal or county codes.

Typical chicken-wire fencing may not protect your birds from predators.

Chicken Tractors

Chicken tractors are lightweight, bottomless shelter pens designed to move wherever grass control and soil fertility are required. They appeal to small-scale raisers, and maintaining a chicken tractor in your garden not only enriches your soil but also allows your birds to supplement their diets with yummy greenery and crunchy bugs, worms, and grubs. There's a place for a chicken tractor (or two) on most every hobby farm!

Choose a well-drained area where storm runoff and melting snow won't make coop floors and outdoor enclosures a sodden mess. Remember also that your coop area should be close enough for you to enjoy your chickens because that is why you wanted them in the first place.

Chicken tractors are engineered to move around your yard or farm, usually by hand, to areas in need of enrichment. Size can range from 3 × 5 × 2 feet tall for three or four chickens to a whopping 8 × 12 × 3 feet tall. Usually, the sides are crafted of wood-framed wire mesh, and a hinged roof protects inhabitants from the elements and allows easy operator entry. If you build compact tractors to fit the width of spaces between your garden rows, the chickens will neatly weed and fertilize your garden without filling up on produce while they do it (provide water and feed inside the tractor, and move the unit once or twice a day). Larger chicken tractors can be set atop spots needing more thorough cultivation over longer periods of time. Chickens are day laborers; remove them from the tractor at night.

Building a Cheaper Chicken Coop

If your desire to keep chickens exceeds your ability to buy or build a standard chicken coop, don't worry: chickens can adapt to simple accommodations. If you keep them dry, safe, and out of drafts, they'll be happy. Consider the following inexpensive options.

You can build chicken accommodations into existing structures, such as your garage or back porch, the kids' abandoned playhouse, or an unused shed. Build coop-style quarters, or house your chickens in cages. Show-chicken fanciers often cage their birds in wire rabbit-style hutches.

It's best to forgo keeping heavy-breed chickens if they must live in cages; continually standing on wire floors will likely damage their feet. Allow 7 square feet of cage floor space for each light-breed chicken or 6 square feet per bantam. Place a 2-foot-by-2-foot sheet of salvaged cardboard in one corner so inhabitants can rest their feet; replace it when it gets soiled. Caged chickens appreciate roosts—affix them at least 6 inches from the floor—and boredom-squelching amusements, such as toys.

A large wooden packing crate fitted with a hinged roof makes a dandy indoor or outdoor coop. Prop the lid open during the day (if you keep flying breeds, you'll have to fashion a screen) and close it at night. Outdoors, install a latchable dog door in one side and attach a small fenced run.

If appearance doesn't matter, fashion a funky, cost-effective walk-in coop out of tarps, a welded wire cattle panel (or two), and scrap lumber. It's amazing what can be done with plastic tarps if aesthetics don't count.

Owls and Weasels and 'Possums, Oh My!

It's no fun—for you or your birds—when hungry midnight marauders visit the chicken coop. The best way to thwart potential predators is to lock your birds inside a safe, secure chicken coop at night. Another approach, if it's legal in your state, is to humanely trap and relocate bothersome nighttime marauders.

After sundown, block or screen in every door, window, and any other crack or portal in the outside walls. To be effective, screening must be small-holed and made of strong material. A mink or weasel

DID YOU KNOW?

If it's possible for a chicken to hurt itself on something, it will. This is important to remember when building a coop. Get down on your hands and knees—at chicken level—and really look at the finished coop to try and spot chicken hazards, like a stray piece of sharp wire poking out.

can easily slip through 1-inch chicken wire, and larger species can simply rip it down. Choose ¾-inch or smaller mesh galvanized hardware cloth for screening windows and building outdoor enclosures to save your chickens' lives.

To discourage chicken-swiping predators, install concrete block foundations into coop floors, set at least two rows high. In addition, bury outdoor enclosure fencing at least 8–12 inches into the earth. Make sure the buried fencing is toed outward, away from the fence line. If winged predators, such as daytime hawks or nighttime owls, pursue your birds, cover the outdoor enclosure with chicken wire; for this purpose, it works nicely.

If your birds range freely and you live where chicken-thieving hawks wreak havoc, think camouflage. For example, don't choose a white-feathered breed. Plant ground cover—bushes, hedges, and flower beds—so that your chickens can hole up in the greenery when predators soar above.

If predators are simply getting into your feed, secure container lids using bungee cords. The key word is *secure* for both your chickens and their feed.

Chow for Your Hobby-Farm Fowl

While your chickens' nutritional needs vary depending on age, sex, breed, and use, their diets must always include water, protein, vitamins, minerals, carbohydrates, and fats in adequate quantities and proper balance. All chicken keepers are in agreement on this point. But when it comes to the question of how best to supply all of those dietary elements, it's a different story. Ask any fanciers how each feeds his or her chickens, and you'll find two distinct camps: those who never feed their birds anything except commercial mixes, and those who never feed their birds commercial mixes

without supplementing them. Opinion runs high on both sides about which approach is better.

Which one you should take really depends on your primary reason for keeping chickens. If you raise birds strictly for their meat or eggs, commercial feed is the way to go. Commercial bagged rations are formulated to serve up optimal nutrition, thus creating optimal production. Supplementing commercial feed with treats, table scraps, scratch (a whole- or cracked-grain mixture that chickens adore), or anything else will upset that delicate nutritional balance.

However, if, like us, you see your chickens as friends and don't care if their growth is slightly slower or if they produce fewer eggs, then consider supplementing their diets. They'll appreciate the variety, and you'll appreciate the much lower cost of a supplemented diet.

Water

Consider this: an egg is roughly 65 percent water, a chick 79 percent, and a mature chicken 55– 75 percent. Blood is 90 percent water. Chickens guzzle two to three times as much water as they eat in food, depending on their size, their type (layers require more water than broilers), and the season—up to two or three cups per day. So whether you use a commercial or home-based diet, your chickens require free access to fresh, clean water.

Chickens need water to soften what they eat and carry it through their digestive tracts; many of the digestive and nutrient-absorption processes depend on water. In addition, water cools birds internally during the hot summer months. If you eliminate water from your chickens' diet, expect problems immediately. Chickens don't drink a lot at any single time, but they drink often.

WASTE LESS

Chickens waste about 30 percent of feed in a trough feeder that's full. If the same trough is only half full, they waste only 3 percent. Save yourself some time and money by spreading less feed for your chickens in more troughs.

Chicken waterers are designed to keep the water clean.

of about nine parts water to one part chlorine bleach.

Commercial Feeds

Whether you buy it or mix it yourself, a healthy, happy chicken's diet should provide the following:

- Sufficient protein based on the age and needs of the bird
- Carbohydrates, a major source of energy
- Thirteen vitamins to support growth, reproduction, and body maintenance: fat-soluble vitamins A, D3, E, and K; and water-soluble vitamins B12, thiamin, riboflavin, nicotinic acid, folic acid, biotin, pantothenic acid, pyridoxine, and choline
- Macrominerals (those needed in larger quantities) and trace or microminerals (those needed in only minute amounts) to build strong bones and healthy blood

Water temperature can affect how much chickens will drink. They don't like to drink hot or too-cold water, so keep waterers out of the blazing sun. When temperatures soar, plop a handful of ice cubes in the reservoir every few hours. In the winter, replace regular waterers with heated ones or add a bucket-style immersion heater to a standard metal version. You can also swap iced-up waterers for fresh ones containing tepid water every few hours. In subzero climates, heated waterers are a must; even a heated dog bowl is acceptable.

Even if one waterer is enough, choose two. Otherwise, bossy, high-ranking chickens in your flock's hierarchy may shoo underlings away from the fountain.

As with the feeders, hanging waterers from hooks or rafters with the drinking surface level with your smallest chickens' backs will give the best results. If you can't hang a waterer, make certain it is level to avoid leaking.

Whichever type of waterer you use, and wherever you hang your waterers, clean and rinse them every day. Scour them once a week (more often in the summer) using a stiff brush and a solution

A D V I C E   F R O M   T H E   F A R M

FEEDING YOUR CHICKENS

My chickens get yummy breakfasts every other day: oatmeal, rabbit feed (for a nice greens-based meal), raisins, scrambled eggs, cat food (protein—and they love it), apples, leftovers, mac and cheese (a favorite!), green beans, and Cheerios (another favorite).

—Jennifer Kroll

Let your chickens graze. My chickens keep our 3 acres almost totally free of ticks. Ticks for eggs—that's a really neat trade!

—Sharon Jones

Clean, roomy accommodations and a balanced diet are essential to chickens' good health.

cells, supporting enzyme activation and muscle function and regulating metabolism; hens require additional minerals, especially calcium, to lay eggs with nice, thick shells

- Fats for energy and proper absorption of fat-soluble vitamins and as sources of fatty acids, necessary for supporting fertility and egg hatchability

Commercial feeds are designed to meet the aforementioned needs precisely. To meet protein requirements, commercial feeds include a variety of high-protein meals made of corn gluten, soybeans, cottonseed, meat, bone, fish, and dried whey. Too much protein can be as bad as too little, so balancing this nutrient is especially tricky. Carbohydrates are much easier; they naturally compose a large portion of every grain-based diet. While some of the thirteen vitamins listed are plentiful in natural foodstuff, commercial feeds cover all bases by adding a vitamin premix. As for fats, commercial feeds contain processed meat and poultry fats in measured amounts. Fats provide twice as much energy as other feed ingredients, making them especially useful in starter feeds and growing rations. Mixing your own commercial-style feed is an option (and often a must for producers of organic meat or eggs), but balancing the nutrition is a complex task.

Common ingredients in commercial feeds include corn, oats, wheat, barley, sorghum, milo, soybean, and other oilseed meals; cottonseed or alfalfa meal; wheat or rice bran; and meat by-products, such as bonemeal and fishmeal. Ingredients are finely ground to produce easier-to-digest mash; sometimes they are pelleted or processed into crumbles so there is less wasted food.

Commercial baby chick food is usually medicated; some feeds for older chickens are medicated, too. Each type of feed, designed for a specific group of birds, contains nutrients in slightly different measures, so choose the correct feed: starter, grower, layer, breeder, or finisher. That information will be printed on the label, along with precisely how much to feed, so always check to be certain.

Commercial feeds also contain ingredients that many fanciers don't approve of, such as antibiotics and coccidiostats for birds that don't need them, pellet binders to improve the texture

of pelleted feed, and chemical antioxidants to prevent fatty ingredients from spoiling. Again, read the labels! If you'd like to offer your chickens commercial feed but want to avoid the questionable additives, ask your county agricultural agent or feed store representative what "natural" commercial feeds are available locally. The Murray McMurray Hatchery sells organic feed and ships throughout the continental United States.

Maintaining Nutritional Value and Freshness

To retain full nutritional value and assure freshness, purchase no more than a two-to-four-week supply of commercial feed. Don't dump new product on top of remaining feed; use up the old feed first or scoop it out and place it on top of the new supply. When storing feed, place it in tightly closed containers and store it in a cool, dry place out of the sun. Plastic containers will work, but if gnawing rodents are a headache, store grain in lidded metal cans. A 10-gallon garbage can holds 50 pounds of feed.

If your chickens refuse commercial feed, examine it closely. Sniff. It may be musty or otherwise spoiled. If it seems all right, you're probably dealing with picky chickens that prefer scratch, treats, and table scraps. Cut back on goodies until they eat the chicken feed, too. Distributing treats only after they've dined on their regular rations will encourage them to be less picky.

The Supplement Approach

According to proponents of supplements, supplemented hens lay better eggs, and supplemented broilers taste better. That's something you'll have to decide for yourself. What we present here are methods that chicken keepers can use to supplement their chickens' diets.

Free-range chickens find greens, insects, and other treats as they forage.

chance. Capture katydids, grasshoppers, and other tasty insects to toss to your chickens. If you do, they'll soon come running when they spot you.

Good Home Cookin'

Chickens happily devour table scraps. Avoid fatty, greasy, salty stuff; anything spoiled; avocados; and uncooked potato peels. Also, strongly scented or flavored scraps, such as onions, garlic, salami, and fish, can flavor hens' eggs. Almost everything else from your table will be fine—even baked goods, meat, and dairy products. Your chickens will love it all.

Many folks assume that free-range chickens will grow healthy eating seeds, weeds, and bugs. They won't. However, if you supplement free-range findings with scratch or commercial feed, your chickens will cheerfully rid your yard and orchard of termites, ticks, Japanese beetles, grasshoppers, grubs, slugs, and dropped fruit. One caveat: they'll also strip your garden clean, so think "fenced garden" if you raise free-range chickens.

Grit and Oyster Shells

Since chickens don't have teeth, they swallow grit—tiny pebbles and other hard objects—to grind their food. If your chickens free-range, or if you use easily digestible commercial feed, you won't need to provide your birds with grit. Otherwise, commercial grit (ground limestone, granite, or marble) can be mixed with their scratch or container-fed to chickens on a free-choice basis.

Ground oyster shell is too soft to function as grit, but it's a terrific calcium booster for laying hens. Feeding oyster shell to hens on a free-choice basis allows the hens to eat it when they wish.

Scratch

Many hobbyists and small-flock owners supplement commercial feed with scratch in measured proportions to not upset the nutritional balance of the feed. Scratch is a mixture of two or more whole or coarsely cracked grains, such as corn, oats, wheat, milo, millet, rice, barley, and buckwheat.

Chickens adore scratch grains. Chickens instinctively scratch the earth with their sharp toenails to rake up bugs, pebbles for grit, seeds, and other natural yummies. Scratch strewn on their indoor litter, anyplace outdoors, or in separate indoor feeders satisfies that urge.

Greens and Insects

Hobby farmers and poultry enthusiasts often grow "chicken gardens" of cut-and-come-again edibles like lettuce, kale, turnip greens, and chard. Chickens of all types and sizes relish greens. Greens-chomping hens lay eggs with dark, rich yolks.

Insects add protein to chickens' diets. Free-range chickens harvest their own bugs, but coop and run-caged birds don't have that

Providing your chickens with scratch grains helps them satisfy their urge to find tasty tidbits wherever they roam.

Domestic chickens belong to the Phasianidae family, as do quail, grouse, partridges, pheasants, turkeys, snowcocks, spurfowl, monals, peafowl, and jungle fowl. Domestic chickens are descendants of the Southeast Asian red jungle fowl (*Gallus gallus,* also called *Gallus bankiva*), which emerged as a species roughly 8,000 years ago. Today, red jungle fowl have disappeared from most parts of Southeast Asia and the Philippines, but a genetically pure population still exists in measured numbers in the dense jungles of northeastern India. In Latin, *gallus* means "comb," and that is how chickens differ from their Phasianidae cousins. While chickens vary widely in shape and size, all have traits in common, including general physiology, behavior, and level of intelligence.

CHICKEN
BEHAVIOR AND HANDLING

Physiology

Chickens see in color; and their visual acuity is about the same as ours. While they don't have external ears, they do have *external auditory meatuses* (ear canals) and can hear quite well. Their frequency range corresponds to ours. Their sense of smell is poorly developed, and they don't taste sweet flavors. They do, however, easily detect salt in their diets. Other important physiological characteristics to be aware of concern the digestive tract, internal and external structure (bones, muscles, skin, feathers), and sexual characteristics.

Digestive Tract

Chickens have no teeth. Instead, whole food moves down the esophagus and into the *crop*, a highly expandable storage compartment that allows a chicken to pack away considerable amounts of food at a time. When packed, it's externally visible as a bulge at the base of the neck. Unchewed food trickles from the crop into the bird's *proventriculus* (the "true stomach") and then to the *ventriculus* (another stomach, more commonly called the *gizzard*) to be macerated and mixed with gastric juice from the proventriculus. The food finally passes to the small intestine, where nutrients are absorbed, and then to the large intestine, where water is extracted. From there it moves to the *cloaca*—the chamber inside the chicken's vent (where its digestive, excretory, and reproductive tracts meet via the fecal chamber)—and finally out of the body via the vent. Food processing time for a healthy chicken is roughly three to four hours. Urine (the white component of chicken droppings) also exits the cloaca, but via the urogenital chamber.

Bones to Feathers

Although chickens have largely lost the ability to fly, some of their bones are hollow (pneumatic) and contain air sacs. Smaller fowl can fly into trees and over fences; when harried, heavy breeds try but usually aren't able to get airborne. Chicken muscles are composed of light-colored (white meat) and red (dark meat) fibers. Light muscle occurs mainly in the breast; dark muscle occurs in the chicken's legs, thighs, back, and neck. Wings contain both light and dark fibers.

Skin pigmentation can be yellow, white, or black. When a yellow-skinned hen begins laying eggs, skin on various body parts lightens in a given order: vent, eye ring, earlobe, beak, soles of feet, shanks. When she stops laying, the color returns in the reverse order.

Day-old chicks are clothed in fluffy, soft down. They begin growing true feathers within days and are fully feathered in four to six weeks. All genus *Gallus* birds, including wild jungle fowl, molt (shed their feathers and grow new ones) annually. Chickens molt from midsummer through early autumn, usually a few feathers at a time in a set sequence—head, neck, body, wings, tail—over a twelve-to-sixteen-week period. Molting chickens are stressed and can be skittish, moody, and irritable. Molting hens will lay fewer eggs or stop laying altogether.

Sexual Characteristics

Growing chicks generate secondary sexual characteristics—including combs and wattles—between three and eight weeks of age, depending on their breed.

Differences in size, color, plumage, wattle, and comb are easy to spot between a hen and a rooster.

All birds in genus *Gallus* are crowned by fleshy combs, and all of them, except the Silkie, sport a set of dangly wattles under their chins; other Phasianidae do not. Cocks develop larger and more brightly colored combs and wattles than their sisters. At about the same age, *cockerels* (young male chickens) begin crowing (pathetically at first) and sprout sickle-shaped tail feathers and pointed saddle and back feathers.

Pullets (young female chickens) reach sexual maturity and commence laying eggs at around twenty-four weeks of age. Although female embryos have two ovaries, the right ovary atrophies and only the left one matures. A grown hen's reproductive tract consists of a single ovary and a 2-foot oviduct, or egg passageway. The ovary houses a clump of immature yolks, waiting to become eggs. As each matures—about an hour after the hen lays the previous one—it's released into the oviduct. During the next twenty-five hours, roughly, the egg inches along the oviduct, where it may be fertilized, enveloped by egg white (albumen), sheathed in a membrane, and sealed in a shell. Because each egg is laid a bit later each day and because hens don't care to lay in the evening hours, the hen eventually skips a day and begins a new cycle the following morning. All eggs laid in a single cycle are considered a clutch.

Behavior

Chickens are easily stressed. Stress seriously lowers disease resistance, and stressed chickens don't thrive. Panic, rough handling, abrupt changes in routine or flock social order, crowding, extreme heat (especially combined with high humidity), and bitter cold can stress chickens of all ages. Labored breathing, diarrhea, and bizarre behavior are the hallmarks of stressed fowl. To keep stress levels low, it is important that you understand chicken behavior.

Pecking Order

In 1921, while studying the social interactions of chickens, Norwegian naturalist Thorlief Schjelderup-Ebbe coined the phrase "pecking order," now used to describe the social hierarchies of hundreds of species, including humans. In any flock of chickens, there are birds that peck at other flock members and birds that submit to other flock members. This order creates a hierarchical chain in which each chicken has a place.

A flock of chicks generally has its pecking order up and running by the time they're five to seven weeks old. Pullets and cockerels maintain separate pecking orders within the same flock, as do hens and adult roosters. Hens automatically accept higher ranking roosters as superiors, but dominant hens give low-ranking cocks and uppity young cockerels a very hard time.

In a closed flock with an established pecking order, there is very little infighting. Each chicken knows his or her place, and, except among some roosters, there is surprisingly little jockeying for position.

The addition of a single newcomer or removal of a high-ranking cock or hen upsets the hierarchy, causing a great deal of mayhem until a new pecking order evolves. Because brawls are stressful, it's unwise to move birds from coop to coop.

Low-ranking chickens are shushed away from feed and water by bossier birds, and they rarely grow or lay as well as the rest. Indeed, low-ranking individuals sometimes starve. If pecked by their betters until they bleed, they may be cannibalized by the rest of the flock. It's important to provide enough floor space, feeders, and waterers so that underlings can avoid the kingpins and survive.

Mating

Like adult roosters, cockerels soon begin strutting, ruffling feathers, and pecking the ground to draw the eyes of nearby hens. This behavior is called *displaying*. Chicken mating behavior is direct and to the point. The rooster chases the hen or pullet, the female crouches when the rooster mounts, and insemination occurs. Cocks tread with their sharp toenails and sometimes rake hens with their spurs while mating, occasionally to the point of shredding the poor biddies' backs.

The pecking order determines who eats first and who has to wait.

Tidying Themselves Up

When their surroundings permit, chickens are tidy birds. They preen by distributing oil (from a gland located just in front of their tails) over and between their feathers. They also dust-bathe. A shake of the feathers after dust bathing sets things right.

Chicken IQ

When they founded their show-biz animal-training business in 1943, operant conditioning (clicker training) pioneers Marian and Keller Breland began by training chickens. Among the couple's first graduates were a chicken who pecked out a tune on a toy piano and another who tap-danced wearing a costume and shoes. These were barnyard chickens rescued from a neighbor's stew pot.

Chickens are as intelligent as some primates, says Chris Evans, animal behaviorist and professor at Macquarie University in Australia. Chickens understand that recently concealed objects still exist, he explains—a concept that even human toddlers can't grasp. Chickens have good memories. "They recognize more than one hundred other chickens and remember them," says Dr. Joy Mench, director of the University of California-Davis Center for Animal Welfare. Dumb clucks? No, indeed!

House Chickens

Thanks to the invention of a nifty little item called the chicken diaper, many chicken fanciers keep a house pet or two. As one house-chicken advocate succinctly says, "What's the difference between a parrot and a chicken except $3,000?" Makes sense, and the chicken may even favor you with eggs! If you live in town, however, it's important to check your municipal chicken-keeping ordinances before moving any chickens indoors. Some cities that allow outdoor hens stipulate that you can't legally keep them inside your home.

Which Breed?

Any breed can make a fine house chicken, although the smaller the chicken, the less poop it generates. Silkies, Seramas, and similar bantams are perennial favorites, although some people prefer more substantial, full-size birds. If you have close neighbors, it's best to stick to hens (even a bantam rooster's crowing can be disruptive). It's logical to raise chicks in the house, but adults adapt to indoor living quite well. Many sick or injured chickens move indoors "just until they're well" and never go back to the coop again. One hand-raised chick that has bonded with humans instead of chickens makes a fine single house pet. However, if you start with non-bonded adults who grew up among other chickens, plan on two—then each has a friend.

Chicken Diapers

If you think you'd like to keep chickens in the house, investigate chicken diapers. Ruth Haldeman, member of the Yahoo! "People with House Chickens" e-mail group and founder of ChickenDiapers.com, is the queen of chicken diapers. She designed them and has custom-sewn them for clients since 2002. Visit her chicken diapers website, described by Ruth as "a clean place for chickens and their humans to sit" and read her chicken diapering FAQs.

Any kind of chicken with a tail knob and stiff tail feathers can wear a diaper (rumpless breeds such as purebred Araucanas and Manx Rumpys can't). Even chicks can wear diapers when they begin growing stiff tail feathers at roughly four weeks of age. Ruth crafts a stretchy, adjustable diaper for chicks, but a typical chick grows through at least three different sizes before it reaches full size.

An adult hen's diaper typically needs changing every two to three hours, but disposable plastic liners mean you needn't change the whole thing. Some people keep house chickens in comfy indoor cages at night and when they're away from home, so diapering is a part-time thing.

Chicken Stories

To some people, chickens are simply meat or egg makers. To others, intelligent, affectionate chickens are the best darn pets on earth. The

The petite Serama is a popular indoor chicken.

following stories were written by members of the "People with House Chickens" e-mail group. Read them, tame a special chicken or two of your own, and share your stories.

Piggy Hero

About three and a half years ago, I was living on the family farm in Brooks, Maine, and attended a local country fair where I fell in love with a fluffy black Silkie rooster who was sitting in a ten-year-old girl's lap, complacently allowing himself to be petted. I went back and forth from that poultry exhibit and finally went home with that roo-boy in a cardboard box.

That roo was Piggy (originally named Suleyman), and since I had only the one, and no chicken house set up, and because I like all manner of living things in my house, he became a very pampered and beloved house pet, and in many ways my very best friend. Piggy has never been constrained in any way, except to the house (though, of course, he does have supervised yard forays) and so has developed his own routines.

At night, Piggy makes his way to his bed, which is the back of the couch (eternally towel-draped), and, generally speaking, he will then sleep through all manner of activity and noise. Now and then, he'll arise with a second wind if the lights are turned on or if he hears activity in a distant room, but generally he sticks to his routine. In the morning, he hops up, crows, gets himself a bite to eat, and seeks out a bed to jump into, where he will greet any sign of movement by gently nibbling a nose or tugging on hair in an attempt to waken a sleeper for a petting session.

One night, Mark was away, and our housemate Paul and I were the only humans in the house. Piggy had gone to bed as usual several hours before, and Paul had retired to his upstairs bedroom. I was in my room at the back of the house, picking sluggishly at a couple of watercolor commissions. Around 11:30, I passed Piggy on the way to the kitchen to put on the tea kettle and then went to my room.

Two minutes later, I heard Piggy hit the floor with a thud. This caught my attention because it was late and there was nothing to wake him. I called his name. Suddenly, he rushed into my room and leapt onto the bed, clearly agitated and making distress noises. Thinking vaguely that perhaps an animal had gotten in and frightened him, I ran through the living room to find the kitchen half engulfed in flames, visible only as a large red glow through thick black smoke. I tried to turn on the light and thought the bulb must have blown.

I screamed for Paul and fumbled with the outside door until it opened, got the outside light on, and tore open the door of the bulkhead to the basement, where the garden hose had recently been stored. As Paul dialed 9-1-1, I turned the hose on full force, aiming it at the door of the kitchen as smoke pumped out like a huge black caterpillar.

By the time the fire trucks arrived, the walls and ceiling were still smoldering, but I had gotten the flames out. We were hustled from the house, and the fire axes went to work. At one point, I sneaked in and checked on Piggy, and he was standing alertly on the bed where I last saw him, so I left him there and went back outside while the firemen finished up.

They said that surely, after another minute, we would have lost the whole house and—they did not need to say it—possibly our lives as well. They asked what had happened, and I told them all, including Piggy's role.

The next evening, Piggy, who when he is not being a hero is just a regular stay-at-home kind of guy, ate a prodigious quantity of his favorite food—a peanut butter, cream cheese, and olive sandwich (you read right)—wiped his beak repeatedly on my leg, and snuggled down beside me for a nap. I intend to propose that a monument be erected in Piggy's likeness on the fire-station lawn!

—*Cindy Ryan*

A Tribute to Eggnus (2003–2004)

Eggnus came into our lives on April 6, 2003. I had placed an order for two chicks at a feed store, and when we picked them up, I saw a box labeled "White Rocks." Inside it were the cutest yellow chicks with a touch of peach on their chests! One chick looked up at me,

HOBBY FARM ANIMALS

SOMETHING TO CROW ABOUT

Everyone knows that roosters crow at daybreak, but the truth is, they will crow around the clock—even in the dead of night—if something arouses them.

Each rooster's call is unique. Some roosters crow louder than others do. Dominant roosters have higher pitched crows than those further down in the pecking order. Roosters inherit their style of crowing, so family members sound somewhat alike.

Hens occasionally crow, too. In the old days, a crowing hen was considered an unfortunate omen. In reality, a crowing hen may be ill, aged, or taking on the leadership of an all-hen flock.

The most remarkable avian crowing machines are roosters of the longcrower breeds. These roos start their calls like other roosters do, but their final note is an eerie, sustained sound that goes on and on until the rooster runs out of breath. Longcrowers of the Drenica breed of Kosovo reportedly crow for as long as sixty seconds at a time. American longcrowers crow for seven to fifteen econds (the typical rooster's crow is two to three seconds at best).

almost pleading for me to pick her, too. I couldn't resist; I had found my Eggnus.

I had a Silkie hen named Emma, who was raising two chicks of her own, and Emma adopted Eggnus, who immediately cuddled under Emma and joined the other chicks. All through Eggnus's life, she remained close to her foster mom.

PUNKY RAINBOW

Spring 2001 to March 26, 2003 Member of the Kroll Family

There was something about her. When she came home, she followed me everywhere. Punky Rainbow became a house chicken for a while. She would sing and follow me around the house. Punky kept her little wings outstretched as if preparing for takeoff as she and our family played tag up and down the hallway. We trusted each other completely. She was so gentle. I could pick her up without a single flap. Punky Rainbow would preen my eyelashes. Many city folk met her at our pet store; Punky was the first living chicken they had ever seen. When she moved to the farm to have babies, Punky and I still sat and talked. She seemed to understand what I said, and chirped and purred back to me. She went with us camping, to the beach, sledding, and to the pet store.

"Just this side of Heaven is a place called the Rainbow Bridge where Punky Rainbow is playing. We'll cross the Rainbow Bridge together."

—Jennifer Kroll, Fluff 'N Strut Silkies

However, Eggnus quickly outgrew her mom and the other large-breed chicks. She would trail behind the others, and when she'd run, we called it the "wiggle waddle." She would hop back and forth sideways, and she had trouble stopping when she started running; once, she even knocked down a rooster who was standing in her path!

When she was four months old, Eggnus went lame. I took her to a vet. Eggnus was the talk of the waiting room—this big white chicken in a dog crate. She made friends with everyone at the vet's office, including the vet's assistant, who told me that Eggnus was a Cornish/Rock cross.

The vet thought Eggnus had injured her back jumping down out of the coop because she was obese. At four months old, she weighed over 9 pounds! She was given vitamin shots and medicine, and I put her on a diet.

After that, Eggnus would go outside in the mornings with the flock, and then she'd come into the house, where she had her own personal fan and lived in a plastic crate in the hallway. She'd flash those big eyes at you and tilt her head sideways. She loved to be petted, and she greeted everyone who passed by.

Eggnus died January 3, 2004, in her sleep. I knew from the research I did on her breed that she wouldn't live to be very old. However, knowing this didn't make her passing any easier. We buried her beside a peony bush she had loved to lie beneath.

I knew Eggnus was special the first time I saw her—I just didn't know how special she really was. She always had a cheerful attitude and the sweetest personality. I will never know why a Cornish/Rock cross was in a box with White Rock chicks, but if she hadn't been, I would never have known my Eggnus.

A gentle giant of a chicken with a heart of gold: Eggnus, I will love and remember you always.

—*Patty Mousty*

It's infinitely easier to keep chickens healthy than to doctor them after the fact, and, in most cases, keeping them healthy isn't at all hard. To keep healthy, happy, bright-eyed chickens, begin with healthy, happy, bright-eyed chickens.

Buy from reliable sources. Don't stock your coop with someone else's rejects or bedraggled bargain fowl picked up at country swap meets. Consider starting with day-old chicks from reputable hatcheries and know what you're working with right from the get-go. Maintain a closed flock. Don't indiscriminately add chickens to your collection. It upsets the flock's hierarchy and causes infighting and stress. It's also the best possible way to introduce disease.

CHICKEN
HEALTHCARE

Provide suitable quarters. They needn't be fancy, but they must be clean, roomy, well ventilated, and draft-free. Feed your chickens what they need to thrive. Keep plenty of clean drinking water available and have enough feeders and waterers so that every chicken can eat or drink whenever it pleases.

Avoid unnecessary stresses. Keep your kids or dog from chasing the chickens, handle them gently, make changes gradually, and don't upset the status quo. Laid-back chickens tend to be healthy chickens.

It's important to recognize problems early, while they're still fairly simple to fix. Consult the "Is My Chicken Healthy?" chart on the next page for general signs of good health and illness.

Maladies: Parasites and Diseases

A savvy chicken raiser knows that early detection is a key to keeping chickens healthy. Most chickens are exposed to wild birds, which commonly spread parasites and disease. Left unchecked, parasites can spread like wildfire through your flock, causing anemia, weight loss, decreased egg production, and even death.

Internal Parasites

Zap internal parasites before they cause problems. However, don't rely on over-the-counter remedies. Instead, collect a community fecal sample using material from a number of chicken plops and take it to your vet to be checked for worm eggs. If worms are present, he or she will prescribe a custom-tailored wormer for your flock. Do this twice a year to keep your birds worm-free.

Communicable Poultry Diseases

It's unlikely that your pet chicken or small flock will succumb to one of the major exotic-poultry diseases. However, these diseases are out there, so you need to be able to recognize their symptoms. If you do suspect that one such disease has a toehold in your flock, it's your duty to report it to the health authorities.

Avian influenza: Avian influenza is a highly contagious respiratory infection caused by type A influenza orthomyxoviruses. Symptoms vary widely and range from mild to serious. Death occurs suddenly; the disease sweeps through a flock in just one to three days. Avian influenza occurs in many forms worldwide and must be reported in the United States.

Fowl cholera: Fowl cholera is an acute, relatively common disease caused by *Pasteurella multocida* bacteria. It spreads rapidly and kills quickly; chickens die within hours after symptoms appear. Humans handling infected birds sometimes contract upper respiratory infections.

Infectious coryza: Infectious coryza is found worldwide; in the United States, most cases occur in the southeastern states and in California. A respiratory infection caused by *Haemophilus paragallinarum* bacteria, coryza often manifests in combination with other chronic respiratory diseases. While contagious, it usually isn't fatal.

A healthy chicken is bright-eyed and alert.

Infectious bronchitis: Infectious bronchitis is a common, fast-spreading, and highly contagious respiratory disease caused by several strains of coronavirus. While chicks often succumb to the infection, adult chickens generally survive, but they remain carriers for life.

Fowl pox: There are two types of fowl pox: wet, sometimes called avian pox; and dry, also known as fowl diphtheria (neither is related to human chicken pox). Caused by the same virus and transmitted in the same manner, the former is primarily a skin disease while the latter affects both the skin and the respiratory tract.

Infectious laryngotracheitis: Infectious laryngotracheitis is a slow-spreading but serious upper respiratory-tract infection caused by a herpes virus. It manifests worldwide and commonly afflicts laying hens during the winter months.

Marek's disease: Marek's disease is a global scourge; it kills more chickens worldwide than any other disease. Marek's symptoms vary according to its victims' ages, but it often culminates in sudden death. Caused by six different herpes viruses, it is virulently contagious. Chicks vaccinated at one day of age are usually immune for life.

Newcastle disease: Newcastle disease manifests in many forms and in varying degrees of severity. It's not generally fatal, and survivors gain immunity for life. Caused by a paramyxovirus,

IS MY CHICKEN HEALTHY?

What to Check	Healthy	Unhealthy
Temperature	103–103.5°F	Higher/lower temperature
Respiration	Easy, even	Labored, rasping, coughing
Posture	Stands erect, head and tail elevated; alert and active	Hunkered down, tail and wings drooping; depressed
Appetite	Eats often	Not interested in food
Thirst	Drinks often	Excessive thirst
Manure	Formed mass, gray to brown with white caps; fecal droppings may be frothy	Liquid or sticky; green, yellow, white, red
Condition	Feels heavy, firm, powerful	Thin, weak, thin-breasted
Feathers	Smooth, neat, clean	Ruffled or broken, dirty, stained
Comb and Wattles	Bright red and firm	Shrunken, pale, or blue
Eyes	Bright and alert	Dull, watery, possibly partially closed
Nostrils	Clean	Crusty, caked
Legs and Feet	Plump, scales clean and waxy smooth, warm joints	Enlarged, crusty scales; hot, swollen joints; crusty, cracked, or discolored soles of feet

Disease can spread quickly. Keep your flock disease-free with care and cleanliness practices.

Newcastle can trigger minor eye infections in humans who handle the live vaccine or infected chickens.

Picking and Cannibalism

Although not diseases, picking (or pecking) and cannibalism are the most vexing aspects of chicken keeping and clearly major threats to the health of your flock. On occasion, chickens literally peck

EXTERNAL PARASITES

Poultry Lice

Although there are many forms of poultry lice, the most common are the chicken body louse (*Menacanthus stramineus*) and the shaft louse (*Manopan gallinae*).

- Fast-moving, six-legged, flat-bodied insect with broad, round head; 2–3 millimeters long, straw-colored (light brown)
- Female lays 50–300 tiny white eggs near base of feather shafts
- Does not suck blood; feeds on dry skin scales, feathers, scabs; gives infested birds a moth-eaten look
- Spends entire life cycle on host
- Primary infestation seasons are fall and winter
- Dust or spray birds and their environment using commercial products such as malathion, permethrin, Rabon, or Sevin (always consult your veterinarian before using such products)

Chicken Mite (also called Red Roost Mite)

Dermanyssus gallinae

- Slow-moving, eight-legged insect; 1 millimeter long (the size of coarsely ground pepper); gray to dark reddish brown
- Lays white or off-white eggs on fluff feathers and along larger feather shafts
- Sucks blood only at night; hides in cracks and crevices in coop or poultry building during the day
- Primary infestation season is summer
- Dust or spray birds and their environment using commercial products such as malathion, permethrin, Rabon, or Sevin (always consult your veterinarian before using such products)

Northern Fowl Mite

Ornithonyssus sylviarum

- Slow-moving, eight-legged insect; 1 millimeter long (the size of coarsely ground pepper); brown
- Lays white or off-white eggs on fluff feathers and along larger feather shafts located on host's vent, tail, back, and neck
- Sucks blood
- Spends entire life cycle on host; feeds day or night
- Primary infestation seasons are fall, winter, and spring
- Dust or spray birds and their environment using commercial products such as malathion, permethrin, Rabon, or Sevin (always consult your veterinarian before using such products)

Scaly Leg Mite

Knemidokoptes mutans

- Slow-moving, eight-legged insect; 1 millimeter long; gray
- Sucks blood
- Burrows into and lives under the scales of the feet, causing lifting and separation from underlying skin; results in swelling, tenderness, scabbing, and deformity, and related joint problems may occur
- Coating the entire leg shaft with petroleum jelly or vegetable, mineral, or linseed oil every two days may help smooth and moisturize scales
- Dust or spray birds, coop, and roosts with carbaryl products such as Sevin (always consult your veterinarian before using such products)
- Ivermectin pour-on (also called spot-on) is a systemic agent used to control both internal and external parasites

Pecking at scratch grains gives chickens something to do.

each other to death. Even worse, once it happens, the habit is readily established. It's important to nip this vile habit in the bud with adults and chicks alike.

Cannibalism usually takes root when one of the flock is injured. The sight of blood from an injured chicken draws her peers because chickens peck at anything red. Unless the injured chicken is removed, the pecking escalates, and she's likely to be pecked until she's dead. Although mildly injured chickens can be left in the flock and treated with

FIRST-AID KIT

Be ready for inevitable avian emergencies by assembling a chicken first-aid kit. You will need:

- A sturdy container. Supplies for a small flock fit neatly in a large fishing tackle box or a lidded bucket. Keep the container stocked; if you use something, replenish the used stock with a new supply. Store the kit where you can find it quickly. Write your veterinarian's phone number on the lid using permanent marker (program the number into your cell phone and post it by the house phone, too!).
- A flashlight with a strong beam. Tuck it to a zipper-seal bag with an extra set of batteries and a fresh bulb.

Into a second zipper-seal bag, place injury treatment materials such as:

- First aid tape in several widths
- Stretch gauze
- Gauze pads
- Cotton balls
- Wooden popsicle sticks to use for splints

To prevent spills and messes, store medicinals in a third zipper-seal bag:

- VetRX Veterinary Remedy for Poultry (for respiratory distress)
- Anti-pick spray or ointment
- Wound powder, herbal wound dressing, antiseptic spray or ointment
- Saline solution for cleaning wounds and injured eyes
- Bach Rescue Remedy
- Any other medicines that your veterinarian recommends

A fourth and final zipper-seal bag for tools:

- Scissors
- Dog toenail clippers (for snipping nails and beak tips)
- One or more 5-cubic-centimeter or larger catheter-tip disposable syringes (for feeding and watering debilitated chickens)

Add a pet carrier for transporting sick chickens to the vet and a quarantine cage, and you'll be set for most any emergency. Put your first-aid kit together now, before you need it. You (and your chickens) will be glad you did.

Various circumstances can result in
a bruised comb, like this rooster's.

Address the problem of overcrowding by removing some birds, moving the flock to a roomier coop, or turning them outdoors. You can also install additional waterers, feeders, and nesting boxes. Dim the lights, install fans, and create more shade to prevent the overly high activity level that results from bright lighting and the edginess caused by steamy heat. If you suspect that diet is the culprit in negative behavior, try different combinations until you find something that works. For instance, confined, solely scratch-fed chickens sometimes don't receive sufficient nutrients from their diet. If they begin pecking one another, switching to commercial feed or a commercial-feed/scratch blend sometimes helps. Conversely, chickens fed a strictly commercial diet sometimes peck out of boredom.

Strewing scratch grains, garden greenery, or acceptable table scraps adds dietary variety, and scratching and nibbling at these goodies gives idle chickens something to do. Flighty, nervous birds such as Leghorns, Minorcas, and many other Mediterranean breeds are more likely to peck; circumvent this problem by ruling them out for your farm. Finally, and most fundamentally, avoid situations that will leave your birds stressed, such as rough handling, temperature extremes, and abrupt changes in routine.

commercial products such as Hot Pick to thwart further pecking, it's better to separate them until they've fully recovered. In general, don't leave wounded, lame, weak, undersized, odd-colored, or otherwise unusual birds in a flock of aggressive peers. To preserve their lives, move them to safer quarters.

Many other factors influence cannibalism within a given flock, including overcrowding, intense lighting and heat, diet, breed-related problems, and stress.

MAJOR CHICKEN MALADIES

Disease	Symptoms
Avian influenza (viral)	Mild form: coughing, sneezing, decline in egg production
Fowl cholera (bacterial)	Oral and nasal discharge, diarrhea, ruffled feathers; high mortality
Infectious coryza (bacterial)	Green diarrhea, darkened head and comb, swollen feet, paralysis, swollen wattles, listlessness; high mortality
Infectious bronchitis (viral)	Watery eyes, putrid nasal discharge, swollen face and wattles, sneezing
Fowl pox (viral)	Coughing, sneezing, gasping, nasal discharge, respiratory distress, depression, marked drop in egg production
Infectious laryngotracheitis (viral)	Dry form: brownish-yellow lesions on unfeathered skin of head, neck, legs, and feet Wet form: labored breathing; lesions in the mucous membranes of the mouth, tongue, upper digestive tract, and respiratory tract; reduced egg production; usually low mortality
Marek's disease (viral)	Partial paralysis, blindness, wasting, tumors

There are few things more enchanting than a clutch of baby chicks. Chances are, you'll want to raise some, and you don't even need a hen to raise chicks. There are three basic ways to start your chick collection. You can take the quickest and easiest route by buying chicks from a hatchery. If, however, you want to be involved from egg to newborn chick, you can choose the most labor-intensive route instead—incubation. Or you may decide in favor of the old-fashioned approach to chick-making—with a hen.

Whether you start with hatchery chicks or incubate your own (in an incubator or under a hen), you'll arrive at the same point: with a clutch of chicks to brood.

RAISING
CHICKS

Hatchery Chicks

Most folks begin with hatchery chicks. They're inexpensive, readily available, and, if you shop around, you'll find scores of breeds and varieties to choose from. Call feed stores in your locale, especially prior to "chick time" (early spring), and ask what they'll be selling this year. Our small-town feed store, Hirsch's in Thayer, Missouri, sells the standard varieties, but it offers exciting rare breeds, too: German Spitzhaubens, Naked Necks, Black Sumatras, Dominiques, and Jersey Giants—even red jungle fowl!

The difference between mail-order and local pick-up hatchery chicks is that the former are more likely to be stressed. However, handled correctly on arrival, healthy chicks from solid hatcheries rebound quickly. Most mail-order hatcheries tuck in an extra chick or two for each twenty-five you've ordered to cover your losses in case a few don't make it.

Incubator Chicks

If you have access to fertile eggs and a sense of adventure, you can incubate your own chicks. You'll need an incubator and its accessories: a quality thermometer, a hygrometer to measure humidity, and a water pan if the incubator you choose doesn't have a built-in water reservoir. You'll need to build or buy an egg-candling unit, too.

Good home incubators are expensive, and incubating chicks is exacting work. If the atmospheric conditions inside the incubator are off for even a few hours, your eggs very likely won't hatch. Factoring in equipment costs, you can buy a lot of chicks for what it will cost you to hatch them at home. However, when eggs in your batch begin pipping (when the peeping chickies peck holes in their shells)—what a priceless thrill! So if you want to try home incubation, here are the bare-bones basics.

Choosing and Maintaining Your Incubator

There are two types of incubators: forced air and still air. A forced-air incubator is fitted with one or more powerful internal fans that continually circulate air around the eggs. The ones used by hatcheries and commercial growers are monstrous things capable of incubating thousands of eggs set in stacked trays.

Tabletop models designed for hobbyists and small-flock poultry keepers hold from forty to one hundred eggs in a single tray and cost in the neighborhood of $100–$600. Most are auto-turning units, meaning that they turn the eggs for you at predetermined intervals. More expensive models come complete with one or more thermometers and sometimes a hygrometer.

It's easier to maintain constant heat and humidity levels—essential for good hatches—in forced-air units. Temperatures are the same anywhere in a forced-air incubator; in still-air models, temperatures stratify, so it's considerably warmer near the incubator's lid than it is on its egg rack or floor. A still-air incubator relies on vents set into its sides, top, and bottom for ventilation, which is not a particularly reliable system. Forced-air incubators are initially more expensive than comparable still units, but they're worth it.

It doesn't cost a fortune to get started with a still-air unit and, with care, they'll hatch a lot of eggs. However, it's very easy to overheat eggs and essentially cook them. It's a great deal harder to regulate heat and humidity in a still-air incubator.

1. The egg is just starting to hatch. 2. The chick is on its way out. 3. And—finally—the new chick emerges!

Both types incorporate see-through lids or transparent observation ports, heating elements, built-in water receptacles or space to set a water pan, and egg racks. Automatic egg turners (which are very nice to have) are optional equipment on all but the most expensive models. Each brand and model differs from the rest. Unless you follow the instructions in the manual provided with your incubator, you're unlikely to have much of a hatch.

It's important to set up the incubator indoors, in a draft-free area, out of direct sunlight, and away from heat vents and air conditioning units. It must be thoroughly cleaned and disinfected before and after every hatch.

Two days before setting your fertile eggs, disassemble, scrub, and disinfect your incubator and accessories, and then put it back together and fire it up. This gives it a chance to come up to heat and also gives you time to adjust the temperature and humidity. Temperature, humidity, and ventilation must all be set properly and monitored to ensure successful development of the embryos.

Temperature

Place a trustworthy thermometer in the incubator at egg height. It should not be too close to the heating element and about 1 inch from the floor. In some units, the egg tray itself will do. Some folks play it safe and set thermometers in two different locations.

If your incubator has a thermostat, set it at 99–100° F for forced-air units and 101–102° F for those without a fan. Inexpensive incubators usually lack thermostats; in that case, follow the owner's manual exactly—you must be able to reliably regulate the heat. A tip: if your incubator is heated with standard light bulbs, use bulbs of different wattages to adjust the heat; a 40-watt bulb is a good one to start with.

Don't add eggs until you've maintained the correct heat and humidity for at least eight hours. The acceptable range is from 97–103° F. Check often, at least twice a day. You'll find that your vigilance will pay off in chicks.

Humidity

To prevent moisture loss from your eggs—and to get them to hatch properly—you must control humidity scrupulously. Relative interior humidity should stay at 55–60 percent from day one through day eighteen. For the last three days of incubation, humidity should be increased to 65–70 percent. Without a boost of additional moisture, pipping chicks stick to their shells. Maintaining that higher humidity level is so vitally important that you must not open your incubator during the final three days of hatching, even to turn the eggs.

You'll need a hygrometer to measure the incubator's evaporative cooling effect. Install the hygrometer so that its wick (not its bulb) is suspended in water. To determine relative humidity, compare incubator thermometer temperature and hygrometer (wet bulb thermometer) readings. The greater the spread, the more evaporation is taking place. Ideally, hygrometer readings will run between 83 and 87. Raise the unit's relative humidity, especially just prior to the three-day hatching period, by increasing the evaporative water surface. You can also use an atomizer to spritz moisture through your incubator's vents.

Ventilation

Oxygen reaches the developing chick embryos via the fifteen thousand or so pores in the average eggshell; carbon dioxide exits the same way. Hatching eggs must breathe, or the chicks inside will die. As embryos mature, they require more fresh air. Incubator vents provide ventilation in various ways; read your unit's manual to learn if, when, and how you should adjust the vents.

Manipulating the Eggs

A setting hen rolls the eggs—either unintentionally, while shifting its weight, or intentionally with her beak—several dozen times a day. If the hen didn't do this, the embryos would stick to their shell membranes and die. With no hen in sight, it's up to you to do the egg turning (as the process is called) necessary for your incubator embryos to thrive.

Turn the eggs first thing in the morning, last thing at night, and at least once during the day; turning them more often is definitely better. Alternate the turns so the side that is down one night will be on top the following night. Use a soft lead pencil to mark one side of each egg with an "x" and the other with an "o" so you can keep track of the turning cycles.

When placing eggs in still-air incubators, position them on their sides with their small ends tipped slightly downward; rotate them

POWER OUTAGES

If your power goes out, don't abandon the hatch. Carry your incubator to the warmest place in your house. If it's a forced-air model, crack the lid to admit fresh air; it's better to keep still-air models closed.

Should the power come on within a few hours, resume incubation, but candle the clutch four to six days later and discard any eggs that appear to have died. Surviving eggs require more time to hatch, sometimes as much as one or two days.

one-half turn at least three times a day. In forced-air incubators, place them small end down and tilted in one direction; tilt them in the opposite direction to turn them.

Do not turn eggs during the last three days before hatching. Chicks will be maneuvering into pipping position, and moving disorients them. In addition, to maintain constant 65–75 percent relative humidity throughout the prehatch period, you don't want to open the incubator for any reason.

Candling Eggs

It's also up to you to check that those embryos are indeed thriving by candling the eggs. Candling eggs is the process of holding an unhatched incubating egg up to a strong light source to examine the contents inside and determine whether the egg is fertile. Infertile eggs and those that have quit growing need to be removed from the incubator before they begin to rot.

Candling doesn't harm the growing embryos as long as you work quickly. Most folks candle eggs at seven days and again one week later. You'll need a dark room and a candling device. Turn out the lights, switch on your candler, and then, one by one, hold the eggs in front of the intense light.

Infertile eggs appear empty, with only a shadow of the yolk suspended inside. You'll spy a spider-like clump of dark tissue in fertile eggs: a glob with blood vessels radiating out from it like spokes. "Dead germs" are eggs that were fertile but have died; in such eggs, a faint blood ring circles the embryo. If you're not certain that you're seeing one, return the egg to the incubator; you can catch it next time.

White eggs are easier to work with than brown or speckled ones, and the second candling is trickier than the first. Is that embryo alive or not? It can be hard for a newbie to tell.

Incubating Timetable: Preparation to Pipping

First, you should buy or collect hatching eggs. Two days prior to setting them, scrub the incubator squeaky clean. Use a weak bleach solution or a commercial product to disinfect the unit and all supplies. Haul the incubator to the spot where it will be parked for the twenty-one-day hatch cycle and then fire it up. Spend the next two days fiddling with heat and humidity levels until everything is perfect.

Six hours before you plan to set them, remove your hatching eggs from storage and allow them to come slowly to room temperature.

Once incubation commences, religiously monitor conditions inside the unit. Make minor adjustments as needed. Turn the eggs at least three times a day, but don't open the incubator any longer or more often than you absolutely have to. Candle the eggs on days seven and fourteen. Discard those you're reasonably certain won't hatch.

Three days prior to the expected hatch (on the eighteenth day of incubation), turn the eggs for the final time and then crank up the heat and humidity. Don't open the incubator again until the hatch is complete. One day before the hatch, clean and disinfect previously used chick waterers and feeders. Set up your brooder and switch on the heat.

When pipping begins, don't help chicks out of their shells. Opening the incubator compromises the rest of the hatch, and a chick that can't break free is nearly always crippled or too weak to survive. Make sure to leave the chicks in the incubator for at least twenty-four hours, certainly until they're fluffy and dry. Taking them out early subjects them to chills. After twenty-four to thirty-six hours, remove the chicks to a warm box and carry them to their brooder.

Hold remaining eggs to your ear and listen closely. If they're going to hatch, you'll hear movement and possibly cheeping; put those eggs back in the incubator and discard the rest of the unhatched eggs. When the hatch is complete, disassemble, scrub, and thoroughly disinfect the incubator and its accessories before packing them away.

If 50 percent of the eggs you set hatch, celebrate! A lot can go wrong during the home incubation process, and a 50/50 hatch is a good one, indeed.

Chicks the Old-Fashioned Way

A cheaper and easier—and often more successful—way is to hatch eggs under a broody hen. Not all hens brood eggs, and none of them do it all the time. For example, Leghorns and Leghorn crossbred superlayers almost never brood. At the other end of the spectrum, some Silkie and Cochin bantam hens set and hatch chicks at the drop of a hat. Most bantam, dual-purpose, and heavy-breed hens (especially Asiatics, such as Brahmas and Cochins) will set; Mediterranean and Continental breeds don't often tend to brood.

The Hen

If a hen lingers in a nest box after laying, or if she ruffles her feathers and sputters when you take the egg, lift the hen off the nest and set her on the floor. Check on her later. If she's on the throne again, you've probably got a broody in the making.

The hen can set on the eggs she's accumulated if you allow her, or on eggs of your choosing, but you shouldn't let her set them in the henhouse because other chickens will pick on the hen. And if she leaves the nest for a daily constitutional and returns to find an interloper on the eggs, there will be a brawl—eggs can be shattered, and the broody might lose. It's best to set the hen up in private lodging.

Hens prefer cozy, secluded cubbyholes in which to set. Build your broody a disposable getaway by slicing ventilation slots that are 1 inch from the tops of all four sides of a lidded cardboard box that is slightly larger than the hen. Generously pad the bottom with chopped wheat straw or shavings.

Hollow out a bowl-shaped depression in the nesting material and then wait until dark to relocate your hen. When you do, wear gloves—the hen will peck *hard*!

Reach under the hen and remove the eggs. Nestle them into the broody-box nesting material and then move the hen. Slide your hand beneath her, fingers facing up, and wait until she quits fussing and settles down on your palm. Then transfer her to her temporary home and close the lid.

Place the box in a predator-proof, dimly lit spot. Get a small, deep container of water and cuddle it into the litter in one corner of the box; place a shallow bowl (tuna cans work well) of feed in another. Don't open the lid again until late the following day. This will give your hen time to acclimate to the new home, and it'll appreciate the solitude during this period. The hen will have food and drink if she wants it (she probably won't, but giving her the option will make you feel better), and it won't mess the nest. If the hen seems content, cut an entryway in one side of her haven and then pat your back—you've successfully resettled a hen!

Once a hen begins setting, her physiology changes. Body temperature drops to around 100° F, and the metabolism slows. The hen molts feathers from her breast and underpinnings, making it easier to warm the eggs and, later, the chicks. She remains on the eggs full time until they hatch, except for half an hour or so every day or two to drink, grab a bite to eat, and make "broody poop" (a memorably stinky gigantic glob of bird doo).

The hen will set until the chicks hatch or until it's obvious that they won't. If she's still setting on unhatched eggs after twenty-three days, she can foster day-old hatchery chicks or another hen's newly hatched babies. Put chicks in the hen's nest at night; by morning, chances are she'll decide that the eggs are her very own.

The New Chicks

Don't allow a hen with brand-new chicks to free-range or immediately rejoin the main flock. Other chickens may harry the hen or the babies, and when the little ones trail mama through wet grass, they easily chill. Chilled peeps are likely to die. Cats, hawks, skunks, and their kind adore chicken appetizers, so if you want your chicks to live, house them and their mama in their own separate coop.

Babies should be fed a chick-starter ration inside of a creep feeder. This is a structure with small openings that only they can enter. An overturned heavyweight cardboard box with chick-size openings carved into its sides makes a fine, free creep feeder. Weight the box top with a brick or flat stone just heavy enough to keep mama from tipping it over.

Make certain the peeps can always reach fresh water. Place a chick waterer in the brooder pen. Drop some marbles into the drinking surface to prevent the chicks from drowning and to keep it clean.

When they're a month old and nicely started, the chicks and their mama can join the flock. But monitor things for a while because some adult chickens are mean to other hens' chicks.

Hatching Eggs 101

Whether you hatch chicks under a real, live mama or inside a "tin hen," you'll want to begin with quality eggs. Following are several requirements for quality eggs.

Fertile. Hens don't need a rooster's input to lay eggs, but if you want fertile eggs, you definitely need the services of Mr. Roo. He shouldn't be too young or too old; he also should be actively breeding hens. You need one prime rooster for every eighteen bantam, twelve for light breeds, or eight if there are heavy-breed hens in your flock.

Clean. A poop-smeared egg can spread fecal-borne disease to the embryo within and to other, cleaner eggs in the clutch. Scrubbing won't help because it removes the protective natural sealant present on

Keep new chicks away from the rest of the flock at first.

newly laid eggs and is likely to force bacteria through the egg's porous shell. Some folks lightly sand away splats with finest-grain sandpaper, while others believe that doing so weakens the shell. Is it worth the risk? Probably not, especially if other, cleaner eggs are available.

Average size. Huge eggs don't hatch well; undersized eggs hatch undersized chicks. Average size is best.

"Egg shaped" and intact. Toss cracked eggs, misshapen eggs, and any with shells that are wrinkled, rough-textured, or thinner than normal.

Promptly gathered. Collect hatching eggs first thing every morning and recheck nests throughout the day. Don't let the eggs get chilled, overheated, or unnecessarily soiled due to being in the nest for too long.

Properly stored. Fertile hatching eggs should be placed in egg cartons, small end down and tilted to one side, and stored in a cool (50–60° F), humid (70–75 percent) place in your house. Never keep hatching eggs in the refrigerator! A simple way to turn them (at this point, you need to do this only once a day) is to slip a 1-inch slice of a two-by-four under one end of the carton and then move it to the opposite end the next day. Keep it up until you're ready to set the eggs. Don't store them longer than ten days—and less time is a whole lot better.

When You Don't Want Chicks

Not everyone wants their hens to brood. Furthermore, not every hen makes for the very best mom. Some abandon their nests mid-incubation, while others are so scattered that they stomp on and shatter their eggs. Less maternal souls sometimes hatch their broods successfully but dislike their chicks, so they'll peck their chicks' little noggins or even kill their babies (and eat them!).

Free-spirit hens gallivant off and leave their babes to their own devices. Whatever causes these hens to behave this way, they shouldn't be allowed to set eggs unless you're willing to snatch the newly hatched chicks and raise the peeps in a brooder.

If you want to turn your broody hen's mind to laying instead of hatching chicks, you have to "break her up." Breaking up a broody sounds easy—just cart the eggs away and off the hen goes—but it's not. Most broodies continue setting, eggs or not. And sometimes this goes on for a good, long (nonproductive) time.

If you've removed the eggs, but your hen stoutly insists on setting, try some proven ploys to get her to stop. First, after nightfall, pull on your gloves, hoist Ms. Broody up out of the nest, and resettle her in a completely new location for three or four days (or more, if needed). Disassemble the nest before you release the hen (toss the nest if disposable). If she was holed up in one of your laying nests and you moved it, scrub and disinfect it, refill it with fresh nesting material, and take it back to the henhouse where it belongs.

If that doesn't work, try temporarily housing the hen in a "broody coop," a wire mesh cage (a small all-wire rabbit hutch works well) suspended from the roof or a rafter, where cool air circulating around the hen's torrid underpinnings is likely to break through her single-minded trance. Just don't hang her in a draft! Three or four days in the broody coop should work like magic. Placing two handfuls of ice cubes under the hen several times a day or bathing her in cold (but not icy) water and then carting her off to short-term incarceration should also speed along the process and cool the underpinnings, too.

Whatever you try, persevere: you can outlast a hen! Provide your hen with plenty of food and water, despite what a few old-timers may tell you. Many seasoned chicken keepers claim that starving broodies makes them surrender sooner, but lack of food will weaken an already stressed bird, and she may even die.

To discourage hens from going broody in the first place, choose a breed that doesn't tend to set. Promptly remove fresh eggs from nest boxes because encountering an inviting collection of eggs triggers the urge to brood.

Brooding Peeps

So now you have chicks. Whether from a hatchery, from your incubator, or out from under a hen, they're yours to raise. It can be tricky, but if you stick to the rules, brooding peeps isn't hard. To have fun doing it and to lose fewer chicks, here's what you have to know.

The Brooder

Most small-scale or urban/suburban chicken raisers don't own brooder houses. Initially, it doesn't matter. Today's favorite chick

brooder is a clean, dry, draft-free semi-topless (leave the flaps on so that you can regulate ventilation) cardboard box stowed in your home in a warm, out-of-the-way spot. When the chicks outgrow it or it gets smelly, simply dispose of it. Transfer the chickies to a new box and shred the old one for the compost pile.

Other ingenious homemade brooders can be fashioned of plastic storage boxes, flexible plastic wading pools, old aquariums, dog crates, or rabbit hutches fitted with cardboard draft shields. Or you can always buy a ready-made brooder. A traditional galvanized steel box-type model with mesh floor, built-in water and feed troughs, and its own heating unit can house fifty chicks for up to fourteen days.

Stromberg's plastic brooder, which has a twelve-chick capacity and is made of high-impact plastic, is a good bet for pet and small-flock owners. It resembles an airline pet carrier, breaks down easily for cleaning, and heats nicely with a 75- to 150-watt incandescent lightbulb.

The Furnishings

Whichever type of brooder you have, you'll need to furnish it with a heat source, litter, feeders, and waterers. Consider the following when choosing which kinds to buy.

Heat Sources

Ready-made brooders usually incorporate their own heaters. If yours doesn't, or if your brooder is the homemade kind, you'll need a reliable heat lamp to warm those tender chickies, or they won't survive.

Infrared heat lamps, which come in clear- and red-bulb versions, provide the constant heat that wee chicks require. Red bulbs throw less light and are said to prevent juvenile picking that sometimes leads to cannibalism. When choosing an infrared lamp, opt for a UL-approved model with a porcelain socket and a lamp guard. If reusing an older unit, make certain that its cord isn't frayed.

Used improperly, heat lamps can burn down barns and homes, so be absolutely certain that the lamp can't fall or overheat nearby flammable surfaces. Hang it by a chain, not by its cord! You'll need one lamp fitted with a 75- to 150-watt bulb per fifty to seventy-five peeps. Storage-box and aquarium brooders can be heated using everyday lightbulbs in the 75-watt range.

Litter

The litter you choose makes a world of difference. It must be insulative, be absorbent, and provide lots of grip. Lodging new chicks on slick sheets of newsprint or flattened cardboard makes

their tiny legs slip to the sides, causing *spraddle leg* (a disability in newborn chicks caused by an inability to properly grip a surface with their feet). Spraddle-legged chicks can sometimes be salvaged by hobbling them with Band-Aids or makeup sponges until their legs correct, but spraddle leg is a problem better prevented than cured.

Sawdust sometimes confuses new chicks, who think it's food. Ingested sawdust leads to *pasty butt*—droppings stuck to a the chick's tush. If you don't soak or pick it off, the chick can't eliminate and will quickly die. Sawdust litter should not be used until chicks are a few weeks old.

First-class litter materials include pine shavings, coarsely ground corncobs or peanut shells, rice hulls, peat moss, sand, and old bath towels weighted down at the corners and laundered whenever they get soiled. Don't use hardwood shavings; some types are toxic to tiny chicks.

Initially bed the brooder area with 3–6 inches of litter, more if it's chilly outdoors and the brooder sits directly on a floor. Stir and fluff litter every day. Scoop water spills and messes as they occur. Add bedding whenever needed to keep things cozy and tidy.

Feeders

Flying saucer-shaped galvanized steel feeders with hole-studded snap-top lids are ideal for tabletop brooders. Folks with more chicks to brood will probably prefer trough feeders (allowing 2 feet of feeder per each twenty-four chicks).

Empty pressed-fiber egg cartons make easily accessible and disposable feeders for tiny chicks, although the chicks will climb on and poop in them, wasting a lot of feed in the process.

If you're raising meat chicks, choose 20–22-percent protein broiler starter for them. If they are pets or future layers, standard 18–20-percent protein chick starter is a wiser choice.

Most starters are medicated with Amprolium to prevent coccidiosis until the chicks develop their own immunity to this common disease. Some feed is laced with antibiotics, too. Nonmedicated feed is available, but you'll have to ask for it—and likely pay a premium price.

HATCHING YOUR OWN

The little incubators do work. I had to keep the room at a steady temperature, though. If the room got warmer, so did the incubator. All I had was the little thermometer that came with it and the little plastic wrap reflectors. It may have been beginner's luck, and there was certainly a lot of prayer involved, too!

—Patty Mousty

When you find a hen who likes to hatch eggs and does a good job of it, guard her with your life! You can set guinea eggs, duck eggs, and even goose eggs under her if she's big enough. Ducks or geese with a chicken mom are such fun. It drives the chicken mom crazy when her chicks make a beeline for mud puddles!

—Marci Roberts

I think it would be better for new chicken people to buy chicks. There are many things that can go wrong when you incubate, and that could discourage a new chicken owner.

—Helen Jenson

Unless you supplement your chicks' diet with scratch, they won't require grit. If you do choose to supplement the food, buy special chick grit or caged-bird grit (available from a pet supply store).

Waterers

If your brood is a small one, you'll need one plastic-based quart-size Mason jar waterer for each dozen chicks. For larger broods, a gallon-size waterer serves fifty chicks. Chicks easily drown in waterers, so whichever type you choose, add marbles or pebbles until only their beaks can get wet. When you place units in the brooder, don't position any near the heat source; chicks don't fancy warm water and might not drink at all if it's hot. Empty and brush scrub waterers every day; rinse them with a weak bleach solution once a week. Make certain that waterers are filled and functional at all times.

Many veteran chicken raisers spike peeps' drinking water with table sugar (one-quarter cup sugar per gallon of water) to give them a needed energy boost. Others swear by vitamin and electrolyte supplements such as Murray McMurray Hatchery's Quik-Chik and Broiler Boosters. If you plan to use water supplements, lace the peeps' drinking water right from the start.

Putting It All Together

At least twenty-four hours before anticipated hatch or delivery, set up your brooder, switch on the heat lamp, and bring everything up to heat. The temperature at chick height (2–3 inches from the floor) must run a constant 95° F for the first full week; use a thermometer to check it twice a day. You'll lower the temperature by about 5° F each week until the chicks are five weeks old. After that, maintaining heat at 70° F (or indoor room temperature) generally does the trick.

Chicks instinctively pick at whatever they see, so for the first few days, carpet their litter with nubby paper towels. Sprinkle a thin layer of chick starter on it to encourage the chicks to pick at feed instead of litter. Tempt slow-learning chicks with chopped boiled egg yolk sprinkled on paper towels or drizzled atop their regular feed.

Listen and Look

Listen to and watch your chicks. Contented chicks converse in gentle cheeps, while frantic, shrill peeping means that they're chilly. Up the heat by moving the lamp closer to the chicks or substituting a higher wattage bulb.

Happy chicks spread throughout the brooder, allowing each other plenty of space. Chilled chicks cluster beneath the heat source, sometimes piling atop one another and suffocating the babes at the bottom of the heap. Overheated chicks extend their teensy wings and pant; they flock to the outer edges of the brooder to flee excessive heat. When the peeps huddle at one side of the heat lamp or another, suspect a draft.

Ongoing Care

Once chicks are eating well, fill feeders halfway (to reduce waste), but never let them run out of water or feed. If feed gets damp or dirty, dump it; rinse and dry the feeder before replenishing it with fresh starter.

Carefully and frequently handle future pets, but only for a few minutes at a time; otherwise, it's best to leave tiny chicks in the brooder. Supervise children and remind them that the peeps are fragile. Make sure that kids wash their hands after handling chicks.

Nip toe- and feather-picking in the bud. Chicks pick each other when they're too hot or too crowded, when their light is too bright, if the air is too stale, if their feed is inadequate, or sometimes simply because they feel like it. Picking leads to cannibalism—not a pretty sight.

If your chicks begin picking one another, add grass clippings to their diet. Strew bits of healthful greenery around the floor and let them pick at that instead. Switch the clear heat lamp or lightbulb for a red one, which should have a calming effect.

Remove picked chicks to safer quarters and dab their wounds with antipick solution, such as Hot Pick, Blue Kote, or pine tar—an old favorite—to heal them and deter further picking.

As chicks mature, provide additional floor space, feeders, and waterers. By the time they're six weeks old, chicks are fully feathered and fit to face the world. It's time to move pullets to the henhouse and meat chickens to quarters of their own.

Whether you keep chickens for pleasure or for profit, you probably either use or sell their yummy eggs. Raising chickens for meat is a sure way to know exactly what does—and does not—go into your bird before it reaches your table. Homegrown poultry and eggs are infinitely fresher and tastier than anything you can buy in a store, and producing wholesome, farm-fresh eggs and meat is cost-effective and relatively easy.

FARM FRESH EGGS AND MEAT

CHICKENS

Egg-Laying Hens for the Job

Among production breeds, Leghorns are the queens of the henhouse; they lay early and often. Their compact, wiry bodies put everything they have into laying eggs, yet they're not many small-flock keepers' first choice. Leghorns are noisy and flighty, and if you plan to eat your spent hens, there's not much meat on those bones when a Leghorn's laying days are through.

A slightly more substantial layer is the Red Star sex-link hen, a reddish brown bird accented with flecks of white. The product of a Rhode Island Red rooster and a Leghorn hen, the Red Star sex-link lays handsome brown eggs and is far less flaky than a Leghorn.

The Red Star's close cousin is the gold-accented Black Star sex-link hen (cockerels are black with white barring), a cross between a Plymouth Rock hen and a Rhode Island Red rooster. The Black Star lays bigger, although slightly fewer, eggs than the Red Star. Like the Red Star, she's a fairly easygoing hen.

Old-time layers and dual-purpose chickens, such as Brahmas, Dominiques, Cochins, and Wyandottes, have a place in today's henhouses, too. They begin laying later and don't lay as many eggs, but they keep it up longer than production-breed hens.

Bantams for eggs? Sure, why not? Bantam eggs are tiny, but fanciers claim that they're the tastiest of all hens' eggs, and bantam layers require less feed and space.

Egg-Laying Timetable

When your home-raised pullets are six weeks old and ready to leave the brooder, move them to their own safe haven, away from aggressive older chickens. Switch their feed to a 15–16-percent protein grower ration and optional supplements such as "big girl" scratch and greens. Don't forget to set out a free-choice grit container.

Around twenty weeks of age, upgrade to a 16–18-percent layer ration and add calcium-boosting, free-choice oyster shell alongside their grit. Never let feeders or waterers run dry. Keeping fresh, pure drinking water in front of your hens is a must! Even a few hours without water affects their lay. If your hens are superlayers, such as Black Star or Red Star sex-links, or from fast-maturing production Leghorn strains, they'll begin laying at between twenty and twenty-four weeks of age.

A pullet's first eggs will be teensy treasures. New layers rarely grasp the concept of nest boxes, so you'll find eggs wherever they land. Tuck an artificial egg, such as a wooden or marble one, a sand-filled plastic Easter egg, or a golf ball, in each nest. Pretty soon, your hens will understand and begin laying in the nesting boxes.

By week thirty-two, most hens will be up to form. They'll continue laying full bore for at least two years and can continue laying for up to twelve years. As they age, hens' eggs will increase in size but decrease in number.

Remember that hens don't lay while brooding or raising chicks, and they may stop laying as winter days grow short. Hens require fourteen hours of daylight to keep producing, so in northern climes, lighting the henhouse is an absolute must. Add extra hours of light before sunrise so your chickens will naturally go to roost at dusk. Use a timer so you don't forget; it's important to be consistent. Just one or two days without additional lighting can throw their production out of whack. Or, give your hens a winter break; they'll ultimately last longer if you do.

Each year, your girls will molt; they'll shed and regrow their feathers a few at a time. Molting generally begins as summer winds down and extends for twelve to eighteen weeks. A fast molt is a fine chicken trait indeed because egg production slows or ceases as Ms. Hen molts. The sooner molting is finished, the sooner the hen will lay.

Stressed hens lay fewer eggs. A whole passel of things can stress chickens: extreme heat or cold, fright, illness, parasites, adding new chickens to the flock, or taking away familiar friends. Business as usual keeps stress down and is good for laying. Strive for peace and serenity in the henhouse if you love fresh eggs.

Who's Been Eating My Eggs?

Everyone loves fresh eggs. Even hens. Egg eating begins innocently enough, when an egg is accidentally cracked or shattered and a curious hen takes a nibble. *Mmm-mmm*, good! The hen keeps an eye out for more golden goodness, and when the other hens notice, their curiosity is piqued. They sample, too. Yummy! One fine day, one of them realizes that if it pecks really hard, it can sometimes serve itself. Pretty soon, your flock is eating more eggs than you are. What to do?

- Revamp your coop's nesting area. Provide more nests for the flock so there is less traffic and, ultimately, fewer broken eggs. After all, a single broken egg can trigger this hard-to-zap habit.
- Relocate nesting boxes away from the fast lane. Install them at least 24 inches from the floor in a secluded corner of the coop.
- Keep plenty of clean, cushiony litter in each box. Protective padding saves many an egg.

- Ban broodies from the henhouse. They're happier off by themselves and are not taking up valuable space in the henhouse.
- Feed high-calcium commercial layer rations with oyster shell served up free-choice to strengthen the eggshells.
- Pulverize eggshells to feed to your hens. They're a dandy source of calcium, and getting used to the taste won't give a hen that "ah-ha!" moment when she realizes she's dining on egg.
- Stressed chickens pick. Keep everything in your hens' environment as low-key as you possibly can. Avoid changes in their daily routine, and never let them run out of fresh feed and water. Don't introduce new chickens to an established laying flock. Keep your chickens reasonably cool in July and warm in February. Absolutely avoid overcrowding.

PICK A COLOR

Brown eggs

Aseel • Australorp • Brahma • Buckeye • Chantecler • Cochin • Cornish • Delaware • Dominique • Faverolles • Java • Jersey Giant • Langshan • Malay • Marans • Naked Neck • New Hampshire • Orpington • Plymouth Rock • Rhode Island Red • Sex-Links (both Red and Black) • Welsummer • Wyandotte

White eggs

Buttercup • Campine • Crevecoeur • Dorking • Houdan • La Fleche • Lamona • Leghorn • Minorca • Polish • Redcap • Silkie • Sultan • White Faced Black Spanish • Yokohama

Tinted eggs (not quite white)

Ancona • Campine • Catalina • Hamburg • Lakenvelder • Modern Game • Old English Game • Sumatra

Colored eggs

Ameraucana • Araucana • Easter Egger

GREAT-TASTING EGGS

Follow these tips for clean, great-tasting eggs:

- Use plenty of cushy nesting-box litter and check it often. Remove muck promptly. Dump everything out and replace it with fresh, fluffy litter at least once a month.
- Collect eggs first thing in the morning and at least once or twice more during the day. The longer the eggs stay in the nests, the more likely they'll become mud-smeared, pooped on, or cracked. For the same reason, supply plenty of nesting boxes.
- Collect eggs in a natural fiber or coated wire basket. To prevent breakage, don't stack them more than five deep.
- Don't clean eggs if they don't need it. Carefully erase minor spots with fine-grit sandpaper. When you must wash soiled eggs, do it before the eggs cool. Cooling causes shells to contract and suck dirt and bacteria into their pores.
- Use water that is 10° F warmer than the eggs themselves. This causes their contents to swell and shove surface dirt out and away from the shell pores. Gently scrub them in plain water or mild egg-cleaning detergent (available from poultry supply retailers). Never soak eggs.
- Dry washed eggs before storing them.
- Refrigerate eggs in cartons, large-end up. To preserve quality (and prevent development in fertile eggs), get them into the fridge as soon as you can. Date the cartons and use up older cartons first.
- Eggs absorb odors from strongly scented foods, so try not to store eggs with such items.

- Handle hens gently when collecting eggs. If a hen breaks eggs in a hasty retreat, clean up the mess right away.
- Identify culprits and cull them to pet homes—or even the stew pot. You'll know them by the dried yolk remnants decorating their beaks and heads.
- Don't assume that your hens are noshing all of the eggs. Predators such as skunks, opossums, and the occasional snake fancy chicken eggs, too.

Raising Chickens for Meat

Start with meat-breed chicks. Chicken-raising newbies who spring for low-priced Leghorn cockerels will be disappointed. Light breed chickens guzzle twice the feed and never flesh out to prime eating size.

Super or Dual-Purpose Birds

Superbroilers convert feed to flesh at lightning speed. They take eight to twelve weeks from hatch to slaughter at 4–5 pounds live weight, or they can be slaughtered earlier (at five to six weeks as Cornish Game Hens) or later (at twelve to twenty weeks and 6–8 pounds live weight) if the raiser prefers.

Old-fashioned dual-purpose chickens make for delicious eating, too, although they mature slower and demand more feed for each pound of weight they pack onto their sturdy frames. Old standbys such as Rhode Island Reds, White Rocks, and New Hampshires need twelve to sixteen weeks to grow to broiler size, but their slower growth spares them the health and structural problems that superbroilers experience during their far shorter lives. Old-fashioned breeds are more flavorful, too. If you're licensed to sell dressed birds, meat from most such breeds can be labeled as Heritage chicken.

Most superbroilers and popular dual-purpose breeds are yellow-skinned, white-feathered chickens, simply because that's the kind of bird Americans prefer. Brown and black semi-superbroilers can be ordered from large hatcheries such as Stromberg's and Murray McMurray. They're more active than white Cornish Rocks, and they bloom a smidge more slowly, but, like dual-purpose broilers, they're less susceptible to the structural problems that plague their faster-maturing kin.

Broilers' Timetable and Requirements

You must feed your meat peeps 20–22-percent protein broiler starter; less protein simply won't do. Figure 100 pounds of starter per twenty-five chicks.

When your chicks leave the brooder at roughly six weeks old, switch to broiler finisher. They'll remain on this ration until they're slaughtered.

Most folks raise broilers indoors, allowing 2 square feet of floor space per bird between six and ten weeks of age and then 3 square feet until slaughter. Continuous, ultra-low lighting encourages nighttime noshing, which is especially important during sizzling summer climates. Some growers limit feed intake, others keep feed in front of their birds all the time. Whichever method you choose, make sure that plenty of clean, cool water is available. Adding a vitamin-electrolyte product to their drinking water is a wise idea, too.

Broiler Don'ts

Unfortunately, astronomical weight gain comes at a hefty price. Cornish Rock broilers' broad-chested, meaty bodies mature faster than their skeletal structures can support. Crippled legs and crooked breastbones are the norm.

Lame birds crouch to ease pain, and they develop breast blisters. Superbroilers are also prone to heart attacks. For best results with swift-maturing meat birds, heed this advice:

- Keep litter fluffy and dry. Crouching on hard-packed litter irritates broilers' keel bones.
- Don't give meat chickens roosts; as they hop to the floor after roosting, they'll damage their legs. Pressing against roosts causes

The Rhode Island Red is a popular dual-purpose breed.

breast blisters, too. Remove anything in their environment that they could leap up on.

- Don't capture or carry meat fowl by their legs.
- Don't startle or chase these injury-prone fowl. Broiler stampedes equal torn muscles, slipped joints, and heart attacks.
- Lodge lame chickens in separate quarters with easy access to food and water. Many will recover within a week. To prevent needless suffering, cull any birds that don't.

Making Meat Down on the Farm

You can butcher all of your broiler chickens at the same time or create a continuous home supply by starting a new brood when the first is four weeks old and then slaughtering one-fourth of the older group at seven, eight, nine, and ten weeks of age. Process roasters at 6—8 pounds live weight. The older they get, the less efficiently they convert feed to meat.

If you've never processed chickens before, have someone show you how. The process isn't tricky, but it should be done with precision. Second-best, print or download a heavily illustrated university publication and follow its instructions to the letter.

Bucks for Clucks

You don't need a huge flock of chickens to earn a tidy amount of egg money with your feathered friends. In fact, there are several tried-and-true ways to make money with chickens. Here are a few to consider.

Sell Fresh Eggs

It seems logical to sell excess eggs to family and friends or maybe from a roadside stand. But first, investigate the egg laws in your community, county, and state to make sure that selling eggs is legal and, if it is, to learn how you're required to store and market your farm-fresh eggs.

Many states' egg statutes are available online, but if yours aren't, your county extension agent is the person to see. He or she can put you in touch with your state's egg-sales regulatory program. Some states impose stiff penalties for simple infractions that you may not even be aware of, so don't omit the important step of familiarizing yourself with the regulations.

Consider my own state's rules. The Arkansas Egg Marketing Act states that people who own less than 200 hens can sell eggs directly from their farms, providing that the following requirements are met:

1. Eggs are washed and clean;
2. Eggs are prepackaged and identified as ungraded with the name and address of the producer;
3. Used cartons are not used unless all brand markings and other identification is obliterated; and
4. Eggs are refrigerated and maintained at a temperature of 45° F or below.

Small-producer laws such as these usually apply to selling eggs from your farm. Selling at farmers' markets or flea markets, through community-supported agriculture (CSA), or directly to restaurants or grocery outlets—even natural food stores—generally requires you to be licensed, and more stringent laws apply.

Eggs for Direct Sales

When you market eggs, go the extra mile to produce a really fresh product. Keep those nest boxes extremely tidy, gather eggs at least twice a day (more often is better, of course) to prevent soiling, and get the eggs under refrigeration as quickly as possible.

KEEP IT CLEAN

To quash disease and avert parasite problems, always strip, scrub, and thoroughly disinfect your broiler quarters between batches of birds.

If you do, and you raise your chickens indoors, you usually needn't vaccinate or deworm them, especially if you feed them medicated commercial rations.

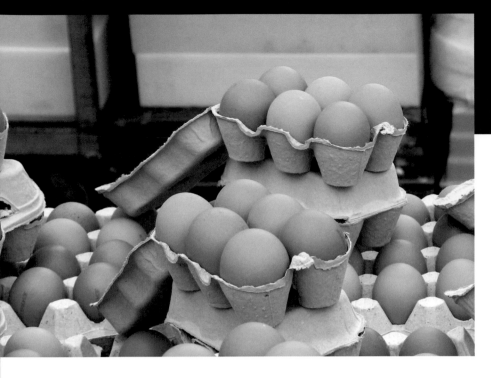

You can often find fresh eggs at farmers' markets.

Choose productive breeds that lay the color of eggs—brown or white—that buyers in your locality prefer. Consider a specialty product such as blue eggs from Araucana hens (the breed is on The Livestock Conservancy's Heritage Chicken list), free-range eggs, or organic eggs. However, if you call your eggs organic, be sure that they really are. Stiff penalties apply to marketing nonorganic foods as organic.

Even if you don't wash eggs for home use, you will be required by most state egg laws to wash eggs that you are going to sell. Wash eggs carefully, place them in clean cartons, and pop them in the refrigerator as soon as you can.

Sell Hatching Eggs

Another good way to earn a bit of egg money is by selling hatching eggs, especially if you keep rare or unusual purebreds that other chicken fanciers would like to own. Hatching eggs and an incubator are the ideal solution for people who want to raise rare-breed chicks but not the twenty-five peeps needed for live shipment.

A good place to sell hatching eggs is on eBay (there are more than 1,000 hatching-egg auctions up for bids as I write this) or Eggbid, the eBay of the poultry world. Ads in chicken fanciers' magazines bring results, as do classified ads in magazines such as *Hobby Farms* and *Chickens*. Don't forget freebie ads in *PennySaver*-type newspapers.

You must be ultra-conscientious when selecting eggs to sell for hatching purposes. Collect eggs at least four times a day, sort them for size and conformity, and store them in a room in neither cold nor hot conditions, away from drafts and sunlight. Turn them at least twice a day. Eggs should be no more than three days old when you ship them. Hatching eggs should be clean but unwashed, fertile, and well shaped. While it's widely understood that a seller

WHAT ABOUT THOSE CARTONS?

Nearly every state's egg laws prohibit the reuse of commercial egg cartons unless all previous markings are fully eliminated. Most states, however, allow you to reuse your own cartons if customers bring them back to be refilled.

Salvage commercial cartons, if you like, by covering original markings using a product such as Diagraph Quickspray Blockout Ink, an aerosol or brush-on cover-up that comes in tan or white. It doesn't, however, look professional, so if you're promoting your eggs as a quality product (and you should), consider starting with new, unmarked cartons. Eggcartons.com and similar companies sell plain or custom-imprinted paper, foam, and plastic egg cartons at discount prices. Mail-order chick hatcheries often sell them, too. Even farm stores sometimes keep new egg cartons in stock. Paper (or pulp) cartons generally cost less, and they're earth-friendly, so many consumers prefer them. Be sure to order some six-egg cartons as well as twelve-egg models; you can market six eggs at a slightly higher cost per egg, and they're convenient for clients who don't consume a lot of eggs.

In virtually every state, you must mark your cartons with your name and contact information. If you're selling specialty eggs (free-range, organic, Heritage), that information should be prominently noted on each carton, too. Make a good impression by printing an attractive label to affix to blank egg cartons. Consider adding tips on the perfect boiled egg or a recipe for French toast. Consumers like this type of information, and it makes your product stand out among the rest.

Hatching chicks to sell may require special licenses and health certifications.

assumes no responsibility for eggs once they're shipped, send the best eggs you possibly can.

There are many secure ways to ship hatching eggs, but this is a good method that has worked for me:

- Obtain free US Postal Service Priority Mail or Express Mail shipping boxes from your post office. The size 4 Priority box (6 × 6 × 6 inches) is great for sending a few eggs; size 7 (12 × 12 × 8 inches) is perfect for shipping a full dozen or more.
- Use lots of bubble wrap and packing peanuts. Cut strips of bubble wrap about 6 inches wide and 12 inches long to ship full-size eggs. Wrap each egg around its sides and tape it, then follow with a second piece of bubble wrap and tape. You can still see both ends of the egg at this point. Wrap a third piece of bubble wrap around the egg lengthwise to protect the ends and then set that egg aside.
- Place a layer of packing peanuts in the bottom of your box; then line it with several layers of bubble wrap and place the eggs close together inside. Pad the sides as needed, leaving several inches at the top in case the shipper sets something heavy atop your box. Fill that space with packing peanuts, too.
- Close the box and check to make sure that the sides don't bulge (if they do, eggs may crack when the box is stacked with others at the post office or in transit), then shake the box lightly to be sure that there is no movement inside. Finally, seal all of the edges using wide packaging tape.
- Mark all sides of the box "Fragile—Hatching Eggs." Also write the buyer's phone number on the box in large letters, along with a note requesting that postal workers call the buyer when the box arrives at the post office.

PRICING EGGS

Don't underprice your eggs. Customers expect to pay more for niche products, so price your eggs accordingly. Visit at least three mainline grocery stores and one natural-foods store to see what eggs are bringing in.

- Ideally, you should send the box on a Monday morning, never on a Thursday, Friday, or Saturday, when it might be left in a cold or hot truck or mail-handling facility over the weekend.

Sell Live Chickens

You can also incubate extra chicks to sell, although you should talk to your county extension agent before you do. Many states consider anyone who sells live chicks a hatchery, requiring the seller to obtain a license and show proof that his or her flocks have been tested for pullorum disease. Swap meets are popular places to sell chicks, but they also present many opportunities to expose unsold chicks to disease.

Another option is to raise those chicks and sell them as ready-to-lay pullets or as young laying hens. Many people are anxious to get into chickens but don't want the bother of raising chicks. In most places, pullets and young laying hens of everyday breeds fetch between $15 and $30 per chicken.

If you raise show-quality chickens or rare-breed fowl, you may choose to sell them to breeders far away from your locale. You can do this, weather willing (chickens can't be shipped during extremely hot or cold weather), although it isn't cheap (figure about $40 to ship an adult heavy-breed chicken).

Breeders frequently ship birds from pullorum-free flocks via USPS Express Mail, and you'll need an approved bird-shipping carton to do so. This is a sturdy cardboard box with a tapered top and holes punched in it, designed to safely transport one or more 6-ounce or larger birds. The tapered top ensures that the box won't have others stacked atop it while the holes allow for adequate airflow. Suppliers, such as Horizon Micro-Environments, and some major hatcheries carry these boxes. The Horizon Micro-Environments boxes are sized to ship up to two full-size

DID YOU KNOW?

The American Poultry Association says that hens of the best laying breeds, such as Leghorns and their sisters from the Mediterranean class, lay between 250 and 280 eggs per year. More conservative sources claim that a good hen lays about 180 eggs per year.

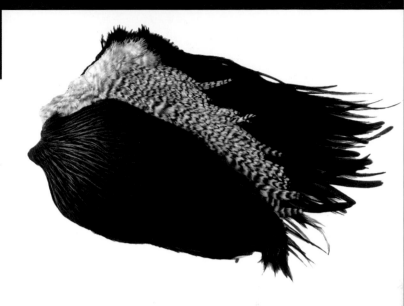

Rooster capes with feathers of various types, to be used for making flies.

or four adult bantam chickens and have mesh inserts under the preperforated breathing holes and even provide space for the buyer's mailing address.

Insurance guarantees that your birds will reach their destination in the quoted time frame, but it does not guarantee that they'll be alive, so it's up to you to take pains to ensure their safe shipment. Make certain that they are healthy, well fed, and well hydrated prior to shipping day. Place a thick layer of absorbent material in the bottom of the box (coarsely chopped wheat straw works well) and supply moist edibles such as orange or apple slices, fresh pineapple slices, or watermelon rind to help hydrate the birds during shipment. As with hatching eggs, write the buyer's phone number on the box and request that the buyer be called as soon as the birds arrive.

Find out when mail pickup occurs, and take boxed birds to your post office no more than an hour before the truck is scheduled to arrive. Save your copy of the shipping form and call or e-mail the tracking number to the buyer. Be sure that the buyer has your phone number, and ask him or her to call or e-mail you when the birds arrive.

Sell Feathers

Fishing-fly tiers and craftspeople treasure the natural, undyed saddle, hackle, and tail feathers of some breeds and varieties of roosters. Fly tiers often prefer feathers from bantam versions because of their smaller, more delicate feathers.

To harvest the feathers for sale, prepare capes from roosters that you slaughter to eat. A cape is a dried skin that comprises the neck, back, and sides of the bird. Remove it in one piece and tack it onto a hard surface, skin-side up. Scrape all of the fat and meat from the skin and then liberally dust the skin with powdered borax (find it in the cleaning section at the supermarket). Let the skin sit for a couple of weeks, scraping off any moist areas and reapplying borax as needed. When a cape is fully dry, store it in its own roomy zipper-seal bag with a little extra borax and a few whole cloves (borax keeps the cape dry, and the cloves improve its overall aroma) in a dry, cool location away from marauding cats and dogs.

If there is any chance that your feathers or finished capes could have feather mites, treat them to a two-week stint in the freezer. Then, package them attractively and sell them from your craft-show or farmers' market booth or online.

Bang Your Own Gong

It doesn't take a lot of money to promote a home business, but promote it you must. Customers are out there, but they have to know where you are and what you sell. There are a lot of little ways to make a big promotional splash, such as creating a website, handing out business cards, posting chicken videos on YouTube, or writing a blog. If you're thinking outside the box, how about giving a demonstration, preparing an educational program for your kids' school, or sponsoring a 4-H poultry project. Offer to write a chicken column for a newspaper for free to get your name and your business out there. Hold free consultations to folks interested in raising chickens. Invite them to your farm, show them your chickens or eggs, and give them good, solid advice—without a sales pitch.

WHAT ABOUT DRESSED CHICKENS?

While it's perfectly possible to sell dressed birds, don't do it before discussing the legalities with your county extension agent, who can put you in touch with the organization that oversees the sale of home-processed meat in your state. If you want to sell meat chickens, it's easiest to sell them as live birds and let buyers take the process from there.

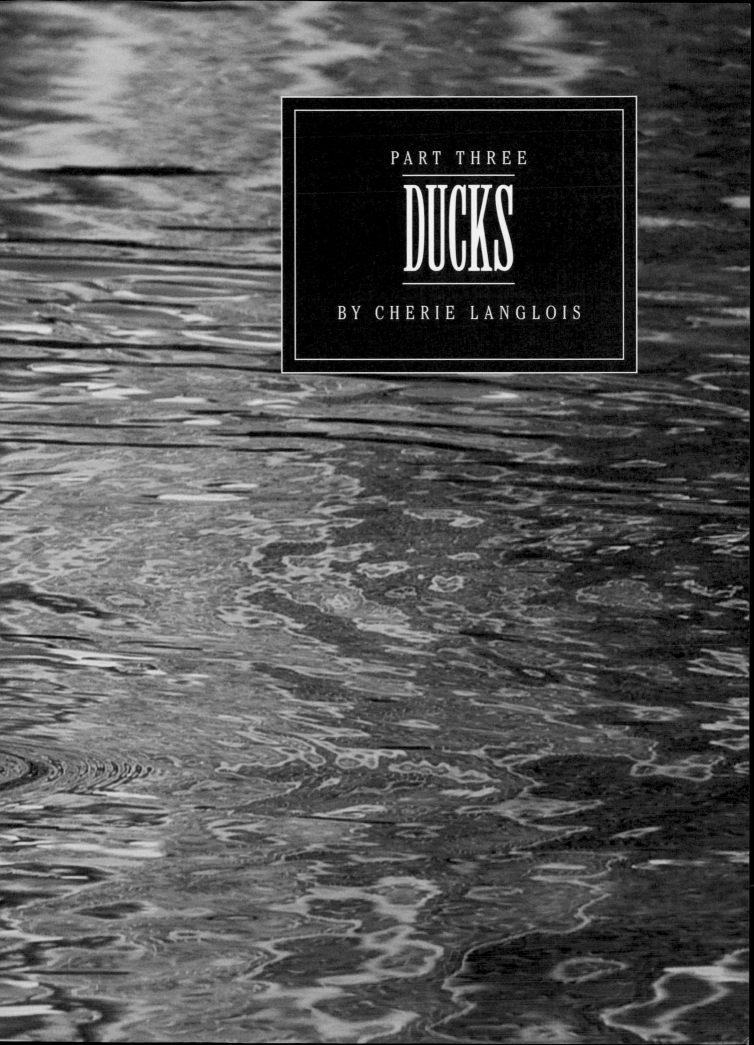

PART THREE

DUCKS

BY CHERIE LANGLOIS

WHY DUCKS?

Well, why not ducks? In the United States, the domestic duck still waddles about in the shadow of the immensely popular chicken, while in other parts of the world—especially in Asia—ducks are just as important as chickens in the lives and diets of humans. What do these water-crazy birds have that make them as much of an asset to farms as landlubbing poultry?

CHOOSING THE RIGHT

DUCKS

For starters, ducks are one of the hardiest, most efficient foragers out there—even more so than their clucky cousins. Properly tended, these birds seldom get sick. Given some freedom to roam pasture, pond, or orchard, they'll glean much of their own feed. Of course, you're welcome to spoil them if you want to, but a small duck flock doesn't have to be babied with elaborate, heated accommodations. Do you live in the frigid North? Lack a pond on your property? Work full time? Not a problem! Ducks will adapt to a wide range of climates and living conditions and thrive on a minimum of daily care as long as you meet their basic needs.

In return, ducks are generous, industrious creatures. Like chickens, ducks on the prowl for their chow provide valuable pest control, weeding, and fertilization services. They efficiently convert food sources into protein-packed meat and eggs, and they give us dreamy-soft duck down for pillows and comforters. Colorful and personable, ducks favor us with intangible gifts as well. They make lovely exhibition fowl and gentle, endlessly amusing pets. Given access to any body of water, they flap their wings, dunk their heads, and splash like playful kids forever on summer break.

The American Poultry Association (APA) recognizes only nineteen domestic duck breeds in its Standard of Perfection. Nevertheless, ducks do differ, and taking time to do some research into duck-breed diversity can save the prospective duck farmer some frustration.

With the exception of the Muscovy, domestic duck breeds descend from the ubiquitous wild Mallard. Through the years, selective breeding by fanciers and farmers has produced a number of Mallard-derivative breeds in a charming mix of colors, shapes, and sizes. Like dog or chicken breeds, each recognized duck breed has certain desirable characteristics, from plumage color to egg output, that distinguish it from other breeds. For example, a big white Pekin duck looks very different from a skinny Runner duck, and the two differ considerably in their egg-laying prowess. They don't even act the same—in general, the Pekin has a more laid-back personality than does the active, true-to-its-name Runner.

Don't expect all individuals of a certain duck breed to be clones of one another, either. Some duck breeds are divided into varieties, usually denoted by plumage color or pattern. Call ducks, for instance, have many colorful types, including buff, snowy, pastel, and white. Various *strains*—birds descended from one flock or breeding farm—exist as well. Breeding and environmental factors such as diet or imprinting (a rapid learning process by which a duckling learns to recognize and become attracted to another duck, animal, or object) can also produce variations in size, color, personality, and more.

Meet the Breeds

Now let's meet a few of the more commonly raised duck breeds from each of the four APA classes: bantam, lightweight, middleweight, and heavyweight. Other duck types have developed in various parts of the United States as well as in other countries, and certainly more will appear in years to come. The Livestock Conservancy also maintains a list of endangered duck breeds. Keep your eyes open—you may stumble upon an interesting breed that is not listed by either the APA or The Livestock Conservancy.

Bantams: Little Ducks, Big Personalities

Bantam ducks, weighing in at a slim 1–3 pounds, make easy-to-handle exhibition fowl, friendly and amusing pets, and even tasty meat birds. Their small size doesn't keep them from consuming their fair share of bugs and slugs, and it enables them to fly very well. Another plus: they eat smaller

DUCK MATCHMAKER

You're looking for	Breeds to think about
A prolific egg producer	Ancona, Appleyard, Campbell, Harlequin, Magpie, Orpington, Runner, Saxony
A hefty meat duck	Aylesbury, Appleyard, Muscovy, Pekin, Rouen, Saxony
A good setter and mother	Bantam breeds, Muscovy
A quieter duck than most	Drakes of the dabbler breeds, Muscovy

amounts of concentrated feed than do larger breeds. Their size is influenced by genetics, environment, and nutrition, so if you plan to raise and breed bantams, you'll have to pay attention to proper selection and management to keep them small. The APA recognizes four bantam breeds: the Australian Spotted (a rare breed), Call, East Indies, and Mallard.

The Call Duck
Ducks don't get any cuter than the compact Call duck, with its plump body, round head, and stubby beak and legs. Weighing less than 2 pounds, these birds come in an engaging variety of colors and patterns, including pure white and magpie. The gray Call looks similar to the wild Mallard.

Calls make friendly pets and beautiful exhibition birds. Females lay sixty to seventy-five eggs a year and tend to be reliable setters and mothers. The ducklings are described as delicate. Be advised that Calls are the toddlers of the duck world: active and noisy! The breed's name derives from the ducks' shrill, carrying vocalizations, which hunters historically exploited to lure wild ducks into traps or within shooting range before federal law banned the use of live decoys in 1934.

The East Indies Duck
This exquisite breed looks as exotic as its name. Dressed in inky black plumage with a shimmering green iridescence, this breed is also known as the Emerald, Brazilian, or Labrador duck. It is thought to have been developed over a span of about 200 years, from the early 1800s to the late 1900s, in North America and Britain.

These ducks are popular for show or as pets. Somewhat quieter than Calls, the females make good setters and mothers. East Indies produce up to seventy-five eggs a year—dark gray-green eggs at the start of the season, then progressively paler eggs as the laying season draws to a close.

The Mallard
Domestic Mallard drakes look just as flashy as their wild counterparts, with vivid green heads, white neck rings, yellow bills, and chestnut breasts. Females have inconspicuous but lovely streaked brown coloring that comes in handy as camouflage when they incubate their buff, green, or bluish eggs. The ducklings sport yellow down with black markings. Domestic Mallards, weighing in at 2½–3 pounds, although different from their slender wild kin in size, are similar in their ability to brood their eggs and forage well. Keepers often raise Mallards for meat, for show, for hunting, and as decorative additions to the farm. Varieties include gray (the wild coloration), white, and golden.

Lightweights: The Superlayers
If you want a super egg-laying duck that's unlikely to go broody on you and try to incubate every egg that's laid, then take a gander at one of the lightweight breeds. At 3½ to a bit more than 5 pounds, these active birds tend to be enthusiastic snail and slug foragers, only so-so flyers, and a bit more land-loving than other breeds. The APA recognizes the Campbell, Magpie, Runner, and Welsh Harlequin (the Magpie and Welsh Harlequin are rare breeds).

KEEPING ENDANGERED DUCK BREEDS

If you've decided to add ducks to your farm, consider raising an endangered breed. Rare breeds—rare because fewer farmers keep and breed them—tend to be hardier than their commercial counterparts and better at foraging. They're beautiful, often important historically, and possess genotypic variations that could become important for agriculture down the road.

The Livestock Conservancy, a nonprofit organization based in Pittsboro, North Carolina, has worked since 1977 to conserve and maintain the genetic diversity of nearly 100 breeds of horses, cattle, asses, sheep, goats, pigs, and poultry. The organization's work embraces education, conservation, and research programs, including a periodic census of livestock breeds and the publication of conservation priority lists.

1. The Call duck is one of the APA's four bantam breeds.
2. The female Mallard is less showy than the male but still eye-catching.
3. A female Campbell duck in the breed's popular khaki "uniform."
4. The Runner's lean, upright body is made to move.

The Campbell

An egg-layer extraordinaire, this breed owes its existence to Ms. Adele Campbell of England, who produced the Mallard-like breed from Runner, Rouen, and Mallard stock in the late 1800s. She eventually selected for an attractive, tan-colored bird that resembled the khaki uniforms of British soldiers. The khaki Campbell is still the most popular type, although white and pied varieties also exist.

Campbells that are selected for high egg production rank as the most prolific layers of all the duck breeds; they produce white eggs and can lay up to 350 a year. They're also hardy birds that can adapt to a wide range of environments and climates.

The Runner

The aptly named Runner duck descended from fowl traditionally herded between home and rice paddy in parts of Asia. The movie *Babe,* about a sheep-herding pig, featured a Runner duck in its supporting cast—not surprising, given this duck's active personality and amusing looks. Weighing around 4 pounds, the Runner has a skinny, upright body, resembling a bowling pin on legs, that enables it to cover ground fast. Their eggs are white, and they can pump out up to 300 of them a year—just don't ask them to sit on the eggs!

Runners come in a wide variety of colors and patterns, including chocolate, buff, black, gray, fawn, and fawn and white. They make fun, pest-consuming pets and popular exhibition fowl.

Middleweights: The Do-It-All Ducks

Weighing in at around 6–8 pounds, these all-purpose ducks fall between the light and heavy classes in terms of size, laying ability, growth rate, and meat yield. They make good pest-gobbling foragers and are generally calm pets. Middleweights recognized by the APA are the Ancona, Cayuga, Crested, Orpington, and Swedish breeds (the Ancona, Orpington, and Swedish are rare breeds). If you want a good all-purpose fowl that does a bit of everything, these birds deserve a look.

The Cayuga

Developed in New York during the early 1800s and named for Lake Cayuga, the Cayuga ranks as one of the most beautiful domestic ducks. In the traditional black variety, its dark plumage glows with an iridescent green sheen, accented by a blue wing speculum. In a year, females lay more than 100 eggs, ranging in color from dark gray to green to white during the season. Although not very many Cayuga are being bred, they are frequent competitors at poultry shows.

An adult white Crested duck.

The Crested

Another easily identifiable breed is the Crested. The APA recognizes two varieties of this breed: the black and the white; the latter looks like a Pekin with a fancy feathered headdress. The eye-catching crest, caused by a dominant mutation, has made these ducks popular as unusual pets and pond ornaments, and they also provide a good supply of eggs—100 or more a year. According to waterfowl raiser Dave Holderread, breeding for crests can be challenging because of health problems, such as premature embryo death and skeletal abnormalities, sometimes associated with this mutation.

Heavyweights: The Mighty Ducks

Heavyweights such as the Muscovy and Pekin, and the rarer Appleyard, Aylesbury, Rouen, and Saxony, weigh 7–15 pounds, which makes them the first choice for farmers who want to raise ducks primarily for meat production. These big birds also make placid pets, and they usually stay close to home because their heft makes flying difficult. Two exceptions are female and younger male Muscovies, whose powerful wings help them take flight with ease.

Heavyweight ducks grow like weeds. This is great if you're raising them for meat, but if you plan to keep them long-term as breeders or feathered companions, you'll need to take steps to prevent overly rapid growth, which can lead to leg deformities and lameness.

The Muscovy

Domestic Muscovies waddle to the beat of a different drummer than the Mallard-derivative dabblers do. In fact, Muscovies, adapted for perching in trees, don't really fit the duck image. For one, they aren't nearly as crazy about water as dabblers are. They're also quackless; the males make a quiet hissing sound, while the females emit breathy squeaks (although they will squawk if frightened).

Muscovies are big, strapping ducks, great for meat production, with males weighing up to 15 pounds and females averaging 8 pounds. Their meat contains less fat than that of dabbling ducks. Both sexes possess sharp claws that enable them to perch in trees and on barns—or even on your house roof.

Muscovies are good foragers. They love slugs and worms, among other garden delicacies. The females take first prize as outstanding setters and marvelously protective mothers. They don't produce a lot of eggs throughout the year, but they do set several clutches of up to twenty waxy, cream-colored eggs for an incubation period of about thirty-five days.

Muscovy drakes have unforgettable faces. The beaks and eyes are surrounded by brilliant red skin adorned with fleshy, wart-like growths called *caruncles*. The females have only a bit of bare skin on their faces. The APA recognizes four colors—black (the wild coloration), blue, white, and chocolate. Fanciers have selected for other colors, such as lavender, calico, blue and white, and chocolate and white, as well.

The Pekin

Chances are good that you've tossed bread to snowy Pekin ducks on a pond somewhere or savored this duck's rich and succulent meat at an Asian restaurant. Probably the best known and most common domestic duck breed, the Pekin traces its history back to ancient China. The first Pekins in the United States arrived in New York City from Peking (now Beijing), China, back in 1873.

Mellow and hardy, Pekins pack on the pounds at lightning speed, making them the commercial duck of choice for meat production purposes. Reaching 9–10 pounds, these birds efficiently convert forage and feed to muscle and also produce a fair number of whitish eggs each year (around 100–175). However, they tend to be poor setters and mothers.

Pekins are easy to identify: their plumage is white, often with a creamy cast, and they have wide bodies, thick necks, and orange bills and feet. Pekin babies—think of the stereotypical Easter duckling—have cheerful yellow down, and they imprint easily on the people who raise them, becoming incredibly tame. Pekins make friendly, amusing pets and look lovely parading across a green pasture or lawn.

The head of a female Silver Appleyard, a rare heavyweight breed.

their outside abode prepared before they arrive on the farm.

Males, Females, or Both?

Although there's no way of knowing beforehand the sex of duckling that will pop out of a hatching egg, you can buy ducklings that are already sexed from a hatchery for a slightly higher price. By the way, if you buy them unsexed, expect that the sexes will be evenly split—at least, that's the theory. Waterfowl are one of the few bird species in which the male possesses a penis, so a duckling can be vent-sexed by inverting the cloaca to see if a penis is present. Do not attempt this without the guidance of an expert.

Fortunately, the sex of mature males and females is fairly easy to determine without an internal exam. In the case of Muscovies, males grow much larger and have more caruncles on their faces than do females. Dabbler drakes often have gaudier plumage than the ducks have—the Mallard is a good example. In breeds with similar-looking sexes, such as white Runners or Aylesbury ducks, the drake has curled "sex" feathers above his tail.

Selecting Your First Ducks

Once you've settled on a duck breed (or two or three), you'll need to make a few more decisions before you have the pleasure of watching your own flock waddle about your farm.

Eggs, Ducklings, or Adults?

You can purchase hatching eggs, ducklings, or adult birds to start your flock. Each option has its pros and cons.

Hatching eggs—fertile eggs that have been carefully selected and handled—may cost slightly less than ducklings, but they could be damaged during transit, and there's no guarantee that every egg will hatch. You'll also need an artificial incubator or a broody bird (either a duck or a hen that will sit on eggs) to incubate them.

Ducklings are adorable and fun to raise, and they do surprisingly well when properly shipped from commercial hatcheries, each duckling traveling complete with its own food supply in the form of its yolk sac. Yet, ducklings require more specialized care (and, thus, more of your time) than mature ducks do. It also takes a while before they produce eggs and meat, breed, or embark on pest and pond-plant control.

Purchasing adult ducks may seem like a good way to go, but depending on the ducks' age—and that could be a big question mark if the breeder hasn't kept good records—they might not have as long and productive lives ahead of them as youngsters would. It may also take some time for them to get used to your farm, and temporary wing clipping might be required for flighted breeds to keep them home at first. If you do choose grown ducks, you will need to have

WILD ORNAMENTALS

Another delightful group of ducks deserves mention, although we won't be focusing on them in this book. These are the ornamentals: wild duck species from around the world kept not for meat or eggs or bug control but simply because they captivate waterfowl lovers with their beautiful plumage and interesting behaviors. Species include the exotic Mandarin, the showy North American Wood, the vivid Cinnamon Teal, and the little Ringed Teal from South America.

Ornamentals generally require more space, especially if you want to breed them. Unlike most domestic duck breeds, they also excel at flying; covered enclosures or wing clipping may be required to keep them from taking off. You need a permit to keep native wild fowl such as the Ruddy Duck and Pintail, so contact your state's Department of Natural Resources first.

Ducks enjoy the companionship of feathered friends.

If you want to produce ducklings, obviously you need both sexes. Some raisers keep their breeder birds in pairs for the breeding season, while others use varying ratios (one drake to multiple females). If you want the ducks as pets, all females or all males will be fine except in the case of territorial, often aggressive, Muscovy drakes. If it's egg production you're after, all females will do. Like chickens, ducks will lay eggs without a drake on the premises. The eggs will be infertile, but that's fine if you just want to eat them and not breed more ducks.

How Many Ducks?

Ducks are social creatures, so unless you plan to be an only duck's best friend and can give it ample attention, you'll want your bird to have some feathered companions.

Before you order your ducks, take a look at your farm or yard and water sources. Too large a flock crammed in too small an area and bathing in one tiny pond will make an odorous, unhealthy mess. Although you don't want to begin with more ducks than you can

manage, you also should avoid starting out with too few; adding birds later can upset the pecking order and also risks introducing disease or parasites. Keep in mind that some fluctuation in flock numbers is to be expected: breeding—both planned and unplanned—will add to the flock; at the same time, predators, accidents, illness, and culling will take their toll. Whatever your purpose for keeping ducks, your best bet is to start with a small flock of two to ten birds until you've learned the ropes of duck keeping.

ADVICE FROM THE FARM

CHOOSING DUCK BREEDS

What are your rearing facilities like? Big duck breeds require much more space per bird. They also eat more than light and bantam breeds do. Calls, East Indies, Mallards, and Mandarins can fly. How will you design your pens or manage your flock to accommodate that quality? Decide before you purchase your ducks what you'll primarily want to keep them for.

—Lou Horton

We tried Muscovies, but they were too aggressive—they tried to kill the other ducks' babies. Now we keep Indian Runners, khaki Campbells, and Blue Swedish. They're all good slug eaters, and the Campbells and Runners lay a lot of eggs, which we like. The Runners are good foragers, but they're indifferent mothers; they don't seem to have the patience to hatch out their ducklings. The Swedish are friendly, and they lay fairly well; they're determined to be mothers.

—Angie Pilch

Not only are Muscovies quiet, but they're very large meat birds and often able to defend themselves. The ducks make excellent mothers—all of my ducklings are raised by their mothers—and our little flock keeps our yard relatively free of grasshoppers in the summer.

—Melissa Peteler

I really enjoy Pekins. I don't like to use poisonous slug bait, and Pekins are great for slug control. They take care of all my flowerbeds, plus they don't dig up the beds the way chickens do. Unfortunately, these ducks grow fast and can have leg problems as a result. All of our Pekins have been very mellow and friendly, with the exception of one "attack drake" that made a great guard duck—he wouldn't let any dogs or people near his girlfriend.

—Trish Smith

ENDANGERED DUCK BREEDS

Here's a rundown of duck breeds listed on The Livestock Conservancy's Conservation Priorities List. To learn more, visit www.livestockconservancy.org.

Critical:

Fewer than 500 breeding birds in the United States, with five or fewer primary breeding flocks of fifty birds or more, and globally endangered.

Ancona: A hardy, all-purpose, medium-size breed developed in Great Britain and capable of laying more than 200 eggs a year. It possesses variable plumage of white mingled with lavender, black, silver, chocolate, or blue.

Aylesbury: A large, tame, fast-growing meat breed hailing from England. It has white plumage, white skin, and a long, pinkish-white bill. It lays fewer than 125 eggs a year.

Magpie: A light, active breed developed in Wales that can produce about 250 eggs annually. It flaunts white plumage with a black back and crown (color varieties include blue and silver).

Saxony: A large, active, all-purpose duck from Germany that lays approximately 200 white eggs a year. Its plumage pattern resembles that of the Mallard, but the drake sports a bluish-gray head and back, and the female is a buff color accented with creamy face stripes, neck, and belly.

Silver Appleyard: A sturdy, calm, dual-purpose breed native to England and developed by Reginald Appleyard. It lays more than 200 white eggs a year and sports silver-frosted, Mallard-like plumage with orange legs and feet.

Welsh Harlequin: A lightweight, streamlined Welsh breed that can produce more than 300 eggs a year and has a complicated plumage pattern similar to the wild Mallard.

Threatened

Fewer than 1,000 breeding birds in the United States, with seven or fewer primary breeding flocks, and globally endangered.

Buff Orpington: A medium, dual-purpose English duck that lays up to 200 eggs a year, grows fairly fast, and displays plumage.

Cayuga: A hardy, calm, medium-size breed developed in New York that lays 100–150 eggs a year. Its lovely feathers are black with a green iridescence.

Watch

Fewer than 5,000 breeding birds in the United States, with ten or fewer primary breeding flocks, and globally endangered.

Campbell: An energetic, lightweight English breed that can produce a whopping 300-plus eggs annually. It comes in four varieties: khaki, white, dark, and pied.

Rouen (nonindustrial): Large, mellow meat duck originating in France that resembles the Mallard. It produces only about 35–125 eggs annually.

Swedish: A stocky, medium utility breed developed in an area that was formerly part of the kingdom of Sweden. It lays 100–150 eggs annually and has slate-blue plumage with a white bib.

Study

Breeds of interest that lack either definition or genetic or historical documentation.

Australian Spotted: A rare, active, hardy bantam breed developed in the United States, not in Australia. It lays 50–125 eggs a year. The complicated plumage patterns of the three varieties include spotting on the body.

Dutch Hookbill: A Dutch breed with a unique downward-curving beak. It is a good layer, with 100–225 or more eggs per year, and is an excellent flier. It has been in the United States only since 2000.

When You're Ready to Buy

Finding ducks isn't difficult. In spring, you'll see Easter ducklings on sale at your local feed store or ads for surplus ducks in the newspapers. Finding the ducks you want—healthy, of good show stock, the right breed, and so on—can be more challenging. For hatching eggs, day-old ducklings, and sometimes adult ducks, check out reputable hatcheries and breeders. You can also purchase ducklings of more common breeds at feed stores and obtain ducklings and mature stock from local farmers. If you are able to be there in person to pick out your ducks, look for active, healthy birds with bright eyes, clean vents, and strong legs. Use special care when picking out ducks for breeding stock.

Most duck raisers start their flocks in the spring or summer, when feed stores carry ducklings and breeders have extra stock to sell. Depending on the hatchery, you may be able to order ducklings and hatching eggs from spring through fall.

Well-protected within their insulating waterproof plumage and down, domestic ducks have adapted to a range of climates around the world. You'll find them weathering Midwestern snows and Southwestern heat waves in the United States, heavy downpours in the British Isles, and stifling humidity in Central American rain forests.

When other wimpy livestock race to the barn during a rain shower, ducks stay outside and revel in it. While chickens remain tucked in their cozy coops on a snowy winter day, ducks dabble blithely about their ice-rimmed pond.

DUCK
ACCOMMODATIONS AND DIET

Mature ducks typically don't require the sort of snug accommodations that many folks fashion for their chickens, especially in areas with fairly mild winters. Like wild waterfowl, most domestic dabblers would be content to spend their nights sheltering beneath a bush or floating on a pond. Flighted Muscovies often prefer taking to the trees or some other lofty perch.

Many small flock raisers, however, do provide their ducks with indoor housing or a covered shelter—and not just because they feel like spoiling their feathered friends. Giving ducks a refuge from weather extremes improves egg and meat production; after all, it takes energy for ducks to keep themselves warm or cool themselves off. Not only that, but when exposed to frigid weather, ducks—especially the cold-sensitive Muscovy—risk frostbite to their feet.

Along with providing a house or shelter for their ducks, many raisers keep their birds partially or completely confined in outdoor duck yards, pens, or fenced pastures. An enclosure of some type comes in handy for preventing these ever-foraging fowl from rooting up your spring veggie starts, sleeping—and messing—on your deck, or wandering out onto the road and over to your neighbor's garden.

Secure housing and pens serve an even more important purpose: protection from the host of varmints that will dine on duck eggs, ducklings, or adult ducks if given an easy opportunity to do so. Most domestic ducks, with the possible exception of some fierce, flying Muscovies, truly are "sitting ducks." As a rule, ducks tend to be noisy, colorful, placid, slow on land, and incapable of swift flight (if they can fly at all)—all traits that make them attractive to wild and domestic predators.

Housing Basics

Along with adapting easily to various climates, ducks adjust well to many different housing arrangements. Keepers of small duck flocks successfully use a variety of accommodations for their fowl, from simple doghouses, revamped chicken coops, and livestock stalls to custom-made wooden abodes and netted aviaries designed especially for ornamental ducks. How large or elaborate your ducks' home should be depends on your climate, finances, time constraints, local building regulations, and more. Before you draw up plans for a duck house and yard or modify an existing structure, keep in mind the following important guidelines.

Space Requirements

At the very minimum, a house for mature ducks that have access to an outdoor pen or pasture should allow 2–4 square feet per bird, depending on the breed's size. So, for a flock of ten ducks, you will need a house or shelter of approximately 20–40 square feet. Birds confined full-time need at least twice that much space—the more, the better.

Remember also that if you don't clean your duck enclosure daily, manure and filth will build up. This accumulation is not only unsightly and smelly but also unhealthy for your birds. Larger, less packed pens will stay cleaner longer than small, overpopulated ones. If you rotate yards or use portable pens that can be moved from place to place, you can probably get by with less roomy digs for your ducks.

Ducks in the wild don't mind a dip in an icy pond.

Where to Place the Shelter

Avoid placing your ducks' new abode in a low spot or smack up against your neighbor's property. Take into account, too, how far away you plan to store your cleaning supplies and feed and where the nearest electrical outlet and water spigot are. A slightly sloped site with well-draining sandy or gravelly soil is ideal. Take advantage of existing trees or shrubs, using them as windbreaks, as sound and visual barriers, or for shade. Plant more trees and shrubs if necessary; ducks require protection from wind, snow, and rain, as well as from hot summer sun.

Think about how you want to access the house and yard for cleaning, feeding, gathering eggs, and watering; easy access will save you much time and frustration. If you plan to keep flighted ornamental waterfowl in an aviary setting, consider doing what many zoos do: establish a safety area—a covered area that has its own entrance and exit door—adjoining the aviary. This allows you to go in and out of the aviary but prevents any birds from escaping.

Ventilation

Air quality can suffer in a tightly closed house, where ammonia fumes rise from manure and dust accumulates from feed and litter, adding to the feather dust and carbon dioxide given off by the ducks themselves. These pollutants can adversely affect a duck's eyes and respiratory tract—and yours, too, if you're working inside the duck house. Windows or slatted vents, located at the top of the building to prevent ground-level drafts, will not only add ventilation but also help eliminate mold-promoting moisture in your duck house. Cover any windows or other ventilation openings with sturdy, well-attached screen to prevent predators from sneaking into the house.

Weatherproofing

The duck house or shelter should have a decent roof to shield against rain and snow. Even water-loving waterfowl enjoy an opportunity to dry off now and then. If you live in an area with mild winters and summers, your duck house can probably do without insulation. But if sweltering summers or freezing winters are the norm, give your ducks access to an insulated house that keeps temperatures steady, and you'll find that they consume less feed and produce more eggs and meat.

Adult ducks don't normally need artificial heating in their houses, especially if they have buddies to huddle up with during cold weather.

In a pinch, pile bales of straw or hay around the house for insulation. During hot weather, you may need to set up a sprinkler or misting system if your birds show signs of heat stress, such as panting. If you keep your ducks in outside yards and pastures, remember to provide them with shade, either natural or constructed, so that they can escape the sun. Trees, shrubs, wood shelters, and tarps can also double as shelters from rain and snow.

Flooring, Litter, and Substrate

Duck house floors commonly consist of cement, packed dirt, or wood. Cement, although probably the costliest option, works well for keeping rodents and other predators from tunneling into the house. A cement floor is simple to clean with a hose and scrub brush, provided it is properly graded to prevent puddle-forming dips. Cover the cement with a soft layer of litter to soak up the moisture in the ducks' droppings (droppings are about 90 percent water). The litter will also protect your ducks' feet, which are smoother and more sensitive than tough chicken feet and easily abraded on a hard surface. Dirt floors and wooden floors also require bedding to absorb moisture. A cushiony litter will help your birds stay warm during winter, too.

Litter should consist of a clean, dry, absorbent material, such as nontoxic pine shavings, quality hardwood chips (other than walnut, which is potentially toxic), crushed corn cobs, and rice hulls. Many farmers use straw, but it's less absorbent than other materials. Steer clear of grass or legume hays for bedding; these can quickly turn moldy.

Some raisers keep ducks in wire-floored cages, particularly for exhibition. Again, because ducks' feet and legs are susceptible to injury, make sure that the wire you use has no sharp edges and that the mesh size is no larger than 1 inch by ½ inch in size. Sprinkling litter in part of the cage will give the birds a comfortable place to sit if the wire hurts their feet.

Ducklings are just as likely to swim in their water as they are to drink it.

Nest Sites

Although domestic dabbling ducks have a tendency to drop their eggs anywhere, some will make nests on the ground. To increase your chances of getting clean eggs, set up nest boxes at ground level along the sides of the house or pen, and bed them with shavings or straw. You can even build a nest-box area into the duck house itself. Provide at least one 12 × 12-inch (or larger, if needed) nest box for every four birds. If building a box from wood, leave one end open and nail a 2- to 3- inch strip of wood along the bottom edge of the open end. This will help keep the nesting material and eggs inside the box. Plastic dog kennels with the doors removed, covered kitty litter boxes, and barrels set on their sides can also serve as nest sites. Wild Muscovies nest in tree cavities or elevated nest boxes such as those used for Wood Ducks; domestic Muscovies will nest just about anywhere, as long as it's somewhat dark and secluded.

Outer pens and yards located on bare soil with poor drainage often require an added substrate to keep the ducks from trampling and dabbling the ground to mud. Mud not only makes for unsightly lodgings and dirty plumage but also teems with bacteria and fungal organisms, some of which can cause health problems. During wet weather, the nutrients, sediments, and bacteria from mud and droppings run off into surface and ground waters, polluting wetlands and drinking water. Good substrate materials include sand, pea gravel, straw, and sawdust or a combination of these materials.

Feeders and Waterers

When outfitting your duck digs with feeders and waterers, choose containers that are durable, easy to clean, and stable enough that the ducks can't flip them over in their enthusiasm to get to the food or water. Feed stores and livestock- or poultry-supply companies carry a plethora of poultry feeders and waterers to choose from. You'll also find round, shallow pans for general livestock use; they are made of rubber or hard plastic and make long-lasting, stable water containers and feed dishes. Check out thrift stores and yard sales for tough stainless-steel pans and bowls if you want to save money. Ducks aren't picky.

While almost anything that works for chickens will suffice for ducks, remember that waterfowl have larger, broader bills than pointy-beaked chickens do, so make sure the feeder and waterer you choose have wide enough troughs. Allow plenty of feeder space so all of your ducks, high- and low-ranking alike, can eat at the same time. Water containers for ducks should be at least three inches deep so the birds can submerge their heads to keep their eyes and nostrils clean and healthy. To help keep litter dry in the duck house, consider placing waterers in an outer yard. If your ducks spend the night in the house and have food available to them, however, they'll also need water to prevent them from choking on the feed. Avoid some of the inevitable mess by setting the containers on mesh or on wooden slats.

Protecting Your Flock with Fencing

Secure fencing is the raiser's first line of defense against a variety of predators. Fencing also keeps your flock from wandering off your property and from invading duck-prohibited zones, such as your newly planted vegetable garden. Take a trip to your local feed or building supply store, and you'll encounter a bewildering variety of fencing material: chicken wire, electric wire fencing, woven-wire field fence with graduated openings, chain-link fencing, welded wire of all gauges and heights, nonclimb horse mesh, light electric poultry mesh, and more.

What does a confused duck farmer choose? Wild predators tend to go for prey that's easily obtained, so think in terms of making it as hard as possible to reach your ducks. For starters, surrounding your property with tight woven-wire field fencing, sturdy horse mesh, or chain-link at least 5 feet in height will help ward off wandering dogs and hunting coyotes, especially when combined with electric hotwire fencing. Dogs and coyotes will gladly crawl or dig under fences, so make sure your fence reaches the ground (you may even need to bury the bottom part of it) and regularly check the fence line for holes. An inside buffer fence, if you can manage one, provides additional security. Avoid leaving gaps around or under gates and pen doors where predators or ducks can squeeze through.

A designated nesting area helps keep the eggs in a safe place.

Inner fences employed only to contain, separate, or exclude your ducks can be shorter (2 feet high would suffice) unless you keep a breed adept at flying. The heftier duck breeds don't jump as high as chickens do. Electrified plastic poultry mesh works well to confine your ducks during rotational grazing on pasture, when you want to easily move the flock from area to area. If you keep ducks that excel at flying, such as bantam or ornamental breeds, cover their enclosures with aviary or game-bird netting. Installed over the top of a duck yard, netting has the added benefit of helping prevent losses to aerial predators such as owls and eagles.

When fencing duck yards, beware of chicken wire: it's not very strong, and it tends to sag, making it a poor choice for predator protection. Combining small-diameter (½-inch) chicken wire with a sturdy, larger-gauge welded wire works well for keeping out rats, weasels, and raccoons that will reach inside to snag sleeping fowl. Whatever fencing you choose, make sure that the wire gauge of your fence is appropriate for the ducks you keep. Field fence may hold Muscovies, but it won't keep a Call duck from slipping through. Ducklings require special caution—they squirm through amazingly tiny openings and often can't seem to find their way back home.

Cleaning and Maintenance

Design your flock's house and yard with ease of cleaning in mind. For instance, it should be located conveniently close to a water spigot and the shed where you stash sanitation equipment and litter. If you have to enter to clean it, the house should have enough head room so you won't knock yourself silly on a roof support. Picking up after your messy flock may not rank as one of the most pleasant chores you'll ever do, but it's an important one that will help prevent disease and parasite transmission. It will also keep your air smelling fresh.

If you have time each day, opt for spot cleaning and then follow up with complete removal and replacement of the litter weekly or monthly. Another popular technique used for ducks and other livestock is the deep-litter system, involving regularly stirring up the old bedding and adding a fresh blanket of litter over the top. The shavings, hulls, or straw mingled with duck manure at the bottom begin to compost, generating heat that helps keep the birds warm during winter. Once a year, usually in the summer, the whole works is shoveled out and left in a pile until the composting process is complete. This nitrogen-rich soil amendment can then be used as mulch for your garden.

KEEPING YOUR DUCKS SAFE FROM PREDATORS

The list of potential predators that may dine on ducks, ducklings, or eggs is depressingly long: fire ants, rats, snakes, snapping turtles, crows, seagulls, hawks, owls, eagles, opossums, raccoons, skunks, weasels, mink, foxes, cats, coyotes, dogs, and more. Declaring all-out war would quickly drain your resources; prevention is your best bet.

- Tuck your birds into a secure house or pen at night, which is when many predators emerge to hunt. Reduce the gauge of your wire if needed—rats and weasels can squeeze through 1-inch chicken wire in a flash.
- Don't leave dog and cat food or uncovered garbage outside, where it attracts raccoons and other varmints. Keep your ducks' feeding areas as clean as possible.
- Keep your property off limits to wandering canines by installing a good boundary fence.
- Conduct frequent security checks of fences and pens to look for signs of digging or holes that would allow a hungry intruder easy access.
- Keep vulnerable ducklings safe in an indoor brooder or completely enclosed pen until they grow large enough that they won't be tempting prey for crows, rats, and cats.

The yearly muck-out is also a good time to give the house a thorough hosing and scrubbing with warm, soapy water to remove dirt and dried manure. Follow this up with a disinfectant or sanitizer used according to directions. Rinse all surfaces well and allow to dry, and you're ready to start piling on the litter again. It's extremely important to clean and disinfect the house when changing over flocks, too.

Duck yards also need cleaning. If you use sand or pea gravel as a substrate, a brisk raking is all it takes to rid the pen of manure and feathers. To conserve sand, drill small holes into a big shovel head to make a nifty sifter that captures droppings and debris while letting the sand fall through. Check feeding and watering areas regularly to see if the sand needs changing; moist sand mixed with feed becomes black and foul-smelling, not to mention unsanitary. The pen's substrate should be changed completely as needed—usually once a year.

Water Features
Duck Ponds

When most of us think of duck ponds, we imagine a pool of sparkling water, ringed with cattails and flowers and alive with bright-eyed, quacking ducks. In the real world, ducks produce copious droppings, flatten vegetation with their big feet, and spend their days dabbling and drilling for yummy worms and other edibles in the dirt, turning it to mud. Set a flock of ducks loose on a little pond that has no inlet, outlet, or pump and filtration system, and you'll have a mucky mess in no time. Build a poorly designed and difficult-to-clean artificial pond, and you'll eventually have the same problem.

A natural lake or pond can work well for ducks, provided there's an inlet and an outlet to refresh the water and the water quality is good. Be careful not to overstock; one source lists 100 birds per acre of water as the absolute maximum. If you choose to build a pond, keep in mind that a permit may be required before you can construct one on your property; contact your local government offices to find out what permits and regulations apply.

Building an Artificial Pond

You can use concrete to construct a small artificial duck pond, or you can use premolded plastic forms or the tough plastic lining sold especially for ponds. The pond should be convenient to drain, clean, and refill, and it should have an overflow pipe to prevent flooding. If you run a drainage pipe or hose from the pond into a planted area that needs regular watering, you'll be killing two birds with one stone every time you empty the pond. Make sure the pond has a gradual slope or ramped area where your birds can easily get in and

MAKING MANURE COMPOST

Transform the duck manure and yucky litter you clean out of your flock's house and pens into crumbly, nutrient-rich compost to enrich the soil in your garden or apply to pastures. Making compost doesn't have to be complicated or time consuming. The simplest (albeit slowest) method is to pile up the duck doo and litter, along with any other organic matter you want to get rid of, such as vegetable scraps, lawn clippings, and old hay, until it reaches a height of at least 3 feet. It helps to confine the pile with pallets or a wire cage so it doesn't spread out all over the place. Cover the pile with a tarp so that rain won't wash away the nutrients and bacteria, but make sure that the pile stays moist, and then wait. After six months or so, you'll have lovely compost for your garden.

out. Ducks can drown, and ducklings are especially at risk if they're unable to climb out of the water.

Think carefully about the pond's location; you'll want it somewhere within reach of a hose and convenient for duck watching, yet not so close to your house or neighbors' homes that noise and smells are a nuisance. If ponds in your area freeze during winter, keep in mind that domestic ducks—particularly Muscovies—are prone to frostbite on their feet. An aerator that keeps part of the pond ice-free can help prevent this. If you keep Muscovies, your best bet is to take away their swimming privileges during freezing weather.

To reduce mud around the pond's edges, try spreading gravel or sand and planting sturdy vegetation such as irises and cattails. Again, keeping duck numbers down and rotating access to the pond helps reduce mud formation and gives waterside plants a chance to recover.

Duck Pools

Contrary to what many people believe, domestic ducks can get along fine without a pond, natural or artificial. What they do need at the very least, however, is ample drinking water in a container deep enough for them to dip their heads in to clean their eyes and nostrils. Shallow, 3-gallon hard plastic containers do just fine; two per flock of four to six birds is usually sufficient. The containers are available at feed stores and are easy to scrub clean and move when the area around them gets too muddy or soiled with droppings. They're also large enough that most ducks can stand in them and take a partial bath.

Your birds will be happier and their plumage will stay cleaner, however, if you can provide them with more spacious bathing arrangements. Plastic kiddie pools make great duck pools. They come in several sizes and are fairly easy to move and clean (you'll need to change the water frequently). To give your ducks a way to get in and out of the water, either install a ramp or place a concrete

I feel that animals are healthier if they can roam relatively freely, eat fresh greens, and catch insects. In summer, my Muscovy ducks run free, using the barn when they want to, but primarily living outside. I recently built a large outdoor pen for peafowl, and in the winter I put the ducks in there.

—Melissa Peteler

After losing a number of ducks to bald eagles, I made a large fenced pen and strung high-test fishing line across it, running the lines about 8 feet apart. Then I attached 10-inch-long tinsel streamers along each line. The eagles and hawks were actually screaming at me—they wouldn't come down. Last year, all of the babies hatched out, and nobody was eaten.

—Howard Carroll

Plastic kiddie pools can be found all spring and summer at any number of retail stores and are relatively inexpensive. The depth works very well for ducks, they're easy to maneuver, and they can be moved around a lot to prevent mud holes. The pool's height provides easy access for most adult ducks, while a brick on the

outside and one on the inside is all that's needed for smaller ducks or ducklings. I use a circular saw to drill 1-inch holes in mine and use a stock tank plug so that when I want to empty them, I simply remove the plug and let the water flow out.

—Cat Dreiling

Allow more square footage in houses and yard space than you think you'll need. Sand makes an excellent drainage medium in pens: it keeps the birds cleaner and prevents some disease problems. Pen security is vital—heartbreak comes to those keepers who underestimate the local predators.

—Lou Norton

Ducks are wonderful because they're creatures of habit, and they have a very strong flocking instinct. They're much easier to herd than chickens, which tend to scatter. We train our ducks to go into their house at night. If they haven't gone in on their own when it's time, we just clap our hands and herd them in. The lead duck runs in and the rest make a desperate dash to follow. They're a little like lemmings.

—Angie Pilch

block up against the outside of the pool and one on the inside, to act as steps.

If you live where winters turn frigid, consider a frostless water hydrant with a short length of hose that you can easily drain, a stock tank de-icer to place in your birds' water source, and/or a bubbler to keep a portion of the duck pond ice-free. All of these items can be purchased at your local feed store.

The Duck Diet

While dining on a slime-coated slug or fat grub may sound nauseating to you, to a duck, those foods rank as *haute cuisine*. Like their wild relatives, domestic ducks are omnivorous and love foraging for a wide variety of creepy-crawly fare, such as slugs, worms, mosquito larvae, beetle grubs, and snails. Ducks relish both aquatic vegetation, such as duckweed, and terrestrial vegetation, such as grass, and crave many of the same crops we do: corn, tomatoes, blueberries, lettuce, grains, and more. Ducks eat fish and frogs and will even snatch a mouse or bird on occasion. It's precisely this ability to scavenge for their own chow and efficiently convert a variety of feedstuffs to meat and eggs that makes these fowl so valuable to farmers around the world.

A small duck flock allowed to roam where there is abundant natural forage costs much less to feed than does a confined flock that needs purchased, concentrated rations. The drawback is that a flock subsisting solely on forage will probably not grow as fast or

produce as many eggs as one raised on a nutritionally complete commercial diet. In reality, most keepers of small flocks, unless they raise their birds in complete confinement, take the middle road when it comes to feeding. Many encourage their ducks to hunt for their own vittles part of the day or year while also providing the birds with supplemental feed to meet nutritional requirements and to keep the flock productive.

There's no one right way to feed ducks that fits all flocks and all situations. In addition to checking out the excellent resources at the end of this book, ask other experienced duck raisers about what they feed their fowl. If you can, consult a veterinarian or extension expert in your area who is familiar with waterfowl. In the meantime, here are some basics about duck nutrition and diets.

Duck Nutritional Requirements

The feed you toss to your ducks must provide them with good nutrition so they will have the energy they need for such essential activities as feeding, digestion, breathing, walking, reproduction, and body-temperature maintenance. The diet must supply nutrients, such as protein, that allow the birds to develop healthy feathers, muscle, bones, and eggs. Given an inadequate diet, your ducks' health, growth, and productivity will suffer. To reach their potential, ducks need to obtain the following nutrients and supplements from their diets in balanced amounts.

Ducks are foragers and can find many tasty and nutritious treats in the grass and soil.

Carbohydrates

Ducks use carbohydrates, found in sugars and starches, as fuel to give them energy for flying, foraging, breeding, egg production, and bodily maintenance. Unlike the chambered stomachs of ruminants, such as the goat, a duck's simple stomach is incapable of digesting large amounts of fibrous material. That's why waterfowl feeds are based primarily on carbohydrate-rich grains such as corn, oats, wheat, and milo—not fiber-packed hay.

Fats

Dietary fat provides a concentrated source of energy and serves as an energy reserve. It contains fatty acids important for vitamin and calcium absorption, nerve impulse transmission, and tissue structure. Corn oil, soybean oil, and other fats contained in formulated feeds also help eliminate dust and improve taste.

Vitamins

Although needed only in small amounts, vitamins are critical to healthy growth, nervous-system function, metabolism, and reproduction. A deficiency of these organic compounds can cause a wide range of health problems, from poor blood clotting and nervous disorders to death. Vitamin overdoses are also bad news. Important vitamins are C, A, D, and the family of B vitamins. Commercial feeds usually contain supplemental vitamins not present in the constituent grains and meals themselves.

Protein

When a duck ingests protein—from some nice plump beetle larvae, for instance—its digestive system breaks this nutrient down into amino acids. Amino acids are critical building blocks in the synthesis of proteins. The duck can use them to make up the proteins it needs for the formation of muscle and nerve fibers, feathers and skin, and eggs. But not all amino acids are alike. Essential amino acids can be obtained by the duck only through its diet, whereas nonessential amino acids can be produced by the animal itself. Protein quality varies from source to source, and one diet source alone won't supply all of the amino acids a duck needs. Consequently, you'll see a number of protein sources, such as fish, bone, meat, and soybean meals, in formulated poultry feeds. Grains such as wheat and milo also provide protein.

GENERAL DUCK FEEDING TIPS

- Buy fresh, quality feed, and avoid storing it too long. Poultry feed is perishable; its nutrients deteriorate, and it can spoil if stored too long. Check the expiration date.
- Store rations properly in a cool, dark, dry place. Put feed in a clean container with a tight lid that will keep out disease-carrying rodents and mold-promoting moisture.
- Clean feeders, waterers, and feeding areas regularly. Keep dining areas clean to discourage rodents and reduce mold formation. Regular washing of feeders and waterers will help keep your flock healthy.
- Provide feeder space for all of your birds. Whether you use chicken feeders or thrift-store bowls and pans, make sure there's space enough at the feeders to allow all of your ducks to eat without being picked on by dominant birds.
- Use treats to train and tame your ducks. Offered in small amounts, a favorite food such as whole wheat bread or scratch works wonders to facilitate and maintain your friendship with your flock.
- Make drinking water available at all times. Ducks offered feed but no water can actually choke. Waterfowl go through more water than do chickens: they not only drink more, they also splash around in it and dip their food- and dirt-caked bills in it.

A good commercial feed provides complete, balanced nutrition.

Minerals

Inorganic dietary nutrients needed by ducks include calcium, phosphorous, iron, magnesium, zinc, iodine, and salt. Calcium and phosphorous are macrominerals, meaning that they're required in larger quantities than are trace minerals such as copper and iron. Like vitamins, minerals must be balanced to maintain good health. For example, calcium and phosphorous in the appropriate ratio contribute to strong bones and eggshells. An excess of calcium, however, can lead to kidney problems (see the "Grit and Oyster Shell" section). Calcium carbonate, iodized salt, copper sulfate, and other compounds supplement the mineral-poor grains in commercial diets.

Water

Too often taken for granted, water ranks as the most essential nutrient of all. Water performs a number of life-giving bodily chores: it ferries feed through the digestive tract, eliminates waste products, regulates body temperature, and makes up 90 percent of blood volume. Offer your flock an ample supply of clean drinking water each day.

Grit and Oyster Shell

Ducks possess a gizzard that does the work of teeth, grinding consumed feed into an easy-to-digest form with the assistance of tiny pebbles and coarse sand called *grit*. Free-roaming ducks will ingest most of the natural grit they need while foraging and dabbling, but you'll definitely need to offer confined birds insoluble granite grit, which you can purchase from the feed store.

Feed stores also sell crushed oyster shell for laying fowl, designed to provide the extra calcium they need to produce eggs with strong shells. If you feed a commercial diet, check the breakdown of nutrients on the bag's label to see if the feed already contains enough calcium for your laying birds (they need approximately 2.8 percent calcium).

Keep in mind that an excess of calcium can be detrimental to a duck's health, so don't offer oyster shell to nonlaying birds unless their feed is deficient in this mineral; they require only about 0.8 percent calcium.

Other Supplements

When used according to directions, a powdered vitamin and electrolyte supplement made for poultry can benefit both new ducklings and ducks that are ill, geriatric, or stressed. Probiotics, the good bacteria that make yogurt such a healthy food, may also help your birds ward off disease. Look for both products at feed stores and poultry-supply companies.

Feed Options

Now that you have an idea about which nutrients and supplements your ducks need to thrive, let's look at the four main feed options available.

Commercial Diets

Chicken folks have it easy: step into any feed store around the country, and you'll find balanced commercial feeds made for chicks, laying hens, and meat birds. Some stores even carry organic chicken feed. For duck raisers, it's a different story. In many areas, finding a nutritionally complete feed prepared especially for ducks can pose a challenge.

Still, it's worth doing some detective work to find a quality commercial waterfowl diet, especially if you're a neophyte duck

---

PELLETS, CRUMBLES, OR MASH?

Commercial poultry feeds come in three forms: extruded pellets of various sizes; crumbles, which look like tiny rock chips; and fine mashes. In general, ducks waste less pelleted feed than they do crumble or mash. Mash in particular can adhere to wet bills and end up wasted in the water container as the duck dabbles and rinses. Ducks tend to eat more and grow faster when offered a pellet diet (most pellets are too big for new ducklings, but you can moisten them with water first). If mash is all that's available, try mixing it with water to make it mushy but not too sloppy before serving it to your flock.

Do your research if mixing your own nutritious feed.

Ducks, which gives detailed formulations for making homemade duck feed, is one such resource.

Grass-Based Diets

Animals reared in pasture-based systems can indulge in natural behaviors, such as basking in the sunshine, breathing fresh air, and foraging for grass and other goodies, and they generally have more room to move about and exercise. Proponents of grass-based livestock farming contend that the meat and eggs from their fowl are tastier and healthier than are those of factory-farm fowl.

Pasture-raised poultry are usually kept in one of two ways: in small, enclosed pens with open bottoms (a.k.a. *tractors*) that the farmer moves once or twice daily to fresh pasture, or in free-range situations, meaning that they roam a fenced pasture by day and stay in a house at night. By using electric poultry mesh or other temporary fencing, the raiser can rotate the birds through smaller pasture sections, allowing grazed areas to recover.

If you provide the right rotations (moving the flock before they mow the grass below 3 inches or so) and proper pasture management (mowing, liming, and reseeding as needed), your ducks can glean much of their diet from pasture as they dine on tender grasses, dandelion leaves, chickweed, and insect treats.

owner. Complete diets take the guesswork out of feeding, plus they're scientifically formulated to provide ducks with the proper ratios of all of the essential nutrients they need to stay healthy and produce well.

If a local feed store doesn't carry duck feed, ask if they can order some. If you're willing to pay a bit more, you can also order waterfowl rations online through a feed supplier or through some hatcheries or poultry-supply companies. Follow the suppliers' instructions on how much feed to offer your flock.

Do-It-Yourself Diets

If you keep a lot of ducks or can't locate commercial waterfowl feed, concocting your own duck ration out of ingredients gleaned from the feed store may save you some money. Wheat, whole oats, cracked corn, meat and bone meal, iodized salt, oyster shell, commercially prepared vitamin premixes, and other components can be combined to form a healthy feed for your flock. Some raisers even have their duck rations custom-mixed, ground, and formed into pellets at a feed mill. Converting the mix to pellets prevents picky ducks from choosing their favorite grains and wasting the rest.

Creating a custom duck diet isn't easy. You can't just throw ingredients together any which way or pick a couple of your flock's favorite food items (such as corn) and leave it at that. Be aware, also, that some feedstuff used for other animals, such as rapeseed meal and peanut meal, can be toxic to ducks under certain circumstances. Your birds need the appropriate nutrients in the right balance; no one food item will provide all of their nutritional requirements.

Keep in mind that ducks' dietary needs change as they grow. For instance, a laying duck must consume plenty of calcium to produce all of those eggs, but an excess of calcium in a growing duckling's diet can harm its skeleton. It's critical to have some knowledge about duck nutrition or to at least know where to go to get the accurate information you need. Dave Holderread's *Storey's Guide to Raising*

DID YOU KNOW?

Although experts warn that adding other foods to a complete ration can cause an unhealthy imbalance of nutrients, in practice, many raisers do treat their ducks to garden and kitchen goodies, from berries to bread heels. Avoid offering your ducks spoiled, sugary, fatty, or—if you keep layers—strongly flavored foods such as onions and garlic, which may impart their flavor to eggs. Similarly, a fish-heavy diet can affect the taste of duck meat. Chopped dark leafy greens and hard-boiled eggs make healthy additions to any duck's diet. Offer extras—particularly bread and other starchy foods—in limited amounts because too much of these items will ruin your ducks' appetite for their more nutritious fare.

Increased carbohydrates during winter months help ducks keep themselves warm.

You'll probably still need to supplement this fresh forage with commercial or homemade feed, particularly as plants go dormant in late fall and winter.

Make sure that your ducks' feeding grounds are free of toxic pesticides, fertilizers, and herbicides. The ducks, for their part, will enrich the soil by depositing their nitrogen-rich droppings as an eco-friendly fertilizer.

Other Poultry Feeds

If you lack pasture, can't find a commercial duck diet, or feel uncomfortable mixing duck rations from scratch, don't panic. Poultry feeds formulated for chickens, turkeys, or game birds can also work for duck flocks, although some dietary modifications and additions may be necessary because these birds have slightly different nutrient requirements. For example, ducklings need more niacin than chicks do, or they can suffer from leg problems. If possible, avoid feeding your flock broiler starter or other feeds not formulated for waterfowl that contain antibiotics and other added medications; some of these can harm ducks.

Nutritional Requirements by Age

To ensure that any type of ration—commercial, homemade, or other poultry feed—is appropriate for your birds, you should understand what your ducks require in the way of nutrition at various ages.

New and Growing Ducklings

Newly hatched ducklings need plenty of calories, more protein than mature ducks need (about 18–20 percent total), sufficient niacin, and

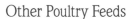

DID YOU KNOW?

To market organic eggs or meat from your duck flock, you must follow the National Organic Program standards established by the United States Department of Agriculture (USDA). This means ensuring that your ducks' certified organic feed is free of antibiotics and other prohibited substances, allowing the ducks outdoor access, and keeping their foraging grounds free of chemical pesticides, herbicides, and fertilizers.

a calcium-to-phosphorous ratio of 1:1. Their feed must be in a form that tiny beaks can handle, such as crumbles, mash, soaked pellets, or small dry pellets (no longer than ⅛ inch). It's recommended that ducklings receive this starter diet for the first two to three weeks.

Fast-growing ducklings from about three to eight weeks of age will experience a slower growth rate if switched to a 15–16-percent protein diet during this time. This lower protein grower diet can help reduce the incidence of wing problems, such as "angel wing," and leg problems, such as lameness, associated with speedy growth. Producers who want fast-growing meat birds often stick with the higher protein fare until the birds are ready for slaughter at around eight to twelve weeks of age.

Mature Ducks

Somewhere around the nine-week mark, a duck's dietary needs change again. From this point on, except during breeding or laying, feed a maintenance diet that contains about 14 percent protein. Don't give them feed meant for layers—it has more protein and calcium than they need. Ducks that you keep as pets, rather than as layers or breeders, can stay on this 14-percent protein diet to maintain a healthy weight.

Nutritional Requirements by Breeding Stage

A duck's nutritional requirements also change during its breeding stages—the times during which the bird mates, lays eggs (if it's a female, of course), and undergoes postnuptial molt.

Breeding Birds

About two to four weeks before breeding season starts, begin feeding breeding ducks a breeder ration with about 17 percent protein. Commercial waterfowl breeder diets are formulated to provide balanced nutrients that promote breeding performance, hatching egg numbers, and the development of healthy ducklings without making ducks fat. Keep feeding this ration until egg-laying has stopped.

DUCK DIET TIPS

Our ducks are picky. They turn their beaks up at regular farm grain mixes. They'll eat all-purpose poultry pellets, and these have a sufficiently high level of protein for them. We mix it with some cracked corn—that's their candy. We feed them leftover lettuce from the garden and cut grass clippings, but they won't eat spinach at all. They love frogs and slugs!

—Angie Pilch

I feed my Muscovies a commercial duck and goose feed and also give them cooked table scraps, including meat, as they're not vegetarians. In the summer, I come upon grasshoppers while working outside, and I'll give them to my ducks. My ducks seem to be pretty healthy, and I think the most important thing is good, clean food and water.

—Melissa Peteler

I chose to go the custom route because I prefer that my ducks have whole grains and fish meal to more closely simulate their natural (and naturally omnivorous) diet. There also seems to be less waste with a whole-grain mix. I feed mash to my ducklings. It's a nonmedicated chick starter from the same mill that mixes my feed, with 5–7 pounds of brewer's yeast added per 100 pounds of feed so the ducklings get the niacin they need.

—Jenifer Morrissey

Only a fool invests good money in breeding stock and then "cheaps out" in feeding them. If at all possible, get feed especially made for waterfowl. The brand of feed is less important than the quality of the ingredients and freshness. Vitamin packages in feeds lose their potency within a matter of a few months or so.

—Lou Horton

Laying Birds

Laying ducks whose purpose is to produce eggs for eating also need plenty of protein and calcium while pumping out all those lovely eggs. Commercial layer diets contain approximately 16–17 percent protein and 2.5–3 percent calcium. Offer these birds some oyster shell if their diet provides insufficient calcium. Start your flock on this ration two to four weeks before laying season commences and keep feeding it until all laying halts.

Nutritional Requirements during Cold and Hot Weather

Because ducks don't hibernate, they need plenty of carbohydrates from their feed to help them stay warm and active on frigid winter days. If you live in a cold climate, you'll probably notice that your flock consumes more chow during winter. Many farmers offer their flocks limited amounts of energy-rich cracked corn or scratch grain mix, both available at feed stores, to help them weather the cold. Be careful, though; too much corn can make ducks fat.

During hot weather in summer, laying ducks often require increased protein in their diets to keep up production because they tend to eat less. Check their water sources regularly because their water consumption may also increase on sweltering days.

Free Choice?

Ducklings and growing ducks should be fed on a free-choice basis—that is, allowed to dine on their rations throughout the day and eat all they want. Whether you offer your mature ducks their rations as free choice or as a set amount of feed at scheduled feeding times (usually twice a day) will depend on how you manage your flock (free range or

in confinement) and whether freeloaders such as pigeons and crows pose a problem.

How much your ducks will eat—or need to eat—depends on many factors, including their size and age, the season, and how active they are. If you use a commercial diet, check the label for suggested feed amounts. Although it isn't necessary to weigh your ducks, pay attention to your flock's general condition and food consumption. Are your ducks so fat that they can barely waddle? Do you notice prominent keels or poor plumage, indications that your birds may not be getting enough food? Is your flock leaving loads of grain at the end of the day for the rats and raccoons to gorge on? If so, consider reevaluating how much chow to dish out.

TOXIC MOLD

Mold produces toxins that can harm a duck's internal organs, muscles, and respiratory system. To keep your flock safe, follow these dos and don'ts:

- Don't feed moldy bread, produce, leftovers, or grain.
- Do store your flock's feed in a container where it will stay clean, dry, and preferably cool.
- Don't give your birds more high-moisture foods (wet mash or table scraps) than they can consume in a day. Remove any leftovers.
- Do check that feed grains have been properly dried before storing them.

All birds—including the captivating duck—possess adaptations that set them apart from most other vertebrates. Of course, the unique avian characteristic that attracts and delights us most of all must surely be the feathers! What else gives these animals their eye-popping range of colors and contributes so much to their enviable power of flight? But birds have more going for them than just feathers; they've evolved some other interesting and useful features that you should know about, too.

DUCK

TRAITS AND BEHAVIOR

How Birds Are Built

The wild ducks from which our domestic duck breeds descend can fly fast, far, and high, thanks to a number of specialized adaptations. A flying bird's skeleton is light and strong, consisting of thin, often air-filled, or pneumatic, bones. The bones that make up the wings evolved from the forelimbs of the birds' dinosaur ancestor. The breastbone has a large protrusion called a *keel*, to which the highly developed wing muscles attach. Most birds have more cervical vertebrae than do other vertebrates, and if you've ever seen a duck preen or a swan arch its graceful neck, you know that most birds also have neck bones far more flexible than ours.

Unlike us, birds have no teeth; the avian jaw is narrow and elongated, forming a horn-covered, toothless beak. Birds' beaks vary in shape and size, each type adapted to handling the specific foods in that bird's diet. In most birds, food travels down the esophagus and enters a handy, expandable storage chamber called a *crop*. From there, it moves into a stomach that consists of two chambers: the *proventriculus*, which secretes gastric juice, and the muscular *gizzard*. Standing in for teeth, the gizzard grinds seeds, grains, insects, and other foods with the help of ingested stone particles called *grit*, which the bird picks up as it forages.

Avian digestive, urinary, and reproductive systems all terminate in one chamber, known as the *cloaca*, where urine and fecal material mix together and then exit the body via the *vent*. As we all know, birds reproduce by laying eggs, a characteristic they share with reptiles and their dinosaur ancestors.

In general, birds have terrific eyesight. The duck, for example, sees colors, and each of its eyes has a visual field of more than 180 degrees, giving it binocular vision to the front, to the rear, and even overhead—a huge plus for spotting sneaky predators. Birds' hearing is also well developed, but their sense of taste is poor, and, with the exception of some species, such as vultures, so is their sense of smell.

Birds have a rapid heart rate, a high metabolism, and an active lifestyle that requires them to consume plenty of food (so much for "eating like a bird"). Avian body weights range from a fraction of an ounce (the bee hummingbird) to more than 300 pounds (the ostrich). A bird's compact lungs connect to air sacs that branch out through its body, an amazingly efficient respiratory system that allows a migrating swan to fly at 20,000 feet in altitude and a Ruby Throated Hummingbird to beat its wings up to seventy times a second. This efficient respiratory system, along with a high metabolism, also accounts for birds' extreme sensitivity to breathing toxic substances. Birds are so susceptible to toxic gases that, historically, coal miners were able to rely on this avian attribute to save their own lives. They took canaries down into the mines with them to serve as an early detection system: the birds' demise warned them of the presence of deadly gases.

In an eggshell, birds—and our ducks—are feathered, flying, toothless dinosaurs.

Specific Waterfowl Traits

Although lots of birds spend time around water, what we normally refer to as waterfowl are swimming game birds in the family Anatidae: ducks and their larger relatives, geese and swans. About 150 species of waterfowl are found throughout the world, occupying every continent except Antarctica. More than fifty of these species inhabit North America, most of them migratory to some degree. In their wild state, these talented birds rule the waters, swimming, diving, and dabbling (that is, feeding in shallow water). But they can fly high in the sky and waddle across land with varying degrees of success as well.

A duck's large webbed feet act as powerful propellers in the water.

All waterfowl, domestic and wild, share certain important physiological traits and behaviors. Here's a speedy overview of the ones that will help you better understand the domestic duck.

Webbed Feet and Duck Bills

Look at a Mallard, and you see a bird built for a semiaquatic life. Large webbed feet propel its streamlined body along the surface of lakes and keep it from sinking into the soft mud of marshes and estuaries. Its short legs sit toward the middle of its belly, allowing the duck to walk on land and achieve an explosive takeoff from water. By contrast, a heavier diving duck, such as the Merganser, has legs situated back near the tail and needs a running start to get airborne.

Now check out that funny-looking bill, yellow in the Mallard male (or drake), orange and black in the female (or duck). Mallards and their domestic descendants are called "dabbling ducks" or "puddle ducks." At times, they submerge themselves completely, but, more often, they bob around on top of shallow bodies of water, using their broad bills to dabble for floating plant material, bugs, and mosquito larvae. They also tip tail-up to scrounge around in the mud, sifting out the edibles with their *lamellae*—comblike plates lining the upper and lower bills. That big bill works on land, too, where the Mallard uses it to tug at tender grass, gobble up berries and seeds, and nab slugs and snails.

Fine Feathers

Like our hair, duck feathers and down are made up of dead cells that are pushed up from the epidermis as new cells grow underneath. Composed mainly of a protein called *keratin*, feathers come in many lovely hues. Two main factors influence plumage color: the type

DID YOU KNOW?

Birds have a poor sense of taste. A Mallard has only about 375 taste buds, while we humans possess a whopping 9,000–10,000. No wonder they like slugs!

of pigment deposited during feather development and the light-reflecting and light-absorbing surface qualities of the feathers.

Adult ducks molt their old, worn feathers—including their flight feathers—once a year as bright new ones gradually come in. This annual casting off of feathers, called the *postnuptial molt*, normally occurs after the breeding season. During this time, which may last from 1–2 months, ducks are unable to fly, making them the perfect lunch for a hungry predator. No doubt, this is where the

DUCK TYPES

National Geographic's *Field Guide to the Birds of North America* divides duck-like waterfowl into the following eight types:

Whistling ducks: Upright and goose-like, these ducks are characterized by their distinctive high-pitched whistles.

Perching ducks: Perchers like the Muscovy and Wood Duck frequent wooded areas, forage on the water surface, and perch in trees.

Dabbling ducks: Dabblers glean food from the surface of shallow bodies of water or by tipping tail-up to snag underwater edibles. Members include the Mallard, the Pintail, and most domestic duck breeds.

Pochards: Heavy-bodied diving ducks in this group include the Canvasback and Redhead.

Eiders: These big northern ducks have a dense coat of down to keep them warm as they dive for food in the frigid sea.

Sea ducks: This ocean-loving group of divers includes the stocky Surf Scoter and the Harlequin duck.

Mergansers: These streamlined waterfowl with thin serrated bills are superb divers and fish-catchers.

Stiff-tailed ducks: The only common species of this type in North America, the stocky Ruddy Duck, uses its stiffly upright tail as a rudder when it dives.

Duck Behavior

Like wolves, bison, and humans, most ducks are social animals. You see them in huge feeding or migrating flocks, courting pairs, or mother-and-duckling families, but you rarely see them alone, except in the case of a duck incubating her eggs.

Flocking behavior has its pros and cons. On the one hand, birds living in a large group are at increased risk of disease. They also run the risk of wiping out their food supply. Another danger is that a quacking, fluttering flock will more likely attract a predator's attention than will a single duck floating quietly among the cattails. On the other hand, individual birds within a flock can spend less time watching for predators and more time foraging—a major advantage when it comes to survival and breeding. It's also possible that the hectic, noisy flight of many ducks could confuse a prowling coyote or attacking eagle long enough for the entire flock to make its getaway. The duck's powerful flocking instinct is a definite plus for duck keepers; it facilitates the task of herding the birds from place to place.

term "sitting duck" originated. For some duck species, such as the Mallard, however, nature has provided an ingenious solution to the problem: as the drakes lose their gaudy feathers in a summer eclipse molt, they take on a camouflaging brown plumage similar to that of females—a lifesaving adaptation when they're incapable of flying. A second molt during fall returns them to their dapper selves.

Feathers are vital to a bird. They enable flight, of course, but they also conserve body heat, protect the skin, and help a duck stay afloat by trapping air. Ditto for the soft down that makes ducklings so cute.

To function well, feathers must be kept clean, and ducks need no bribing to take regular baths. While domestic ducks can survive without water to bathe in, they feel and look their best when provided proper bathing facilities. In addition, watching waterfowl bathe is one of the chief joys of having them on your farm or in your yard. They happily dip their heads beneath the water over and over, beating their wings to make a glittering rainbow spray. Ducks know how to have fun!

All birds spend time each day preening their feathers, but for water-loving birds such as ducks, feather preening is much more than just a lengthy beauty routine—it's a matter of life and death. Feathers and down must keep a duck's body warm and dry even when the bird dips for pond plants on a frigid winter day. This means that the bird must groom each feather meticulously into place and regularly distribute oil from its *uropygial gland* (or *preen gland*, as it is also called) over the plumage so it will repel water.

Watch any duck emerge from a bath, and you'll see a whole lot of tail-, body-, and wing-shaking going on. The duck will get busy, scratching at its feathers with its bill and toenails to reposition and relock the barbules and trap in an insulating layer of air, ruffling its feathers to expose the oil gland above the tail, and repeatedly rubbing its head and bill against the gland to release the oil. Then you'll see it distributing this oil throughout the wing and tail feathers by combing them through its bill.

DUCK TRIVIA

· Not all ducks quack. Muscovies make hissing and whistling sounds.

· Ducks can drown. Ducklings are especially vulnerable to drowning if they are unable to climb out of a pool or water container.

· Ducks can be half-awake—literally. A duck can sleep with one eye open and the other shut. One hemisphere of the bird's brain sleeps while the other one (controlling that open eye) remains alert and watchful.

· A duck can fly up to 60 miles per hour. Normal cruising speed is about 20–30 miles per hour.

· Ducks can live fifteen to twenty years, but most ducks' lives are cut short, either killed by predators or slaughtered for meat.

· A flock of ducks goes by other names, including a raft, a paddling, a flush, and a team.

Humans used to think that all birds were birdbrains. Nowadays, we're more enlightened, thanks to a slew of studies showing our feathered friends to be far more intelligent than once believed. So where do ducks rank? I know they can learn to come when called, and they also learn from the actions of their flockmates. I've observed a very shy young Muscovy watch her tame mother take treats from my hand and soon start doing the same.

Ducks living together establish a pecking order just as chickens do, with one duck at the top, ruling the flock; one duck at the bottom of the hierarchy; and everybody else in between. During mating season, wild ducks usually form pairs, performing elaborate courtship rituals that culminate in nest building, egg laying, and incubation of the eggs by the female duck. Domesticated drakes have apparently had this pairing instinct bred out of them; they are more likely to mate indiscriminately with any females in the flock.

Downy ducklings are *precocial*, meaning that once they recover from the strenuous hatching process (usually within a day or so), they're up on their little webbed feet, ready to leave the nest and search for their own food. Compare these independent babes with the *altricial* young of some other birds, such as the robin, that emerge naked and helpless. It's a good thing ducklings are so self-sufficient because although mother ducks do form an attachment to their young that lasts for almost a month, they're not nearly as attentive and protective as mother geese are.

Duck Ancestry

There are thirteen species of perching duck and thirty-nine species of dabbling duck that can be found throughout the world. Domestic ducks, the kind you commonly see on farms today, descend from two species only—the wild Muscovy and the common Mallard. The Muscovy's large size no doubt made it a natural choice for domestication as a meat bird. As for the Mallard, this species may have been singled out for domestication because it was common and because it easily adapted to living near humans.

Wild Muscovy (*Cairina Moschata*)

Tentatively classified as a perching duck, this large, strange, long-tailed duck of the tropics is the ancestor of the domestic Muscovy. Characterized by black plumage with an iridescent sheen and flashy white wing patches, the species inhabits wooded wetland areas from the Rio Grande in southern Texas through Mexico and all the way down to Central and South America. The male has black and red facial skin around the beak and eyes, and both sexes possess powerful claws to help them perch in trees. These nonmigratory waterfowl congregate in pairs or small flocks and prefer to nest in tree cavities and boxes. Their eclectic diet includes aquatic and terrestrial plants, insects, crustaceans, small fish, and reptiles.

According to some sources, the Incas of Peru domesticated Muscovies centuries ago, keeping them as pest-controlling pets and suppliers of feathers, eggs, and meat. Spanish conquistadors took

these hefty ducks back with them from Columbia to Spain during the 1500s; from there, Muscovies were eventually imported to Africa, Asia, Australia, and North America. Along the way, this bird acquired many names, including Barbary duck, Brazilian duck, Turkish duck, Pato, and Guinea duck.

Common Mallard (*Anas Platyrhynchos*)

All other farm ducks descend from the wild Mallard, thought to have been domesticated in China about 4,000 years ago. The Mallard is a common dabbling duck that breeds on and around shallow wetlands throughout North America and other parts of the Northern Hemisphere. Chances are you've encountered these adaptable birds at a park, a zoo, a pond, or a farm field near your home.

Even neophyte duck watchers have little problem identifying this popular game species. Mallard males flaunt shimmering green heads, white collars, chestnut breasts, and orange legs and feet. When they explode into flight, the drakes flash vivid blue wing bars, bordered with white and black, called *speculums*. The female duck, which also brandishes speculums, has subtle brown-penciled plumage that helps camouflage her as she sits on her eggs.

Mallards dine on a wide variety of foods, including acorns, grass, duckweed, fruits, algae, tadpoles, frogs, tiny fish, leeches, mosquito larvae, and crayfish. The female usually makes a shallow nest in cloaking vegetation near the water, where she lays a clutch of seven to ten eggs.

Are Ducks Right for You?

Now that you've been properly introduced to the amazing duck, you probably feel that it's high time you zipped off to the feed store and nabbed some powder-puff ducklings. Not so fast! While a small duck flock can be a terrific addition to a hobby farm, these birds have

DUCKS NEED WATER

One disadvantage of having ducks is that they do need a steady water source. People have reported being able to manage them without adequate pools for bathing, but that's not something I condone. Water is vitally important to ducks [because] it helps them stay healthy and keeps their feathers in good condition. Ducks that are given steady access to fresh, clean water rarely suffer from external parasites.

—Cat Dreiling

some characteristics that can drive an unprepared keeper
to distraction.

Ducks Are Messy

There's a reason Martha Stewart raises chickens and not ducks.
Ducks consume a lot of food and water, and it all has to go
somewhere. To put it plainly, these fowl produce copious amounts
of wet droppings that have a truly unpleasant odor. During molting
periods, their feathers and down fly far and wide. Ducks also love to
dabble messily in their food dishes (thus attracting rodents), in their
water containers, and in the puddles around their swimming holes.
Too many ducks occupying too small an area can bring stinky ruin
down on your farm and negatively impact groundwater and wetlands.
On the bright side, proper management can keep the mess to a
minimum, and all of that duck manure makes great fertilizer!

Ducks Can Be Noisy

Some folks love the quacking of a flock of ducks because it reminds
them of the country; others prefer that ducks be seen and not heard.
If you or your neighbors are sensitive to quacking, you might want
to steer clear of gabby Call, Pekin, or Mallard females. Drakes of all
dabbling duck breeds are less talkative. Muscovies are completely
quackless: the drakes make a hissing sound, and the females emit
soft whistles and squeaks unless they're upset about something,
in which case they might squawk. You can also mute the cacophony
somewhat with landscaping or a solid fence.

Ducks Can Destroy Gardens

In their enthusiastic search for snails, slugs, and other creepy crawlies,
patrolling ducks will uproot young plants—or stomp them flat with
those big, flapping feet. Ducks also relish tender vegetation, some
flowers, and berries, which can lead to conflicts with humans who want
to harvest these crops. Again, management techniques can help.

Ducks Can Carry Zoonotic Diseases

All farm animals are capable of carrying *zoonotic* diseases—
animal diseases that can be transmitted to humans.
Commonsense practices, such as washing your hands and
cooking meat and eggs thoroughly, will help you and your family
avoid contracting salmonellosis or other illnesses from your
flock. It also helps if you make sure that your ducks stay in the
peak of health by keeping their home as clean as possible,
providing a proper diet, and paying attention to biosecurity
issues, such as quarantines and pest control.

Ducks Need Protection

Ducks may be super swimmers, super healthy, and super self-
reliant, but they aren't indestructible. These typically nonaggressive
fowl, along with their eggs and vulnerable young, can succumb to a
variety of hungry predators, from snapping turtles and bald eagles
to raccoons and coyotes. Domestic dogs and cats will also kill ducks
and ducklings. It's up to us to defend our fowl with secure fences,
nighttime housing, and other protective measures.

Local Laws May Prohibit Ducks

Finally, don't forget to check your local zoning laws, even if you live
in what appears to be a rural area. Some places may have limits on
the number of ducks you can keep, while others may prohibit poultry
keeping altogether.

DUCKS AS PETS

Ducklings raised with kindness and plenty of attention
make friendly, fascinating, and responsive pets. Even
people-shy older ducks can learn to overcome their fear
of humans, especially when coaxed with treats. Duck
lovers describe their pet ducks as lying at their feet like
dogs, tapping at the door for handouts, and following
them faithfully around the farm. Some people keep ducks
as house pets—you can even purchase diapers made
especially for ducks. Although my tame Muscovies dislike
being picked up, they seem to enjoy having their back
feathers stroked and being scratched about the head, neck,
and chin. Some pet ducks like to play: one family I spoke to
has a Rouen duck that chases tennis balls. Take that, Fido!

A flock of healthy, happy ducks is a joy to behold: tails a-wagging, they greet the day with enthusiasm—cavorting in their swimming pool; preening their sleek, glossy plumage; searching bright-eyed through the grass for delicious bugs. When your animals are the picture of good health, life on the farm feels right.

For those of us who have become completely enamored of our little flocks, that feeling of well-being changes drastically when a duck falls ill. You notice one morning that Daisy doesn't come speed-waddling over for breakfast with her companions. Instead, she sits in the back of the house, looking listless, feathers ruffled. How sick is she? What's wrong with her? Will the rest of the flock catch whatever she has?

HEALTHCARE FOR YOUR
DUCKS

Happily, waterfowl tend to be hale and hearty creatures when given proper care. But even the most conscientious management won't guarantee that your birds will never get sick or injured. In this chapter, we'll look at proper ways to handle ducks in sickness and in health. We'll also cover basic information on some of the health problems your flock can experience.

Disease-Prevention Basics

Let's begin our discussion of flock health with an important question: How do you prevent illness and injury from ruining your ducks' day—and yours—in the first place? Let's look at some savvy strategies for minimizing problems.

Choose Good Stock

Start with healthy stock from a reputable source. Whether you choose to build your flock with ducklings or with adult ducks, try to obtain your birds from a good source that you know carries healthy stock. If you are picking out the ducks yourself—at a feed store or a farm, for example— choose ducks and ducklings that appear vigorous and active. Avoid those with pasty rears, discharge from the eyes, or leg problems as well as any listless-looking birds that just sit around with their eyes closed.

Maintain a Good Diet

Feeding your birds a healthy, balanced diet with ample fresh water is critical. An inappropriate or unbalanced diet ranks as one of the leading causes of health problems in domestic ducks. Ducks can't thrive on a diet of bread alone—or corn or lettuce or grass—any more than we can. Overdoses, deficiencies, and imbalances of specific nutrients can be bad news for your flock's health. Waterfowl also require an abundant supply of clean drinking water to remain healthy. With ducks' dabbling and bathing habits, keeping their water sparkling clean all the time is impossible, but do change it at least twice a day and scrub the containers regularly to keep the water from turning foul and stagnant.

Maintain Proper Hygiene

Avoid crowding your ducks. Ducks are social animals, but they don't enjoy being packed like sardines. Overcrowding causes filthy conditions, promotes the spread of disease, and increases aggression and stress. So ditch the factory-farm mentality and give your ducks plenty of space to breathe, bathe, eat, preen, interact, and basically behave like ducks.

Keep your flock's housing, pens, and swimming pools or ponds as clean as you can. We've said it once (or twice), and we'll say it again— ducks are messy creatures, especially when confined in small houses and pens rather than given a larger area to roam. Regular cleaning of your flock's abode, dining spot, and enclosures will cut down on parasites, disease-carrying flies and rodents, and harmful molds and bacteria. It will also prevent your ducks from experiencing the sheer stress—and thus lowered immunity—that comes from living in a filthy, foul-smelling environment. You should also clean or flush your flock's bathing facilities regularly to decrease health risks associated with stagnant, dirty water.

Keep your ducks' enclosure and anywhere else they may roam safe.

Protect Ducks from Other Animals

Control pest populations and keep your ducks away from wild birds. Rats, mice, crows, and pigeons drawn to a leftover buffet courtesy of your ducks can bring disease and parasites. So can the migratory waterfowl alighting on your pond and the pretty wild birds flitting about feeders. If possible, use fencing or enclosures to bar your ducks from areas frequented by wild waterfowl; these birds pose one of the biggest disease risks to your flock. You should also prevent access to the ground around your bird feeders, where droppings accumulate. Control rodents as much as you can, using common sense and caution, because you don't want your ducks or any of your other animals ingesting poison or getting caught in a rat trap.

Keep a Closed Flock

A closed flock means that no new birds enter until it's time to replace the flock completely. It also means that all of the resident birds stay put—no going back and forth to shows or fairs to mingle with other waterfowl and poultry. Closed flocks are standard procedure on large commercial poultry farms, where biosecurity is a big, big deal. Obviously, not every small duck farmer can—or even wants to— keep his or her flock completely closed. Still, think carefully before saying "sure, why not?" when your neighbors offer you their last two ducks , or when you're tempted to take a few free surplus fowl sitting

in a cage at your local feed store. Some birds come with stuff you didn't bargain for, such as infectious diseases and parasites.

Quarantine New or Sick Birds

If a closed flock isn't an option for you—for example, you regularly purchase or exchange breeding stock, or you love to show ducks— then put new or returning ducks through a quarantine period before allowing them back into the flock. Even fowl that look completely healthy can harbor disease, and some illnesses have long incubation periods. Keep new birds isolated from the rest of your flock for at least four weeks. Ducks back from the fair or show should stay in quarantine for a minimum of two weeks. Have a separate set of cleaning equipment, feed dishes, and other paraphernalia for each group of birds (or sanitize well before exchanging supplies), and take care of your resident flock *before* tending to the quarantined birds. Any fowl showing signs of an infectious disease should be promptly removed and isolated from the rest of the birds for observation, diagnosis, and possible treatment.

Be Biosecure

With highly pathogenic avian influenza a looming threat, biosecurity is definitely not just for commercial operations anymore. Biosecurity encompasses measures that keep infectious diseases off your farm, including those previously mentioned. Other sensible measures include restricting visitor (especially bird-owning folks) access to the area where you keep your ducks; using a disinfectant foot bath; changing your clothes and footwear after you've visited another farm, a fair, or an auction; disinfecting poultry cages and equipment that have been to another farm or the fair; and disposing of any dead birds promptly and properly.

Keep Their Environment Safe

A strand of wire dangling loose from the fence, a sharp nail protruding from the duck house wall, shards of broken glass, a bottle

DID YOU KNOW?

Vaccination of small flocks is needed only if you've had disease problems in the past, if outbreaks have occurred in your area, or if you frequently show your ducks or acquire new stock. Before embarking on a do-it-yourself vaccination program, call your state diagnostic lab veterinarian or a local poultry extension specialist to find out what vaccines—if any—are recommended in your area. If you must inoculate your flock, ask your veterinarian or an experienced duck keeper to show you how to administer the vaccinations correctly.

cap, a hole in your ducks' enclosure large enough for a raccoon or mink to squeeze through—all of these hazards can spell disaster. When caring for your ducks each day, survey their environment. If their bin of feed smells moldy, toss it. If a bald eagle is hunting overhead, evaluate your duck yard and management techniques to see if you need to make modifications to keep your flock safe. If you spot a clump of baling twine in the grass, pick it up immediately.

Recognizing Symptoms and Warning Signs

Although duck keepers should know the warning signs of ill health or injury in their birds, learning each duck's normal appearance and behavior is just as important. Ducks may not be as obviously individualistic as dogs or horses, but they often have different personalities, even within the same breed. You may discover that your birds' behavior changes during the course of the year: a normally active, vocal duck becomes subdued while molting; a friendly quacker turns aggressive when she hatches her brood.

Be a good observer of your ducks when they're healthy, and you'll find it easier to quickly identify any abnormal behaviors or signs that indicate ill health or injury. Give each of your birds a bill-to-tail inspection at least once a day, really focusing on appearance and behavior. In general, a healthy duck has smooth, glossy feathers, except during molting periods; bright, discharge-free eyes and nasal openings; a clean vent area; and a good appetite. A vigorous duck will take baths throughout the day and spend time preening, foraging, drinking, and sleeping.

Many domestic animals, like their wild kin, instinctively take great pains to hide sickness or injury so they won't be seen as easy prey, and ducks are no exception. Signs of illness in birds may be subtle and difficult to spot, so vigilance is important. The following symptoms should set off warning bells in your head: coughing, sneezing, gasping, nasal or eye discharge, sluggishness, depression, ruffled feathers, watery or bloody diarrhea, poor appetite, an unusual loss of feathers, abnormal drop in egg production or soft-shelled eggs, swelling or twisting of the head and neck, purple discoloration of exposed skin, tremors, drooping wings, lack of coordination, and any sudden deaths within the flock. Indications of injury are usually more obvious: limping, a drooping or oddly bent wing, a closed eye, the presence of blood.

What to Do When a Bird Is Ill

Death or any of the illness symptoms could indicate infectious disease. Your first step should be to isolate the affected duck from your other birds right away. Place the patient in a warm, dry, well-bedded cage, dog kennel, or other small enclosure with food and a water and electrolyte mix, which you should always have on hand as part of your duck first-aid kit (see sidebar). Keep noise and stress to a minimum.

Although it may be tempting to try to diagnose the disease and treat the bird yourself with feed-store medications, your duck will have a better chance of recovery if you seek professional help. In some cases, the survival of your entire flock may depend on it.

Duck diseases can either be noncontagious or contagious (infectious). Noncontagious diseases, such as aspergillosis, will usually affect only a few ducks and won't spread among birds. Contagious diseases, such as avian influenza, can sweep rapidly through a flock; for such cases, seeking help in determining the cause is especially critical.

A phone consultation with the vet can help you avoid some costly mistakes. For instance, using an antibiotic will not cure a viral disease, may cause bacterial resistance when used improperly, and could even kill your duck if it's the wrong kind. Also, if only one or two ducks take ill, don't automatically treat the entire flock unless your veterinarian instructs you to do so.

Other Health Threats

Ducks can also become ill from poisons and parasites. The following sections provide some preventive measures to help keep your flock healthy.

Ducks usually avoid toxic plants, but it's better to be safe than sorry.

Poisons

Foraging ducks seem to either instinctively steer clear of poisonous plants or avoid consuming enough of these plants to do them any harm. But when given no other edible options, ducks may eat toxic plants. For a list of such plants in your area, contact your local extension agent or visit.

Ducks can also accidentally ingest other types of poisons. For example, rodent poisons and insecticides can be highly toxic to ducks; read directions carefully before using these in areas frequented by your flock. Take steps to restrict animal access to these poisons or, better yet, find safer alternatives.

Botulism, another deadly form of poisoning, occurs when ducks dine on toxins produced by the anaerobic bacterium *Clostridium botulinum*. Stagnant bodies of water or decaying plants and carcasses that provide optimum conditions for this bacterium are usually the culprits. The disease causes paralysis, and infected ducks die within one or two days. Your best bet for prevention is to bar your ducks from stagnant water and keep pools and ponds clean.

Aflatoxin poisoning occurs when ducks eat grains or seeds that have been contaminated by several types of mold, usually the result of wet harvest conditions. Never give your flock moldy feed, bread, or bedding. Castor-bean poisoning and poisoning by high levels of erucic acid in rapeseed (canola) meal have also been reported in ducks. Ducks that frequent ponds at or near hunt clubs where lead shot is used often suffer from lead poisoning. For cases of suspected poisoning, contact a veterinarian as soon as possible to discuss treatment options.

Parasites

Parasites are organisms that live and feed off other organisms, providing no benefit in return—freeloaders, basically. Whether we like it or not, these organisms are a fact of life for our birds and for us. Parasites affecting livestock usually fall into two categories: *ectoparasites* (external parasites), such as mites and lice, and *endoparasites* (internal parasites), such as roundworms and *Coccidia*, which spend part of their life cycle inside the animal's body.

Although many parasites are fairly benign, others can cause health problems in birds, ranging from itching to severe anemia. Creepy-crawly lice plague poultry by biting, sucking blood, or dining on skin scales; they can cause intense itching, weight loss, and poor growth. Tiny relatives of the spider, mites lay eggs inside the skin or at the base of feathers; they can also cause weight loss and lead to death from anemia.

DUCK FIRST-AID ESSENTIALS

When injuries or illness strike, time is usually of the essence. Keep the following first-aid items on hand so you'll be prepared for emergencies:

- Antibiotic ointment and antiseptic spray
- Gauze, first aid tape, and cotton balls
- Insecticidal poultry dust for mite or lice infestations
- Pet carrier for transport
- Probiotic powder
- Quarantine cage or enclosure
- Scissors

- Styptic powder to stop bleeding
- Veterinarian's phone number
- Vitamin and electrolyte mix for poultry
- Wooden Popsicle sticks or tongue depressors (for splinting)
- Veterinarian's phone number
- Plastic storage or tackle box to store these items

A duck's wings, legs, and webbed feet are at risk of certain injuries.

Common Leg, Wing, and Foot Problems

Duck legs may look sturdy and strong, but their legs, feet, and wings are actually their weak links. While many chicken farmers will nab their birds by the legs, grabbing a duck this way can result in injury. Ducks can also injure their legs when chased by predators or people. Niacin-deficient diets can cause leg problems in ducklings, and rations with excessive calcium can do the same for growing birds.

Good hygiene will help control ectoparasites, as will providing your flock with proper bathing facilities. You may need to use a commercial insecticide formulated for poultry, such as an insecticidal dust, if your birds have a heavy infestation. Always read labels and follow the manufacturer's directions.

Internal parasites of ducks and other poultry include gapeworm, large roundworm, tapeworm, capillary worm, and *Coccidia*. Mature, healthy animals generally develop a degree of immunity to internal parasites, whereas young and old animals and those weakened by stress are more likely to suffer ill effects from parasitism.

Signs that internal parasites are running amok include weight loss, poor growth, reduced egg output, coughing, diarrhea, lethargy, weakness, head shaking, pale membranes, and even death. Deworming agents, or *anthelmintics*, to combat various parasites can be obtained from your feed store or poultry-supply company, but consult a veterinarian before using these products. Some dewormers are effective against only one type of parasite, and few are approved for use in treating waterfowl. Avoid using dewormers as a preventive measure when your birds don't have a problem; instead, ask your veterinarian to check a fecal sample. Unfortunately, the indiscriminate use of dewormers has already created dewormer-resistant "superparasites."

A group of protozoan parasites called *Coccidia* can cause illness or death in waterfowl, particularly in ducklings. Often present in the soil, these organisms kill cells in the duck's digestive system, leading to watery or bloody diarrhea and a depressed attitude, evident by ruffled feathers and decreased food consumption. Overcrowded conditions and wet, dirty bedding are major contributors to the problem. While coccidiostat medications are commonly employed in chickens and turkeys, caution must be used when treating ducks with these medicines. In general, small duck flocks rarely require medicated feeds.

Bumblefoot, a condition in which the footpads become cracked and infected, usually affects adult ducks kept on hard, dry surfaces, such as concrete and gravel, or on wire. It can also occur when ducks constantly walk and lie on wet, dirty bedding. A reaction between water and the uric acid in duck droppings creates ammonium hydroxide, a weak, lye-like solution that can produce burning ulcers on the feet. Prevent bumblefoot by layering hard surfaces with cushy dry litter, by allowing birds access to grassy terrain, and by keeping your flock's accommodations as clean and dry as possible.

While we're on the subject of feet, remember to keep an eye on your ducks' toenails and trim them if they seem too long or hinder your duck's waddling. If the nails are clear, you'll be able to see where the blood vessel ends to avoid cutting it, but keep some styptic powder on hand just in case. Styptic powder works as a clotting agent to rapidly stop bleeding from nail-trimming injuries.

Twisted wing, also called *angel wing*, occurs when the primary feathers on one or both wings grow up and away from the duck's body. More a cosmetic flaw than a health problem, the condition can be either inherited or caused by a diet too high in protein. If you catch twisted wing early, try correcting it by realigning the wing and feathers into a proper folded position and then taping them into place with masking tape. Be careful not to tape too tightly or block the vent opening, and release the wing after about two weeks. With any luck, your duck will still be able to fly, although probably not as well as a bird that has never had this condition.

Some domestic duck breeds are talented flyers, so you may wish to ground these birds to ensure that they stay on your property or in their pens. Wing clipping is easy to do, and unlike cutting off a wing's pinion, it doesn't permanently disable the bird. While an assistant holds the bird and stretches out the wing, use a sharp pair of scissors to cut the eight or nine long primary, or flight, feathers on one wing at

FLOCK HEALTH

Believe it or not, adult ducks can suffer from hypothermia. They become slow and listless and can't walk very well. We always have a large cage, brooder light, food, water, and electrolytes ready, since one has to work quickly to save [an affected] duck. I place the hypothermic duck's feet on my hands to gradually start warming the bird—it seems that the feet are their thermostats. I alternate putting my warm hands under its wings and feet, until the duck starts moving more. I have syringes to gently wash some electrolyte water over the inside of its mouth if the duck shows no interest in drinking right away. They're soon drinking their water after that.

—Dawn Turbyfill

Some duck breeds physiologically need more niacin to prevent weak legs. Two sources are niacin powder added to the ducklings' water and brewer's yeast added to their feed. The faster this deficiency is caught and treated, the faster they will recover. You'll need to keep an eye on them until the legs are fully developed at about five months.

—Dawn Turbyfill

the level of the smaller secondary feathers. Watch that you don't snip thick, blood-engorged pin feathers that grow in to replace old feathers during the molt. If you are unsure about how to clip wings, ask an experienced duck raiser to show you how to do it. Remember to repeat the procedure when new feathers grow in after the annual molt.

Many raisers prefer not to clip their birds' wings; they believe that their ducks have a better chance of avoiding predators if they can fly. Once they become accustomed to your farm and know where their food and shelter comes from, most domestic ducks prefer to stick around.

Handling Ducks

Once in a while, you'll need to catch a duck to treat an injury or health problem, transport it to a show, check its weight, or perform some other task. Untamed ducks can be quite elusive, so make the effort to befriend your birds, bribing them with yummy treats so they'll permit you to approach them closely. Teaching the ducks to enter an enclosed area each day for their meal will make capturing them much easier. A long-handled, medium-size fishing net comes in handy for nabbing shy, fast, or flighted birds in more open areas (such as your neighbor's backyard!). However, extricating ducks from a net is a tricky procedure that almost always messes up their feathers. Whenever possible, it's best to catch them by hand.

Dress appropriately for duck catching: wear long sleeves to

protect against scratches and prepare for the likelihood of being pooped on. Remember that legs and wings are off-limits for grabbing. Instead, pin the duck's wings against its body with both of your hands or, if you can't do this, catch the bird gently around the neck (don't squeeze, and don't do this with small ducks whose necks are more delicate) to hold it in place until you slip your other hand and arm underneath its body from the front. A captured duck may paddle furiously, so it's important to get control of the feet with one hand while cradling the bird firmly against your side.

If you're dealing with a big duck that might peck at your face, it may be best to carry the bird facing backward. Ducks are not ferocious animals, but they will peck and bite if provoked. Everyone who handles ducks must wash their hands well with soap and warm water afterward. Use an alcohol-based hand sanitizer if water isn't available. Even a minor bout of salmonellosis is nobody's idea of a good time.

Duck Diseases at a Glance

The following are the most common culprits affecting domestic duck flocks.

Aspergillosis

Moist bedding or feed makes a perfect medium for the growth of mold, including the fungus *Aspergillus,* whose spores can sicken ducks by promoting plaques in the air sacs and lungs. Aspergillosis is usually seen in ducklings; symptoms include lethargy and breathing troubles. Prevent this disease by tossing out moldy straw or feed, changing bedding regularly, and sanitizing your incubators as needed. Watch out for hardwood chips that contain bark.

Avian Influenza

Most of these virus strains cause little or no illness in affected birds. Highly pathogenic forms, however, can pass through a domestic poultry flock like lightning. Ducks often survive but then become carriers, infecting chickens and other poultry. Ward off this disease with good biosecurity measures. Be especially careful to prevent contact between your flock and wild waterfowl.

Avian (or Fowl) Cholera

Avian cholera can spread through a flock in an explosive fashion. Symptoms of this bacterial disease include diarrhea, discharge from the mouth, appetite loss, labored breathing, and sudden death. Necropsy may reveal internal hemorrhaging of the heart muscle and abdominal fat; an abnormally large, copper-colored liver; and an enlarged spleen. Unsanitary conditions and stagnant water

Handle fragile ducklings gently and carefully.

contribute to this disease. Control rodents, which can also spread cholera, and dispose of dead animals promptly and properly.

Colibacillosis (or *E. coli*)

A bacterium called *Escherichia coli* causes this common poultry infection, which can infect the yolk sac or cause an infection of the bloodstream in ducklings. Good sanitation procedures will help prevent this disease.

Duck Viral Enteritis

Also known by the frightening name of *duck plague*, this highly contagious disease caused by a type of herpes virus can infect waterfowl of any age. The virus is passed on by direct contact with infected birds or contaminated water. Symptoms include extreme thirst, nasal discharge, watery or bloody diarrhea, lethargy, drop in egg production, and ruffled feathers. Sadly, this disease can cause sudden death. Necropsy will reveal internal hemorrhages and necrosis. Prevent contact between your ducks and wild waterfowl, which can carry the virus, and change footwear after being around wild waterfowl. Vaccines are available, but only on a limited basis.

Duck Viral Hepatitis

Like duck plague, this is an extremely contagious viral disease with a short incubation period that shows up suddenly and results in high mortality rates. The virus, which causes an enlarged, discolored liver, primarily strikes ducklings less than seven weeks old. Signs include lethargy, loss of balance, and spasmodic paddling movements of the feet; the duckling typically dies within an hour. To prevent this disease, keep ducklings isolated from older birds during the first five weeks, restrict access to wild waterfowl and their habitats, practice good hygiene, control rodents, and vaccinate breeder ducks to impart immunity to their ducklings. Knowing the breeder you acquire ducks

from and his or her management methods can help you keep this disease off your farm.

Riemerella (or Pasteurella) Anatipestifer Infection

Also known by its much simpler name, *new duck disease*, this bacterial illness mainly infects ducklings that are two to seven weeks of age. Signs may include discharge from the eyes and nasal openings, sneezing and coughing, sluggishness, head and neck tremors, weight loss, and a twisted neck. The disease can also cause sudden death. Providing clean drinking water and general good hygiene will help prevent the spread of this disease. A vaccine is also available.

TRANSPORTING DUCKS

If you're transporting ducks, follow these guidelines from the Virginia Cooperative Extension to help them reach their destination safely.

- Transport ducks in escape-proof, well-bedded carriers (tough plastic dog crates, for example), with openings small enough to prevent the ducks from sticking their heads through.
- Ensure that the carrier has sufficient ventilation and air flow during warm weather, but avoid exposing your birds to drafts on chilly days.
- Don't crowd birds into crates; this can cause overheating and fighting.
- In cool weather, a blanket or towel tossed over the crate helps keep birds calm.
- Take along a spray bottle with cool water or some ice packs in case your ducks show signs of overheating. Offer them water to drink during rest stops.
- Secure the crates within your vehicle so they won't move during transit.
- Never leave ducks unattended in a parked car or trailer during the summer, and don't leave them in a dark-colored crate or unshaded cage out in the sun.
- Avoid subjecting ducks to extreme changes in temperature (from an air-conditioned vehicle to the hot outdoors, for example).

Ducklings are one of the biggest perks of raising ducks. They're soft, fuzzy, and too cute to believe. Just toss some ducks of the opposite sex together, wait for eggs to start appearing and ducks to start sitting, and a mess of ducklings should be the end result—right?

Not so fast. First of all, getting eggs is one thing (like chickens, female ducks will lay eggs whether a male is present or not), but getting *fertile* eggs and successfully hatching them out is another thing entirely. The strategy just described may work with some broody breeds, such as the Muscovy and the Mallard, but not with others. The Indian Runner, for instance, may lay loads of eggs for you to eat, but ask her to sit on those eggs for a whole twenty-eight days? Forget it!

BREEDING AND

DUCKLING

CARE

Second, even easy-to-breed ducks shouldn't be allowed to reproduce out of control, with no thought on your part as to what you'll do with those adorable babies—which, by the way, grow up fast! Will you raise the ducklings for meat or sell them to others to eat? Will you market them as pets or as breeding or show stock? Will you supply ducklings to feed stores? Please don't assume that you'll just give the ducklings away to friends and neighbors who want them on a whim. And whatever you do, *don't* plan to release your surplus stock at a pond or lake, thinking they'll take care of themselves. They won't.

It makes good sense, before embarking on a breeding program with any form of livestock, to decide on your objectives and what traits are important to you. Knowing your goals ahead of time is critical to your success as a breeder.

Breeding-Flock Basics

Whatever your breeding objectives, try to start out with breeder ducks that show good vigor and—if they're purebreds—appropriate breed conformation. Obtain specifics on conformation and breed standards by checking out the APA's Standard of Perfection and the websites of breeder associations.

The birds should be healthy and neither too fat nor too thin. Avoid breeding any ducks with genetic flaws, and make sure that your birds are fully grown before breeding; an immature duck's first eggs are small and less likely to be fertile. In general, lightweight and middleweight ducks reach sexual maturity at about seven months of age, while heavier ducks can take up to a year. Ducks are at their breeding peak from one to three years of age, although they can successfully breed for longer.

Domestic ducks readily interbreed, so if you keep more than one breed and want to avoid producing hybrids, separate them at least three weeks before the onset of the mating season. Mating will typically start in the early spring, although the timing may vary with climate and latitude. A breeding pair will continue to mate even while the female lays her eggs, so you should prevent the breeding groups from mixing until the ducks complete their clutches and begin to incubate.

Male-to-Female Ratio

The ratio of drakes to ducks in a breeding flock is an important consideration. Breeders of ornamental "wild" ducks, such as Mandarins, Wood Ducks, and Mallards, generally keep these birds in pairs within separate enclosures. Usually monogamous in the wild, these ducks tend to be territorial. With domestic ducks, it's a different situation. Raisers of domestic ducks often keep one drake for each two to six females. Not only are more males unnecessary, but too many can lead to fighting and possibly life-threatening "raping" of the females. Muscovy drakes are especially prone to getting into vicious brawls over their girls. Avoid crowding your birds to keep competition for mates and nest sites to a minimum.

Encouraging Breeding

To maintain your breeders in good condition and increase their reproductive success, offer your mating ducks a breeder feed ration about one month before egg production starts up.

WHO LET THE DUCKS OUT?

Visit any public park pond or lake, and you're bound to encounter domestic ducks scrambling for bread with the wild Mallards and Canada geese. They're probably not farm escapees—most domesticated ducks, even those that fly, tend to stay close to home when offered plenty of food and protection from predators. It's possible that some well-meaning people decided that they didn't want ducks anymore and released them, feeling confident that the birds would revert to an independent wild state or survive off treats tossed by visitors.

Sadly, the free life isn't always a bed of roses for feral ducks (domestic ducks that have reverted to a wild state). They face dangers seldom encountered on the farm, such as winter starvation, abusive humans, and predators. Worse still is the impact that feral ducks have on other ducks and the environment. Breeding out of control, they foul water bodies and urban lawns with their abundant droppings. They hybridize with wild ducks and transmit diseases such as duck plague and fowl cholera.

If you have too many ducks, deal with your problem responsibly. Control overbreeding with diligent egg removal and by finding appropriate homes for your birds.

Keep pens as clean as possible, and make sure that your ducks have their choice of secluded, protected nest boxes littered with straw, wood shavings, or other soft materials. Fresh litter, changed as needed, will help keep the eggs clean and also insulate them from the cold ground on chilly days. Sometimes you can persuade the females to lay their eggs in the nest boxes by placing a "dummy egg," such as a golf ball, hard-boiled egg, or sand-filled plastic egg, in each nest. But don't be surprised or offended if your ducks drop their eggs just about anywhere; many domestic duck breeds and individual ducks no longer have the instinct to nest and brood.

If you allow your ducks to roam free part-time, they may nest almost anywhere on your property that offers a sheltered, secluded site. Muscovy females, hole-nesters in the wild, excel at hiding their nests in hard-to-find spots. If you suspect that a duck has a hidden nest and would like to find it, try playing detective: watch for when she takes a break from egg laying or incubation to feed or bathe, then nonchalantly follow her back to her nest.

Finally, provide your breeding ducks with plenty of fresh drinking water and, if possible, a bathing pool deep enough for them to float in. While some domestic duck breeds can mate on the ground without a problem, heavyweight ducks often need the buoyancy provided by water to facilitate mating. Water also allows the ducks to cleanse away the sticky, bacteria-rich saliva left on their heads during mating.

Hatching Options

Once your ducks start laying eggs, you have two hatching options available: natural incubation by the mother duck or another broody bird and artificial incubation in an incubator machine. Each method has its pros and cons.

The simplest way to hatch out ducklings—and the method that yields the highest egg hatchability—is to let a female duck incubate the eggs for you (if she's willing, that is). After breeding commences and a broody female duck finds the ideal nesting spot, she scrapes and hollows out a simple depression in the litter with her beak, body, and feet. Every day or so, usually during the night or early morning hours, the duck will lay one smooth-shelled egg. For the duration of the laying period, which will depend on the duck's breed and the size of the clutch, the female will still step out and about quite a bit, engaging in all of the usual activities. Each egg, if it's fertile, will contain a cluster of embryonic cells existing in a state of suspended development. Development will continue once the process of incubation is set into motion by the mother duck, a substitute broody bird, or an artificial incubator.

How many eggs can you expect? The final clutch size varies among duck species and domestic breeds: a Mallard will lay a clutch of about seven to eleven eggs, while a domestic Muscovy may lay fifteen to twenty eggs. You'll know that your duck has nearly finished laying when she lines the nest with soft down from her breast. If the duck has produced more eggs than she can completely cover, you'll need to remove some to either discard or incubate elsewhere. The eggs will have a better hatch rate if they go through incubation in a single layer. According to veteran raiser Dave Holderread, good broody hens will frequently hatch out every fertile duck egg they incubate; by comparison, with artificial incubation, the average hatch rate is 75–95 percent of the fertile eggs set.

Setting Ducks

Once the duck begins sitting tight, making only occasional, brief forays for food and water, incubation has begun. If conditions are right, the duck will naturally provide the proper humidity, air flow, and temperature for her egg-encased embryos to develop normally. She instinctively knows when to stand up and allow them to cool off on a hot day and when to hunker down, keeping them toasty warm during a chilly night. Throughout the day, she'll regularly turn the eggs and push them about the nest so they receive an even amount of

DID YOU KNOW?

Increasing day length stimulates mating and egg laying, so some raisers will bring the birds into egg production by using artificial lights in the duck house or pen to extend the "day" to fourteen hours.

A good setter monitors the position and temperature of her eggs.

eggs' expected hatch date, starving herself in the process. Aside from this, try to avoid disturbing the duck. Consider separating the duck from the rest of the flock so they won't bother her, either.

Keep in mind that a setting duck and her treasure trove of eggs are an enticement to many predators, so it's crucial to ensure that your brooding bird has protection. If the duck chose to nest somewhere other than in a secure house or pen, try placing a covered portable pen, electrified poultry mesh, or a large cage around her, nest and all. Attempts to move a duck and its nest, unless it's nested in an easily transported box with a solid bottom, are usually unsuccessful.

heat from her body. When the duck leaves the nest to feed, drink, or bathe, she will tuck a blanket of down and bedding over the clutch to keep the eggs insulated and hidden.

Breeds that make good setters, or brooders, include Muscovies and bantam ducks, followed by the Appleyard and middleweights like the Cayuga and the Orpington. Pekins, Runners, and Aylesbury ducks are generally poor setters, but don't rule them out completely. For Mallard-derivative breeds, the incubation period generally lasts for 24–29 days. Muscovy ducks sit for about thirty-five days or so.

Don't forget about your patient duck during the incubation period; she still needs care and attention. Make sure that she has food and water close by so she won't be forced to leave the eggs unattended for too long, but not so close that the food will soil the nest and eggs or attract rodents.

Nesting ducks normally take occasional breaks. If yours obsessively sticks to the nest, you might need to carefully remove the duck once a day so she has a chance to feed and drink. Be alert, too, for the bird that seems committed to setting an infertile clutch. You'll need to remove the eggs, or the duck may sit well past the

Artificial Incubation

It isn't always feasible for the mother duck to incubate the eggs. As I've mentioned, some ducks lay plenty of eggs but are poor setters, and others tend to flunk out in the mothering department. Where nest-raiding predators pose a problem, even good brooders may have difficulty bringing their eggs to the point of hatching. In such cases, you may need to use artificial means to incubate the eggs.

Here's another scenario to consider: A setting duck can cover only so many eggs, but you may want to hatch out more eggs than that from a particular duck—a rare breed, perhaps, or an exceptional exhibition bird. You'll obtain a greater number of fertile eggs if you collect the eggs daily from the laying duck's nest (she will keep on laying, thinking the clutch is not yet full.) Rather than leave the nest empty, set one or two dummy eggs in it; otherwise, the duck might think the clutch has been raided, and that could cause her to abandon the site for a new one. The eggs you gather will need incubating, and this is where an incubator can prove its worth.

EGG CANDLING

Verify that the eggs are fertile and check for dead embryos by candling the eggs at about seven days. Most breeders do this, and some repeat the procedure once a week until hatching. Candling involves holding and gently rotating the egg against a candling machine or a flashlight in a dark room or closet; this will illuminate the inside of the egg. Your hands should be clean and dry when you touch the eggs, and avoid jarring or shaking them. Do not dawdle while candling—you don't want the eggs to cool off too much—but don't rush, either.

During candling, an infertile egg will look pretty clear; you'll see the air space at the large end and a shadowy yolk but nothing else. An egg with a dead embryo shows a dark line or circle of blood and gives off a foul odor as time goes by. Remove the bad eggs and discard them, but handle them gently so they don't burst. A fertile egg will have a dark spot in the center with a web of blood vessels radiating out from it. At about two weeks, the egg beneath the growing air space will look very dark, but you may be able to make out the tiny heart beating and see the embryo moving. If all goes well, you'll have a baby duck.

A female Muscovy makes a good broody for her own eggs or a surrogate setter for other ducks' eggs.

A third instance in which artificial incubation comes in handy is if you want ducklings without having to breed ducks. In this case, order fertilized hatching eggs through the mail and hatch them in an incubator. Hatching eggs can be purchased from some hatcheries and breeders for less than the cost of a live duckling.

Incubators commonly used by small flock raisers come in a number of sizes and styles with varying price tags. They range from inexpensive plastic-domed miniature incubators to large cabinet models that accommodate several hundred eggs and can cost close to $1,000.

Basically, incubators come in two types: still air and forced air. Still-air machines heat the eggs by convection; they lack fans to move the air inside the unit. They're often less expensive and simpler to operate than forced-air units, but it can be much more difficult to adjust heat and humidity levels, especially for novices. Forced-air incubators come with fans that circulate heated air evenly throughout the interior, helping keep the unit at a constant temperature and humidity level that promotes successful hatching. Many have automatic egg turners and multiple tiered shelves to hold the eggs.

Follow the manufacturer's instructions for setting up the incubator, and be sure to get it running several days before adding the eggs to ensure that it works properly and to establish the correct settings. The settings may need to be altered slightly depending on your climate, elevation, and other factors; it takes practice and experience to be successful. Even under ideal conditions, don't expect a 100-percent hatch rate with an artificial incubator.

Set up the incubator in a room where the temperature stays relatively constant. Avoid cold or drafty areas or spots where the temperature and humidity fluctuate, such as near a heater, a sunny window, or an air conditioner.

About six hours before setting the eggs, transfer them from their cool storage spot to a place where they will gradually reach a room temperature of about 70° Fahrenheit. This will keep them from experiencing temperature shock when placed in the heated incubator. Set the eggs on the incubator racks (if provided), with the large end elevated and the small end down. As with natural incubation, candle the eggs at seven days to identify infertile eggs or dead embryos and discard them so they don't contaminate the good eggs.

Preparing Eggs for Incubation

How you collect, handle, and store the eggs before setting them in an incubator has a big impact on their hatchability. Checking for eggs twice a day contributes to egg cleanliness and will help you beat predators to the punch. When the weather is colder, bringing in the eggs will also keep the embryos from being damaged or killed by extreme cooling. Have clean, dry hands when you collect the eggs, and handle them with care. Eggs that have thin or cracked shells or that look abnormally small or large are best discarded.

Despite your efforts, some eggs may still be soiled with dirt or droppings. Cleanse them gently to wash away egg-damaging bacteria before you store or incubate them. If they're only slightly dirty, brush the soiled spots lightly with a clean, rough, dry cloth or a steel-wool scrubbing pad. Avoid rubbing away the waxy cuticle, or bloom, that covers duck eggs and protects them from bacteria. Although it's possible to wash really grubby eggs with a sanitizing solution, some sources advise against it because it can reduce hatchability. Before storing eggs, mark the shells with the date on which they were laid, along with the breed and individual duck if needed, using a dull pencil or wax crayon.

CLEAN IT UP

It is vitally important to clean your incubator between hatching sessions to prevent the spread of disease. If you're using a new incubator, also clean it before its initial use to remove any dust or chemicals. If it's a used incubator, clean and sanitize it before you use it for the first time. Follow the cleaning instructions that come with the incubator or the cleaning procedure offered by the Mississippi State University Extension Service: Remove all organic matter (eggshells, feather dust, down, and so on) from the incubator with a vacuum or broom and then wash it out well with warm water and detergent. In the case of a particularly dirty incubator, rinse it with a disinfectant solution such as quaternary ammonia.

Muscovy females typically make good mothers, too.

until hatching is completed because opening it causes temperature dips and moisture loss. Many breeders use a separate hatching incubator at the temperature and humidity level required by the emerging ducklings.

Soon the duckling will use the hard egg tooth on its bill to push and peck a hole into the eggshell. Look for this telltale sign of imminent hatching, called *pipping*, on the rounded end of the egg. Over the next day or so, the duckling continues to work its way out of the shell, finally emerging wet, bedraggled, and exhausted. After resting and drying off within the incubator for a few more hours, your precocious duckling will be all fluffed out and ready to head to its new home in the brooder with its fuzzy friends.

If the duckling has managed to peck a hole in the egg but can't seem to make any more progress over the next few hours, you can help it by very carefully peeling bits of eggshell away. If you see blood on the shell membrane, stop, wait a few more hours, and then try again. Unfortunately, ducklings lacking enough vigor to complete the hatching process on their own often don't do as well as their stronger brothers and sisters.

While you wait to accumulate enough eggs to set under a broody or in the incubator, store the eggs in a cool place such as a basement, cellar, garage, or pantry where the temperature stays a constant 50–60° Fahrenheit. Moisture loss can negatively impact hatchability; the humidity during storage should ideally be 75 percent. Place the eggs rounded side up and small side down in egg cartons with their lids cut off. If possible, store the eggs no longer than a week before setting them; the hatch rate declines the longer the eggs sit in storage. Propping up one end of the carton a few inches with a block of wood or a brick and switching the block from one end to the other each day will help keep the embryos from adhering to the sides of the eggs, especially if you have to store the eggs for longer than a week.

Hatching Time

If you're incubating eggs from a domestic dabbling duck, expect the eggs to begin hatching at about twenty-eight days. Muscovy eggs will be ready to hatch at about thirty-five days. Warmer-than-normal temperatures in the incubator can shorten incubation time, and cooler-than-normal temperatures can lengthen it.

About three days before the eggs hatch, you may hear the wonderful sound of the babies softly chirping inside. Double-check that the eggs are in proper position for hatching, with the blunt end tilted up, because this is where the duckling will peck into the air space and take its first breaths. You can stop turning the eggs now. Watch the temperature carefully because the ducklings will generate their own heat while they struggle to emerge, and you may need to reduce the temperature to keep it at the correct setting.

During hatching, waterfowl eggs require more ventilation and a slightly higher level of humidity (88 on your hygrometer, or wet-bulb thermometer, which measures humidity levels within the incubator). If your incubator unit contains a hatching tray, this is the time to carefully transfer the eggs onto it. Try to avoid opening the unit again

Caring for Mother-Reared Ducklings

Ducks vary in mothering ability, but ducklings left in the care of an attentive mother duck or substitute maternal duck won't require a lot of specialized care from you. Of course, the new ducklings need protection from a host of predators, so it's important to house the mother duck and the young in a maternity pen of some sort that's secure enough to keep large and small varmints out. The pen must also be duckling-proof because adventurous ducklings can squeeze through surprisingly tiny gaps and mesh openings. For the first week or so, the mother duck will hover over the babies with wings slightly spread,

DID YOU KNOW?

Direct contact with poultry and their droppings can spread the bacterium *Salmonella*, a common cause of foodborne disease that produces diarrhea, fever, abdominal cramps, and sometimes more severe illness. Make sure that all who handle ducks or ducklings wash their hands well afterward in warm, soapy water for at least twenty seconds. And no kissing of ducks/ducklings allowed!

keeping them warm and sheltered. Still, it's important for the maternity pen to provide the family with shelter from heavy rains, wind, and sun.

Offer the ducklings a waterfowl starter diet that meets their nutritional needs. Baby ducks, with their tiny beaks, need mash, crumbles, or moistened pellets to eat. Place their food, and plenty of it (the mother duck will help itself to some!), in a shallow dish that they can easily access. Provide the mother duck with food in a steep-sided dish to keep the little ones out. Ducklings also relish and thrive on tender grass cut into tiny pieces (no long strands) or finely chopped dark leafy greens. Some rotten fruit hung in a mesh bag out of reach of the ducks will lure fruit flies for the hungry ducklings to snack on. Of course, the mother duck benefits from these menu items, too.

The ducklings and their mother will require an ample supply of water to drink, but use caution. Ducklings can drown in water containers if they have no easy way to climb or jump out. What's more, the mother duck can accidentally crush or drown the ducklings if she climbs into a large pan with them to bathe. Until the ducklings are large and nimble enough to jump out of a water pan, use large, narrow trough-style chicken waterers and avoid giving the ducklings

a container to bathe in. If you can supervise them, however, older babies enjoy going for a swim in a kiddie pool or on a pond where there's ample room and that they can easily exit. Ducklings can quickly become chilled, so schedule swims for fair weather.

Allowing the family out into a safe, grassy area early on will give the mother duck a chance to show the ducklings how to forage. Watch out for farm dogs, cats, or other livestock (including some Muscovy drakes) that might harm the babies.

Caring for Brooder-Reared Ducklings

Ducklings purchased by mail, bought from a feed store, or hatched out in an incubator will need a substitute for their absent mother. A brooder unit provides a warm, nurturing environment for the ducklings until they feather out sufficiently to keep themselves warm. It also gives them a safe haven until they grow big enough to hold their own against other ducks and not be picked off by crows, cats, or rats.

You can purchase a brooder built especially for chicks or ducklings or easily fashion one from a metal stock tank, a sturdy cardboard box, a wooden box, a plastic storage box, or a hard-plastic kiddie

---

INCUBATOR POINTERS

For duck embryos to successfully develop in an incubator, the machine (or the duck raiser) has to regulate the following:

Temperature. Proper temperature is critical: too much heat, and the embryo will cook to death; too little, and development slows and stops. Most incubators come with thermometers for monitoring the interior temperature, and many have thermostats to set the temperature. In a still-air machine, position the thermometer so it will show you the temperature at mid-egg level because temperature will vary at different heights (it will probably be warmer at the top). A forced-air machine doesn't have this temperature variance. Experts generally recommend that the temperature for incubating waterfowl eggs in a forced-air incubator be 99.5° F. For still-air incubators, some raisers advise a higher temperature of 102° F. It's wise to monitor the temperature several times a day and keep a record to refer back to at the end of the hatch.

Ventilation. Developing duckling embryos get oxygen through thousands of tiny pores in the eggs' shells. All incubators come equipped with vents that allow fresh air to enter; the vents can also be used to regulate humidity. Forced-air units have fans that increase air flow, dispersing oxygen, heat, and moisture throughout the incubator.

Humidity. It's normal for an egg to lose water as the duckling develops and the air cell expands. An incubator must provide enough moisture to prevent the egg from drying out but, at the same time, allow a certain amount of moisture loss so that the duckling can successfully hatch. Different incubators come equipped with various types of water reservoirs that can be used to supply moisture and regulate humidity. You'll need a hygrometer, or wet-bulb thermometer, to measure humidity levels inside the incubator. The hygrometer should read between 84 and 86, although the exact humidity level varies depending on climate.

Egg turning. Nesting ducks turn their eggs frequently during the course of the day, an action that prevents the embryo from sticking to the side of the egg and increases hatching success. If your incubator has an automatic turner, the incubator will do the mother duck's work for you. If not, you'll need to gently turn the eggs yourself at least three or four times a day. Marking an X on one side of each egg and an O on the other (with a blunt pencil or wax crayon) can help you keep track of turning the eggs.

BREEDING DUCKS SUCCESSFULLY

I have a two-stage duckling transfer from incubator to brooder because my eggs never hatch all at once. I have a 1-square-foot cardboard box with a burlap bag on the floor and its own heat lamp, plus a very small waterer and feeder. I transfer the dried duckling from the hatcher to the box, dipping its beak in the water and food. After about a day, I move the duckling to the brooder. This staged approach helps the youngest ducklings get their feet under them before they have to cope with their older flockmates.

—Jenifer Morrissey

We put the ducklings in a high-sided bin with a terry towel on the bottom, a brooder lamp, organic chick starter, and fresh water. The towel prevents the ducklings from straddling. After two weeks, we allow the ducklings to swim supervised in a kiddie pool for about fifteen minutes. We set up another kiddie pool (without water) in the house with the brooder lamp, canvas on the bottom, food, and water. They stay as our house guests until about eight weeks of age.

—Dawn Turbyfill

I keep only one [Muscovy] male around, as multiple males will fight over one female and will even drown her if they try to breed her in water. One male gets on top of her to breed, then more pile on, and she doesn't stand a chance. I once pulled a dead female out of a stock water tank.

—Melissa Peteler

pool. Set up and outfit the brooder before your ducklings arrive or hatch. For starters, place the brooder in a convenient, dry, draft-free spot where predators or pets won't be able to reach the new babies. Depending on the container's height, you may need to cover it with a mesh top if your ducklings excel at jumping and climbing. If your brooder is fairly small or if you have a lot of ducklings, you may have to eventually move your brood into larger accommodations to avoid unhealthily crowded, dirty conditions.

Outfit the brooder with a thermometer so you can monitor the temperature down where the babies hang out; 90°F is the usual recommendation for the first week, and then you can reduce the temperature by 5°F per week until it reaches 70° F. To supply the proper temperature, use a heat source such as a hanging metal reflector and light bulb or other type of brooder lamp. A large brooder may need two heat sources. Simply raise the heat source(s) to reduce the temperature. Remember that any heat source can pose a fire hazard if it comes in contact with something flammable, so use caution and common sense.

Allow enough space in your ducklings' abode for them to escape the direct heat of the lamps if needed; however, don't let any ducklings get chilled in the cooler outskirts of the brooder during those first few weeks. Discourage them from wandering too far from their heat source by using a draft guard or shield, which is a roll of cardboard 12–18 inches high that opens out into a circle and lines the inside of the brooder. It eliminates the corners that ducklings tend to crowd into and cuts down on drafts. The ducklings' behavior will tell you even better than a thermometer would whether the brooder is too warm or too cold. If the babies noisily cram together beneath the light, they're probably too cold. If they constantly stay at the outer reaches of the brooder or, even worse, you see them panting, it's too hot for their comfort.

Some raisers recommend initially lining the brooder not with shavings but with an absorbent layer of paper towels, cloth towels, or sheets because newly hatched ducklings may ingest shavings along with their food as they learn to dine on their crumbles or soaked pellets. Never keep ducklings on a slippery surface, such as newspaper, because the constant slipping and sliding can give ducklings a condition known as *spraddled leg*. Within a few days, however, lay down a thick layer of litter material, such as shavings, chopped straw, or sawdust. Be prepared to change their litter often to keep it clean, relatively dry, and free of mold.

Your ducklings will need shallow feeders large enough to allow them all to eat at the same time. Their feed, offered free choice, should be a proper starter ration that includes sufficient protein and niacin with grit provided on the side. Warning: ducklings gobble a lot of food— definitely more than chicks do—and grow fast! Like ducklings raised by their mother, your brooder babies will benefit from some finely chopped grass or dark greens in addition to their commercial fare.

A constant, ample supply of water is also essential; as mentioned, plastic chick waterers with narrow troughs work well. Save swimming for when they're older and can do it outside of the brooder. The warm, moist environment of a waterfowl brooder leads to the rapid growth of mold and the proliferation of bacteria, so, again, clean the feeders and waterers frequently.

When the sun shines, try to get your sheltered ducklings out into a safe outdoor pen where they can graze on young grass, hunt for bugs and worms, swim under supervision, and maybe even see the big ducks. They'll love the exercise and environmental stimulation, and you'll love watching their antics. By the time they're one month old, the ducklings will have grown in enough feathers to keep them warm outside in all but the most inclement weather. Of course, they'll require shelter and protection from predators, just as adult ducks do. Before you know it, your ducklings will have reached their two-month birthday and will essentially be adult ducks.

This chapter will focus on what ducks give us in return for all of the attention we lavish on them and how to make the most of it. Ducks, as I've already mentioned, are useful and generous creatures. Allowed to roam your property, they help control slugs, snails, mosquitoes, and other pests, providing eco-friendly fertilization services as they go. They trim grass and weeds and control aquatic weeds in waterways.

They delight us with their happy-go-lucky attitudes and antics and often make wonderful, responsive pets.

It's doubtful, however, that humans living 4,000 years ago decided to domesticate ducks primarily because they made cute, friendly companion animals. Rather, they were attracted to these birds for the life-sustaining gifts they offered: eggs and meat.

DUCK

EGGS, MEAT, AND DOWN

Duck Eggs for Consumption

You won't find duck eggs sitting next to chicken eggs at most grocery stores, and that's a shame. Like chicken eggs, duck eggs are a rich source of protein, choline (an essential nutrient that may help prevent heart disease), and two important nutrients found in the yolk that may guard against vision loss: lutein and zeaxanthin. Eggs contribute to countless dishes, and they make a terrific convenience food all by themselves. Protected within their shells, eggs store well under refrigeration for up to two months.

Eating eggs in moderation won't clog your arteries with cholesterol, either. According to a press release from Harvard Health Publications, only a small amount of the cholesterol in the food we eat enters our bloodstream. Saturated fats, and those evil trans fats, are the biggest culprits affecting blood cholesterol levels. Dining on one egg a day shouldn't pose a problem for most of us, particularly if we cut back on trans and saturated fats at the same time. If you're still worried, you can always eat the cholesterol-free egg white, or albumen (which consists mainly of the protein albumin and water), and skip the yolk. However, the yolk contains most of the good stuff. You may also find it more difficult to separate yolk from white in a duck egg—they adhere to each other more strongly than do the yolk and white of a chicken egg.

Anyone who keeps laying fowl knows that freshly gathered eggs taste and even look better than store-bought eggs, especially if the birds have been allowed to range and eat a natural, varied diet. The yolks glow a more vivid orange-yellow, the albumen stands up instead of oozing all over the pan, and the eggs scramble to firm perfection. Depending on the bird's diet, fresh duck eggs taste much like fresh chicken eggs, but they do differ in a number of other ways.

For starters, the thick, smooth shell of a duck egg is harder to crack than the shell of a chicken egg; prepare to use more force. Duck eggs also have a yolk that's larger relative to the white and thicker and stickier; this can put some people off. When scrambled or fried, a duck egg has a firmer, chewier texture than a chicken egg does. With the extra protein they contain, duck eggs add fluff and loft to baked goods. Lots of people enjoy eating duck eggs, but if you're considering keeping ducks primarily to provide eggs for your own table, try them first to make sure you like them.

Nutritionally, duck eggs contain slightly more protein, calcium, and B vitamins than chicken eggs do. They also have more fat and cholesterol because of their larger yolks. Interestingly, some people who have allergies to chicken eggs say they can eat duck eggs without a problem (but people can be allergic to duck eggs, too). If you allow your ducks to range, their eggs may be even better for you: some studies have shown a higher level of healthy omega-3 fatty acids in free-range eggs compared with eggs from confined poultry.

Getting Good Eggs

As you know by now, laying ability varies by duck breed. Prolific layers include the Magpie, Ancona, Harlequin, Runner, and the queen of them all, the khaki Campbell, which can pop out more than 300 eggs a year. Once you've settled on a laying breed, getting plenty of good eating eggs comes down to following a few simple guidelines, including offering the proper diet, providing good nesting sites, and confining your flock until nine o'clock in the morning. Ducks normally finish egg-laying sometime during the night or early morning. If you coop your free-roaming ducks until nine o'clock or later, you won't have to embark on an extensive and sometimes fruitless egg hunt every day. It will also keep egg-loving varmints

A duck egg's yolk is a vivid orange-yellow color.

from devouring the eggs before you find them. Gather eggs promptly in the morning so they don't have a chance to get dirty, and keep the duck house and nests clean with regular litter additions and changes.

Cleaning and Storing Eggs Meant for Eating

As noted earlier, eggs have a protective coating called *bloom* that keeps bacteria from entering the porous shell. Scrubbing and washing removes this coating, increasing the risk of contamination and spoilage. Still, ducks are not the neatest creatures, and dirty eggs happen even with preventive measures. Badly soiled eggs are best discarded, but you can clean a slightly soiled egg by gently brushing it with a dry cloth or steel wool. If needed, you can wash a dirty egg with warm water and a sanitizer made for that purpose, but store the egg separately and use it sooner than the others because washed eggs won't keep as long.

Cool eggs as soon as possible. Store fresh eggs pointed side down in breathable cardboard egg containers in the refrigerator. They keep up to two months but will taste much better if used sooner.

Ducks for Meat

It's not only ducks' eggs that have been underutilized here in the United States, it's the meat as well. Chicken and turkey products dominate the poultry scene, while duck meat, unfairly stereotyped as laden with fat and calories, is generally shunned by much of our diet-obsessed populace.

But humans have enjoyed duck meat dating back to ancient times. To this day, domestic duck remains popular in the cuisines of many countries—and for good reason. If you've ever sampled expertly prepared duck in a Chinese, Thai, or French restaurant, you'll know that duck is fabulously flavorful, rich, and tender. Ducks multiply quickly, and they're highly efficient at converting forage into meat. Duck meat is packed with high-quality protein as well as significant amounts of certain B vitamins and important minerals such as iron, zinc, selenium, phosphorus, and copper. Although considered a white meat along with other poultry, duck breast meat looks darker in color than does chicken or turkey breast. This coloration is the result of the increased presence of red blood cells carrying oxygen to fuel the duck's flight muscles.

FOOD SAFETY

Any raw or undercooked meat can harbor bacteria that proliferate under the right conditions and cause foodborne illness in humans. Adopt these safe handling tips to stay healthy:

- Thaw frozen duck in the refrigerator, in the microwave, or in cold water—never on the counter. Cook duck immediately after defrosting it.
- Use refrigerated duck within one to two days or freeze it in its original packaging.
- When storing or handling duck or raw eggs, avoid cross-contamination with other foods. Use a separate cutting board for meats, and clean up and disinfect spills immediately.
- Never partially cook duck and refrigerate it to finish cooking later.
- Cook duck to an internal temperature of 165°F (as measured with a meat thermometer). Refrigerate leftovers promptly.
- Cook eggs thoroughly before you eat them.
- Wash your hands with soap and warm water after handling raw meat or eggs.

Source: USDA Food Safety and Inspection Service

Muscovies are known for
yielding lean meat.

As far as duck fat goes, most of it resides in and under the skin, so if you remove the skin, you'll have a portion even leaner than chicken. A duck of any breed that's allowed to forage and exercise will produce leaner meat than will a confinement-raised duck. Muscovies and their hybrids are particularly known for their lean meat.

Benefits of Home-Raised Meat

Raising your own ducks for the table has some distinct advantages. You'll know exactly how the birds were raised: what they ate, where they lived, and how they were treated, slaughtered, and processed. You can choose what diet—organic or conventional—to feed your meat flock and whether to let them graze in grassy pastures or search for slugs around your garden. You can give them ponds or pools to bathe in and clean, uncrowded houses to sleep in securely at night. You can ensure that their final hour comes with a minimum of stress and pain. What's more, concern about how the animals we eat lived and died is shared by an increasing number of consumers, which is something to think about if you plan to market your ducks for meat.

If you intend to slaughter ducks yourself and have never done so before, ask an expert to come to your farm and show you how to do this necessary but unpleasant chore in the most humane way possible. Of course, not all of us duck raisers can stomach the thought of butchering. Finding a custom slaughterer who will handle small batches of poultry is one option, but locating one who deals with ducks, which pose more of a plucking challenge than do chickens, can be difficult. You might want to find a pragmatic farmer or duck hunter who'd be willing to do the job in exchange for a share of the harvest.

Getting Good Meat

Popular heavyweight breeds developed with meat production in mind include the Muscovy, Pekin, Rouen, and Aylesbury. Hybrids of Muscovies and Mallard derivatives, called *mule ducks* or *mulards*, are also valued as meat birds, particularly in Europe. Ducks of the medium, lightweight, and bantam classes produce delicious meat, too—just not as much and at a slower rate. And these slower-growing birds tend to have less body fat as well.

Ducklings raised for meat usually receive a starter or grower diet containing about 18–20 percent protein to promote speedy growth. They need ample water to drink and water to bathe in. A meat duck such as the Pekin gains weight at lightning speed—nearly a pound a week—and will be ready for butchering in about eight to ten weeks, when it weighs approximately 5 pounds. Muscovies take about fourteen weeks to reach slaughter weight. Rapid growth in meat ducks often causes health problems and leg abnormalities; ducks kept as pets, show birds, layers, or breeders should be switched to a lower protein diet after the first few weeks of life.

Remember that all ducks—even short-lived meat birds—benefit from having a spacious outdoor area where they can exercise, forage, and soak up the sunshine.

Selling Organic Meat and Eggs

The demand for organic foods, including poultry products, has seen amazing growth as consumers become more health-conscious and concerned about the ethical treatment of farm animals. In fact, the demand is so great that farmers making the switch to certified organic production often receive a premium for their products. If you plan to sell duck eggs and meat, making the leap to organic is worth considering.

As defined by the USDA, organic food is food that is produced in an environmentally responsible fashion, without the use of antibiotics, animal by-products, hormones, most conventional pesticides, synthetic fertilizers, or bioengineering. Producers planning to market their duck meat or eggs as organic must adhere to standards outlined in the USDA's National Organic Program. Organically raised poultry, for instance, must receive an organic diet from the second day of life onward and must have access to the outdoors.

Organic poultry production has its pros and cons. Certified organic feed costs more than conventional feed, and it can be difficult to locate. The road to organic certification involves

Ducks' downy feathers are used in pillows, comforters, and jackets, to name a few.

submitting an application and farm management plan (plus a substantial fee), undergoing assessment and review by a certifier, and receiving an on-site inspection of your farm—a process that has to be repeated each year. Organic farmers cannot resort to as many quick fixes as conventional farmers can, so organic production tends to require more work, planning, and management than does conventional farming. For more information on the National Organic Program, including regulations and guidelines, visit www.ams.usda.gov and click on "National Organic Program."

But for those who have embraced it, organic poultry production has a payoff that goes beyond the higher prices many consumers will pay. Raising ducks organically requires organic feed, and the more the demand for organic feed grows, the more farmers will put additional land into organic production. That translates into fewer chemicals contaminating the environment and affecting its inhabitants—including us. For organic food fans, it is enough to know that organic poultry receive a diet free of added inorganic

arsenic (a cancer-causing agent) and antibiotics (which promote the development of drug-resistant bacteria).

If converting to organic isn't a viable option for you, consider pasture or free-range duck farming. As mentioned, eggs from pasture-reared poultry tend to taste and look better than eggs from confined birds. Meat from pastured birds usually has less fat and a firmer texture. Some studies have shown lower cholesterol and higher levels of omega-3 fatty acids, vitamin E, and beta carotene in meat products from pastured birds. And, from your duck's perspective, what could be better than having space to waddle around outside, gobbling slugs and snails, nibbling dandelions, and doing all of the other things ducks are meant to do?

Down and Feathers

Duck down and feathers are the unavoidable—and incredibly useful—by-products of the butchering and plucking process. (And while on the subject of plucking—never live-pluck your ducks; it is painful for them.) Duck and goose down have long been used as warm and fluffy fillers for pillows, comforters, feather beds, ski jackets, and vests. If you've ever snuggled beneath a down comforter on a winter night, you'll know that when it comes to insulating you from the cold, nothing can outdo this remarkable natural product— as long as you keep it from getting wet.

Small flock raisers should consider this fluffy by-product of slaughter for personal—not commercial—use only; it would take years to collect enough down and small, soft feathers to create even one queen-size comforter. A pillow might be a more reasonable goal!

As for those gorgeous duck feathers, you can clean, bag, and save them to use as cat toys, earrings, and quill pens, as well as for fly tying, painting, or other crafty projects.

FOIE GRAS

Foie gras (pronounced *fwah-grah*) is a French term for the fatty liver of geese and ducks, a delicacy first savored by the ancient Egyptians and still served in fancy restaurants today. According to the Humane Society of the United States, however, the commercial production of this expensive luxury food requires the cruel force-feeding of confined ducks and geese with a pressurized pump two to three times a day. This abnormally huge quantity of food makes the liver enlarge up to ten times its normal size, enough so that the birds find it difficult to move. Essentially, the organ becomes diseased. That's not a particularly appetizing thought, especially for those of us who try to treat our feathered friends as humanely as possible! The good news? Some countries have banned the production of foie gras.

Exquisitely colored duck feathers have many practical and decorative uses.

have to put all your eggs in one basket, either: duck raising is perfectly compatible with other farm business enterprises, such as chicken operations, market gardens, community-supported agriculture, and orchards.

Don't leap into the duck business before you look, however. The demand for duck products is still much lower than the demand for chicken and turkey products. Selling any food product for human consumption is a complicated affair, governed by a plethora of rules and regulations that vary from state to state. To find out what insurance, permits, and licenses are required or recommended for your intended business, start by checking with your state Department of Agriculture, your local health department, and your county extension office.

Devise a business plan, research your markets carefully, and start small, growing with your markets. Do you plan to sell eggs from your farm, at farmers' markets, or via both venues? Would you like to supply organic duck breasts to fancy restaurants or to grocery stores? Where will you advertise your hatching eggs and rare-breed ducklings— on your website, in the newspaper, or at local feed stores? Internet marketing has become the norm for many small businesses.

No matter how you plan to market your products, keep detailed records, or you'll never know whether your business enterprise is making or losing money. Whatever you do with your little flock, above all, have fun! You can bet that, given good care, your ducks will surely enjoy themselves in everything they do. After all, they're ducks!

Duck Businesses

Wouldn't it be great to create a business around something you love, such as raising ducks? Raising ducks on a small scale won't make you rich. But if you live in the right area and have a knack for marketing, it's possible to turn a profit either by selling duck products—duck meat, eggs for eating and hatching, feathers, and down—or by selling ducks as pets, breeding stock, or exhibition stock. You don't

---

ADVICE FROM THE FARM

BUSINESS POINTERS

When processing our Muscovies, rather than go through the difficult process of waxing and plucking and dealing with the little dark pin feathers, we skin the breasts and use those. Muscovies have scrawny legs, which tend to dry out when you roast them, and most of my recipes call for duck breasts anyway. Skinning is so much faster and easier!

—Terry Ann Carkner

Our duck eggs are in high demand; we sell out very often. All of our ducks are pasture-ranged and organically fed. We scrub the eggs with baking soda to remove stains, debris, and odor, and inspect them for any cracks or pinholes made by the duck's toenails. This scrubbing—or any washing of the eggs—removes the natural protective oil layer on the shell's surface, called the bloom. When the eggs are dry, we re-oil or rebloom them with certified extra virgin olive oil. This extends the freshness of the eggs by several weeks.

—Dawn Turbyfill

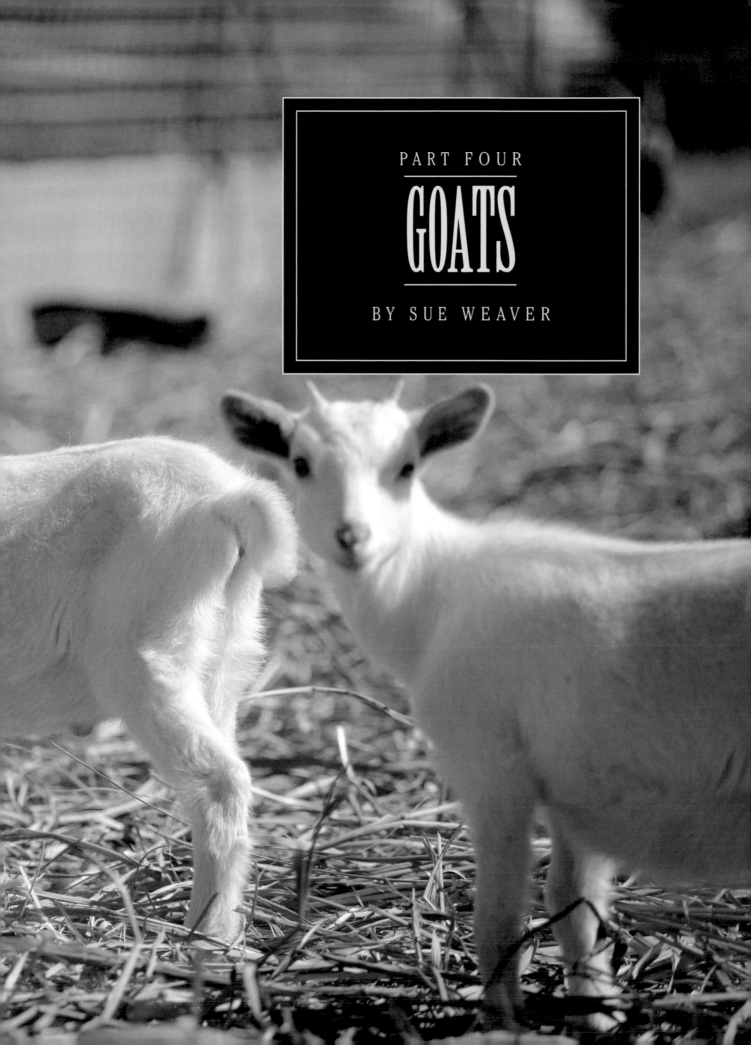

PART FOUR

GOATS

BY SUE WEAVER

WHY GOATS?

Goats were the first domesticated livestock; we've had 10,000 years to get things right. Today's goats provide tasty milk, delicious meat, attractive pelts, and two kinds of renewable fiber. They clear pasture for other livestock by grazing and destroying weeds and brush, they pull carts (goats are amazingly strong), and they pack along the tents and grub when folks go camping.

GOATS:
A PRIMER

It costs little to buy and maintain goats, and only a modest land plot is required to raise them. Goats are naturals for today's hobby farms. Curious, intelligent, agile, and friendly, goats provide hours of entertainment for their keepers. Everyone who has goats loves them. Whether you want to turn a profit with goats or keep a few for fun, we're here to show you how to get started.

How long have goats been around? Where did the first ones come from? Are there many different kinds? What are they like? Who raises goats? Here's a brief look at goats through history and a glance at types, breeds, and traits.

From the Beginning

Goats were domesticated around 8,000 BC by the people of Ganj Dareh, a Neolithic village nestled in the Kermanshah Valley of the Zagros Mountains in the highlands of western Iran. According to archaeologists, goat meat had graced the human menu for more than 40,000 years prior to this. However, toe bones recovered from Ganj Dareh middens are the remains of young bucks, the ones not needed for breeding purposes, and some aged does, females too old to have kids; this tells us that people had begun keeping goats rather than just hunting them.

After a 100–200-year occupation, the good people of Ganj Dareh packed up their families and possessions, including their goats, and traveled south into the arid Irani lowlands. They resettled away from the wild goat's natural range at a place called Ali Kosh. With a movable food supply—goats and two newly domesticated cereal grains, wheat and barley—humans could abandon their longtime roles as hunter-gatherers and take up the mantle of nomadic herders and tillers of the soil. Archaeological excavations at Jericho unearthed mounds of domestic goat bones carbon-dated to 7,000–6,000 BC.

Early domestic goats served their human masters exceedingly well. They provided milk and meat, fiber for tent covers and clothing, skins for leather, hair-on pelts for robes and rugs, and kids to sacrifice to the gods. Goats packed belongings on their backs and drew travois-type sledges. They were friendly and small, thus easily handled, and required minimal care. Best in arid, semitropical, and mountainous countries, goats survived under conditions in which horses, sheep, or cattle would starve.

Goats spread east from the Fertile Crescent across continental Europe and thence to Great Britain. As elsewhere, goats there became "the poor man's cow," thriving in mountain and moorland crofters' fields and folds, from which they sometimes escaped. Their feral descendants still thrive in remote and isolated pockets along the western coast of Ireland, on Snowdonia in Wales, on Lundy Island and the Isle of Rum, and in the Mull of Kintyre, Galloway, and Loch Lomond in Scotland.

During the 1500s, goats came to the Americas with Spanish conquistadors, settlers, and sailors. It was their custom to salt uninhabited islands with breeding stock, allowing them to harvest future meals on subsequent trips. Historians believe that the Pilgrims carried goats on the *Mayflower*'s 1620 maiden journey to the New World. The Pilgrims considered goat's milk a restorative medicine. In the coming centuries, goats accompanied settlers as they pushed westward across North America. By browsing as the party traveled, goats furnished their own eats while providing meat and milk on demand.

Two young Myotonic, or Tennessee Fainting, goats.

By the mid-nineteenth century, generic Spanish goats (also called scrub, brush, hill, briar, and woods goats) could be found in most southeastern states and throughout the Southwest and California. The year 1849 saw the arrival of North America's first purebred goats: seven Angora does and two bucks imported to South Carolina. One of North America's few purely native breeds first made an appearance in the 1880s. An itinerant stranger named John Tinsley came to Marshall County, Tennessee, accompanied by four slightly peculiar goats. When they were startled, their muscles would seize, causing the animals to freeze and sometimes fall over. From these four goats, many believe, emerged the Myotonic goat, a heavy rump breed with a tendency to topple, popular for meat production and ease of handling.

The 1904 World's Fair in St. Louis, Missouri, sponsored the first North American dairy goat show. The *Missouri Historical Review* noted, "This first provision made at a World's Fair for a display of milch goats brought to the Exposition some choice and home bred specimens." At the same World's Fair, Hagenbeck's Wild Animal Paradise displayed two striking Schwartzwald Alpine does in a lavish diorama depicting the Alps. This same year, the United States formed its first goat registry, the American Milk Goat Record, now the American Dairy Goat Association (ADGA).

In 1906, Mrs. Edward Roby crossed Swiss dairy goats with common stock to develop the American goat. With them, she strove to supply tuberculosis-free milk to the children of Chicago at a time when many cows were infected. Although she was moderately successful, parents who had never tasted goats' milk refused to give it to their children.

During the early 1900s, the first Anglo-Nubians (now simply called Nubians) were shipped from Britain to North America. Between 1893 and 1941, 190 Toggenburg dairy goats were imported; between 1904 and 1922, 160 Saanen. During 1922, the first documented purebred French Alpines, twenty-one in a single importation, arrived by ship, followed in 1936 by five Oberhaslis (then called Swiss Alpines). The first documented modern Pygmy goats arrived in North America during the 1950s, originally as novelties to be displayed in zoos. In 1993, the first purebred Boer meat goats, developed in South Africa in the early 1900s, set hoof on American soil. Boers took America by storm, as did Kiko meat goats developed in New Zealand and imported at about the same time.

Goats at a Glance

Domestic goats belong to the Bovidae family, along with other hollow-horned, cloven-hoofed ruminants such as cattle, thence to the Caprinae subfamily, in the company of their cousins, the sheep. Goats are further classified by their genus, *Capra*, and fall into one of six species: *Capra hircus* (today's domestic goat), *Capra aegagrus* (the wild Bezoar goat, ancestor of *Capra hircus*), *Capra ibex* (the wild ibex), *Capra falconeri* (the markhor of central Asia), *Capra pyrenaica* (the wild Spanish goat of the Pyrenees), and *Capra cylindricornis* (the Dagestan tur of the Caucasus mountains). (Some scientists divide goats into as many as ten species.)

Roughly 100 breeds and documented varieties of domestic goats exist in the world today, but fewer than two dozen are available in North America. The world's estimated 1 billion goats have many traits in common, including social structure, flocking instincts, and breeding traits.

Goat Classifications

For the goat keeper, goats fall into three basic categories—dairy goats, meat goats, and goats raised for fiber. Subcategories and crossovers certainly exist. Goats are sometimes used to pull carts and pack supplies recreationally and to clear land.

Dairy Goats

Dairy goats are lithe, elegant creatures developed for giving lots of luscious milk. However, excess kids (bucklings not needed for breeding) are often marketed as *cabrito* (the meat of young kids). Some dairies routinely breed their does to Boer and Kiko bucks to produce a meatier product. Recreational goat aficionados claim that dairy goat wethers (castrated goats), particularly Saanens and Alpines, make the best harness and pack goats bar none.

Dairy breeds readily available throughout North America include the Swiss breeds (Saanens, Sables, Oberhaslis, Toggenburgs, and

WELCOME TO THE GOAT WORLD

Dairy goats are a lot of work when they're lactating because they have to be milked twice a day if they're not raising their own kids, and you can't just milk a goat when you feel like it. It must be done on a regular schedule.

Angora and cashmere goats require shearing—twice a year in the case of the Angora. If you're a hand spinner or if you want to market mohair or cashmere on a commercial basis, that's good; otherwise, it's a lot of work!

I have pasture-run meat goats of no specific breed. They're relatively easy to take care of. These goats would make good pets if you don't like the idea of selling your goats to the butcher or eating them yourself.

—Glenda Plog

Goats are great, and you will find the goat world is a very friendly place. Figure out what you're looking for in a goat, then find a breed (or mix) that best matches what you want. If you are looking for any old goats, see if there is a rescue near you. Farm animal rescues can be hard to find, but they do get very nice goats that the old owners just couldn't keep anymore. They also take in abused animals, so talk to the rescue to see which would be the best match for you.

—Michelle Wilfong

Alpines), the LaMancha (a distinctly American breed), the Nubian (known in its British homeland as the Anglo-Nubian), and the pint-size Nigerian Dwarf from West Africa. Miniature versions of all but Nigerian Dwarfs are out there, too.

An uncommon midsize combination dairy and meat breed, the Kinder goat, was developed by crossing full-size Nubian does with meaty Pygmy bucks. Although Pygmy goats are primarily raised for pets, the does give a surprising volume of high butterfat-content milk.

Meat Goats

Primary purebred meat goat breeds are the immensely popular Boer from South Africa; the all-American Myotonic (also known as the fainting goat) and their selectively improved counterparts, Tennessee Meat Goats; and the New Zealand Kiko goat. Several exciting combination breeds such as the TexMaster (Boer/Tennessee Meat Goat) and Genemaster (Boer/Kiko) have been developed, while generic Spanish meat goats form the nucleus of many commercial herds. All are bred for muscle mass, hardiness, adaptability, and exceptional feed-to-flesh conversion ratio. Pygmies are meat goats, too.

Fiber Goats

The backbone of North America's fiber-goat industry is the traditional white mohair-producing Angora, but the fleece of colored Angoras is in high demand for hand spinning, too. A more diminutive fiber producer is the midsize Pygora, developed by crossing Angoras and Pygmies. Cashmere goats are the Rolls-Royces of the fiber-goat industry, and the American cashmere goat population is growing rapidly.

Recreational Goats

Goats have frequently been driven in harness, sometimes as serious work animals but frequently for recreation. Shortly after their father's presidential inauguration in 1861, Willie and Tad Lincoln were presented with cart goats named Nanny and Nanko. On one occasion, Tad harnessed Nanko to a rocking chair and drove at breakneck speed

through a White House reception, causing many a dignified gent and hoop-skirted lady to leap to safety. Most recreational goat buffs prefer wethers, but does, too, can work in harness or under packing gear. A bonus: a lactating pack doe provides fresh, whole milk on the trail.

Brush Goats

Because goats willingly browse weeds and saplings that other animals won't touch, many people keep them for clearing land of scrub and brush. Dairy does can do the job, but because of potential damage to their large udders, goat keepers prefer not to use them for this particular task. The hands-down champions are hardy generic Spanish goats—they aren't called brush, scrub, and briar goats for nothing.

A Buyer's Guide to Goats

Don't rush out to buy goats. It's a bad idea when purchasing any type of livestock but especially risky with goats. Goats require specialized handling and feeding—and keeping goats contained in fences is never a lark. Goats are cute, personable, charming, and imminently entertaining. They can be profitable, particularly in a hobby-farm setting. But goats are also destructive, mischievous, sometimes ornery, and often exasperating. Be certain you know what you're getting into before you commit.

Find yourself a mentor. Most experienced goat producers are happy to teach new owners the ropes. To track down a mentor, ask your county extension agent for the names of owners in your locale, join a state or regional goat club, or subscribe to goat-oriented magazines and e-mail groups to find goat-savvy folks in your area. A mentor or extension agent can talk with you about which breed will meet your needs, what to look for when buying your goats, and what happens once you do. You need to educate yourself as well.

Choosing the Breeds

Before going goat shopping, know precisely what you want. Make a list of the qualities you won't compromise on as well as which ones

There's no denying that goats are cute, but don't make an impulse purchase.

you're willing to forgo. Some breeds fare better than others in certain climates. Certain breeds are flighty. Some make dandy cart goats, whereas others are too small for harness work unless you plan to drive a team. If you want a goat who milks a gallon a day, a Pygmy doe won't do. However, if you're looking for a nice caprine friend, and you don't want to make cheese or yogurt, a Pygmy doe (or two) could prove the perfect choice. Consider availability as well and whether you're willing to go farther afield to get exactly what you want.

Purebred, Experimental, Grade, or American?

Registered goats generally cost more to buy than do grade (unregistered) goats, but you might not need to spring for registered stock. It depends on your goals. If you plan to exhibit your animals at high-profile shows or to sell breeding stock to other people, you probably do want registered stock. If you want a pack wether, a 4-H show goat, or a nice doe to provide household dairy products, registration papers aren't essential.

A registration certificate is an official document proving that the animal in question is duly recorded in the herdbook of an appropriate registry association. Depending on which registry issues the certificate, the document will provide a host of pertinent details, including the goat's registered name and identification specifics such as its birth date, its breeder, its current and former owners, and its pedigree. Dairy-breed papers also document milk production records in great detail.

The four categories of dairy goats in terms of registration are purebred, experimental, grade, and American. *Purebreds* are registered goats that come from registered parents of the same breed and have no unknowns in their pedigrees. *Experimentals* are registered goats that come from registered parents of two different breeds. A goat of unknown ancestry is considered a *grade*. However, several generations of breeding grade does to ADGA-registered bucks (always of the same breed) and listing the offspring with the ADGA as *recorded grades* eventually results in fully registerable *American* offspring. For example, a seven-eighths Alpine and one-eighth grade doe is an American Alpine; a fifteen-sixteenths Nubian and one-sixteenth grade buck is an American Nubian. However, ADGA terminology doesn't apply to meat goats.

To qualify as a registered *full-blood* in the American Boer Goat Association herdbook, all of a goat's ancestors must be *full-blood* Boer goats. Registered *percentage* does are 50–88 percent full-blood Boer genetics; *percentage* bucks are 50–95 percent Boer. Beyond that (94 percent for does, 97 percent for bucks), they become *purebred* Boers. *Purebreds* never achieve *full-blood* status.

The International Kiko Goat Association registers *New Zealand full-bloods* (from 100 percent imported New Zealand bloodlines), *American premier full-bloods* (of 99.44 percent or greater New Zealand genetics), *purebreds* (87.5–99.44 percent New Zealand genetics), and *percentages* (50 and 75 percent New Zealand Kiko genetics). To avoid making costly mistakes, learn your breed's registration lingo before you buy!

Pets, cart and pack goats, brush clearers, and low-production household dairy goats needn't be of any specific breed. Mixed-blood goats cost less to buy and may be precisely the animals you need.

Availability

If you're seeking Nubians, Pygmies, or Boers, you'll probably find a plentiful supply of good ones close to home. Less common breeds, such as Sables, Kinders, and colored Angoras, may be a different story. If you don't want to travel long distances to buy foundation or replacement stock, pick a common breed or at least one popular in your locale. Conversely, although it takes more effort to start with something out of the ordinary, it also assures a market for your goats.

Purchasing goats from a distance has its pitfalls because you may not be able to visit the sellers and inspect potential purchases in person. If this is the case, buy only from breeders whose sterling reputations (and guarantees) take some of the gamble out of long-distance transactions. The transportation of distant purchases is also an issue, but it needn't be a major one. Livestock haulers and

Saanens occasionally produce colored offspring, called Sables (pictured), which have come to be recognized as a distinct breed.

Don't discount the importance of good conformation; you'll pay more for a correct foundation goat, but it's worth it. Even if you never show your goats, buyers will pay higher prices for your stock.

Health

Never knowingly buy a sick goat! Carefully evaluate potential purchases before bringing them home. A healthy goat is alert and sociable—even semiwild goats show interest in new faces. A goat standing off by itself, head down and disinterested in what's going on, is probably sick or soon will be.

A healthy goat is neither tubby nor scrawny. It shows interest in food if it's offered; when resting, it chews its cud. Its skin is soft and supple; its coat is shiny. Its eyes are bright and clear. Runny eyes and a snotty nose are red flags, as are wheezing, coughing, and diarrhea (a healthy goat's droppings are dry and firm). Unexplained lumps, stiff joints, swellings, and bare patches in the coat spell trouble. Avoid a limping goat because it could have foot rot (or worse).

If in doubt and you really want a particular animal, ask the seller if you can hire a vet to take a look, and consider it money well spent.

Horns

If you don't like horned goats, don't buy a goat that has them! The cores inside a goat's horns are rich in nerves and blood vessels. Dehorning, even done by a veterinarian and under anesthesia, is a grisly, dangerous, and ultimately painful procedure that leaves gaping holes in an animal's skull. With dedicated follow-up care, these holes will eventually close, but why expose an animal to this kind of torment?

Dairy-goat kids are routinely disbudded when they're a few days to a week or so old. This is accomplished by destroying a kid's emerging horn buds by burning them with a disbudding iron. Although it's painful and not a procedure best performed by beginning goat keepers, disbudding is far more humane than exposing a goat to full-scale dehorning later on.

Meat- and fiber-goat producers and recreational goat owners are far less likely to eschew horns, but all goats exhibited in 4-H shows—even the ones that are shown in 4-H meat-goat, fiber-goat, driving, and packing classes—must be hornless or shown with blunted horns.

Should horns be a problem? It depends. You probably don't want them if you confine your goats (they'll butt one another, probably causing injuries), if they'll be expected to use stanchions or milking stands, if you have small children, or if your other goats are polled

some horse transporters carry goats cross-country for a fee. Kids and smaller goats can be inexpensively and safely shipped by air.

If you're buying close to home, locate breeders via classified ads, notices on bulletin boards at the vet's office and feed stores, and word of mouth. Or place "want to buy" ads and notices of your own. To get a feel for breeders and to learn what sort of goats they have for sale, visit breed association websites or subscribe to print and online goat periodicals. Peruse the ads and breeder directories.

Goats auctioned through upscale production sales and consignment sales hosted by bona fide goat organizations are generally the cream of the caprine crop. Never buy goats at generic livestock sale barns. Run-of-the-mill livestock auctions are the goat farmer's dumping ground. Most animals run through these sales are culls or sick, and the ones who aren't will be stressed and exposed to disease. A single livestock sale bargain can bring the likes of foot rot, sore mouth, and caseous lymphadenitis home to roost, sometimes to the tune of thousands of dollars in vet bills and losses. Buy your goats through high-profile goat auctions or from private individuals.

Selecting the Goats

The cardinal rule when buying goats: start with good ones. Choose the best and the healthiest foundation stock you can afford.

Conformation

Acceptable conformation—defined as the way an animal is put together—varies among dairy, meat, and fiber goats. It's important to study a copy of your breed's standard of excellence, available from whichever registry issues its registration papers, before you buy.

(naturally hornless) or disbudded. However, science theorizes that horns act as thermal cooling devices, so if you have working pack or harness goats, or you live where it's hot, they're a boon.

Teeth

A goat has front teeth only in the lower jaw. In lieu of upper incisors, there is a tough, hard pad of tissue called a

COMMON GOAT BREEDS IN BRIEF

Here's a brief look at the different dairy, meat, and fiber breeds as well as some additional choices.

Dairy Goats

Alpine (or French Alpine): Alpine goats originated in the French Alps. They are medium to large, with does at least 30 inches tall and 135 pounds, and bucks at least 34 inches and 170 pounds. Friendly, inquisitive Alpines come in a range of colors and shadings. Because of their productivity and good nature, Alpines are popular in commercial dairy settings.

LaMancha: The almost-earless LaMancha (at least 28 inches and 130 pounds) is an all-American goat developed in Oregon during the 1930s. Fanciers claim that LaManchas are the friendliest of the dairy-goat breeds. They can be any color. Two types of ears occur: gopher (1 inch or less in length, with little or no cartilage) and elf (2 inches or less in length, with cartilage). LaManchas produce copious amounts of high-butterfat milk.

Miniature Dairy Goats: The Miniature Dairy Goat Association registers scaled-down (20–25 inches tall, weight varies by breed) versions of all standard dairy goat breeds. Miniatures have the same standards of perfection as those of their full-size counterpart breeds.

Nigerian Dwarf: Nigerian Dwarfs are perfectly proportioned miniature dairy goats, capable of milking three to four pounds of 6–10-percent butterfat per day. Gentle, personable Nigerians can be any color. They breed year-round, and multiple births, from the average of four to as many as seven per litter, are common. Does are typically 17–19 inches tall and bucks 19–20 inches tall; 75 pounds for both sexes.

Nubian (or Anglo-Nubian): Nubians were developed in nineteenth-century England by crossing British does with bucks of African and Indian origins. Noisy, active, and medium to large in size (does at least 30 inches and 135 pounds; bucks at least 35 inches and 175 pounds), Nubians are known for their high-butterfat milk production, sturdy build, long floppy ears, and aristocratic Roman-nosed faces. All colors and patterns are equally valued.

Oberhasli: Alert and active, Swiss Oberhaslis are medium-size goats (minimum for does is 28 inches and 120 pounds; for bucks, 30 inches and 150 pounds). They are always light to reddish brown, accented with two black stripes down the face, a black muzzle, a black dorsal stripe from forehead to tail, a black belly and udder, and black legs below the knees and hocks.

Saanen: These big (30–35 inches and 130–170 pounds), solid white, pink-skinned goats from Switzerland are friendly and outgoing heavy milkers with long lactations. They are popular commercial dairy goats, often called "the Holsteins of the goat world."

Sable: Sables are colored Saanens, more recently recognized as a separate breed. Because their skin is pigmented, they don't sunburn as Saanens sometimes do.

Toggenburg: Toggs are smaller than the other Swiss dairy breeds. They are some shade of brown with white markings: white ears with a dark spot in middle of each, two white stripes down the face, hind legs white from hocks to hooves, forelegs white from knees down.

COMMON GOAT BREEDS IN BRIEF

Meat Goats

Boer: The word *boer* means "farmer" in South Africa, land of the Boer goat's birth. Big (does weigh 200–225 pounds and bucks 240–300 pounds; height can vary greatly), flop-eared, Roman-nosed, and wrinkled, the Boer is America's favorite meat goat. Boers are prolific, normally producing two to four kids per kidding, and they breed out of season, making three kiddings in two years possible. Boer colors include traditional (white with red head), black traditional (white with black head), paint (spotted), red, and black.

Genemaster: Genemaster goats are 3/8 Kiko and 5/8 Boer and developed by New Zealand's Goatex Group company, the folks who pioneered the Kiko goat. Pedigree International currently maintains the North American Genemaster herdbook.

Kalahari Red: The Kalahari Red is a breed that has developed in South Africa since the 1970s and looks like a large, dark-red Boer. Although a few American producers are breeding true South African stock, most North American "Kalahari Reds" are simply solid red Boers.

Kiko: *Kiko* means "meat" in Maori. Kikos were developed in New Zealand by the Goatex Group. Beginning with feral goat stock, breeders selected for meatiness, survivability, parasite resistance, and foraging ability and, in doing so, created today's ultrahardy Kiko goat.

Myotonic: Today's Myotonic goats (also called fainting goats, fainters, wooden legs, Tennessee peg legs, and nervous goats) are believed to be the descendants of a group of Myotonic goats brought to Tennessee around 1880. When these goats are frightened, a genetic fluke causes their muscles to temporarily seize up; if they're off balance when this happens, they fall down. Myotonic goats come in all sizes and colors (black and white is especially common). They don't jump well, so they're easy to contain, and they're noted for their sunny dispositions.

Savanna: Big, white, and wrinkled, South African Savanna goats resemble their Boer cousins, but with a twist. South African Savanna breeders used indigenous white-goat foundation stock and natural selection to create a hardier-than-Boer breed of heat-tolerant, drought-and-parasite-resistant, extremely fertile meat goats with short white hair and black skin. Savannas' thick, pliable skin yields an important secondary cash crop: their pelts are favorites in the leather trade. A small number of North American breeders offer full-blood Savanna breeding stock, but interest in the breed is skyrocketing. Pedigree International maintains the official Savanna herdbook.

Spanish: *Spanish* is a catchall term for brush goats of unknown ancestry, so no breed standard exists. Spanish goats can be any color, although solid white is most common. Both sexes have huge, outspreading horns.

Tennessee Meat Goat: Suzanne W. Gasparotto of Onion Creek Ranch developed the spectacular Tennessee Meat Goat by selectively breeding full-blood Myotonic goats for muscle mass and size. Pedigree International maintains the Tennessee Meat Goat registry.

TexMaster Meat Goat: The TexMaster Meat Goat, another Onion Creek Ranch development, was originally engineered by crossing Myotonic and Tennessee Meat Goat bucks with full-blood and percentage Boer does (meaning they are a only certain percent Boer, not 100 percent). Pedigree International keeps its herdbook as well.

Fiber Goats

Angora: The quintessential fiber goats, Angoras produce long, silky, white or colored mohair. Angoras are medium-sized goats (does are 70–110 pounds, bucks 180–225; height varies). They aren't as hardy as most other breeds. Multiple births are relatively uncommon. Angoras must be shorn at least once a year.

Cashmere: Cashmere goats are a type, not a breed. Goats of all breeds, except Angoras (and one class of Pygoras), produce cashmere undercoats in varied quantities and qualities. High-quality, volume producers are considered cashmere goats.

Pygora: Pygoras were developed by crossing registered Angora and Pygmy goats. They're small (does at least 18 inches tall and 65–75 pounds; bucks and wethers at least 23 inches tall and 75–95 pounds), easygoing, and friendly, and they come in many colors. Some Pygoras produce mohair, some cashmere, and others a combination.

Other Breeds

Kinder: The Kinder goat (does 20–26 inches, bucks 28 inches; weight varies) is a dual-purpose milk and meat breed developed by crossing Nubian does with Pygmy bucks. Prolific—most does produce three to five kids per litter—and easygoing, Kinders make ideal hobby-farm milk goats and pets.

Pygmy: Nowadays, Pygmy goats (does are 16–22 inches, bucks 16–23 inches; weight varies) are usually kept as pets, but they developed in West Africa as dual-purpose meat and milk goats. Pygmies are short, squat, and sweet-natured. Lactating does give up to two quarts of rich, high-butterfat milk per day, making Pygmies respectable small-family milk goats.

A Saanen kid (left) and a Sable kid.

dental palate. For maximum browsing efficiency, the lower incisors must align with the leading edge of the dental palate, neither protruding beyond it (a condition called *monkey mouth* or *sow mouth*) nor meeting behind the dental palate's forward edge (*parrot mouth*).

Beginning at about age five, a goat's permanent teeth begin to spread wider apart at the gumline and then break off and eventually fall out. A goat with missing teeth is said to be *broken-mouthed*. When his last tooth is shed (around age ten), he's a *gummer*. Aged goats with broken teeth have difficulty browsing, so unless you're willing to feed soft hay or concentrates, check those teeth before you buy.

Sex-Specific Factors

No matter what class of stock you raise—be they dairy, meat, or fiber goats—buy does with good udders. A goat's udder should be soft, wide, and round, with good attachments front and rear. The two sides should be symmetrical. Avoid lopsided, pendulous udders with enormous sausage teats, especially in dairy goats, and reject goats with extremely hot, hard, or lumpy udder because these are telltale signs of mastitis involvement.

Dairy goats should have two functioning teats with one orifice apiece. Deviations from the norm are rare but are serious faults. Dairy kids are sometimes born with additional vestigial teats, but they're usually removed when doelings are disbudded.

CONSERVATION PRIORITY LIST BREEDS

The Livestock Conservancy (www.livestockconservancy.org) includes seven goat breeds on its Conservation Priority List. Two critically endangered breeds require immediate help: the island-bred Arapaw and the San Clemente, of relatively pure Spanish stock. In the "Watch" category is the Spanish goat, while the Myotonic and Oberhasli are listed as "Recovering" and the Golden Guernsey is listed as a "Study" breed.

Meat goats, especially Boers, are often graced with more than two teats. In Boers, up to two adequately spaced, functional teats per side are acceptable. However, nubs (small, knoblike lumps that lack orifices), fishtail teats (two teats with a single stem), antler teats (a single teat with several branches), clusters (several small teats bunched together), and kalbas or gourd teats (larger roundish lumps that have orifices) frequently occur. A blind teat (one lacking an orifice) can be dangerous if newborns consistently suckle on it in lieu of a functional one; the kids literally will starve. Most of these irregularities disqualify a doe from showing.

Male goats have tiny teats, too; they're situated just in front of the scrotum on a buck. Although they aren't important in and of themselves, check for the same irregularities in breeding bucks as you would in does. Bucks with unacceptable teat structure may sire daughters with bad udders. Bucks with more than two separated teats per side generally can't be shown.

Bucks must have two large, symmetrical testicles. When palpated, the testicles should feel smooth, resilient, and free of lumps. An excessive split separating the testicles at the apex of the scrotum (more than an inch in most breeds) is unacceptable. When choosing a buck, size matters. The greater his scrotal circumference, the higher his libido and the more semen he'll likely produce. A mature buck of most full-size breeds should tape 10 inches or more, measured around the widest part of his scrotum. Boer bucks must tape at least 11½ inches (American Boer Goat Association) or 12 inches (International Boer Goat Association) by maturity at two years of age.

When buying a wether, ask when the goat was castrated. Since castration abruptly halts the development of a young male's urinary tract and affects adult penis size, early castration predisposes male goats to *water belly*, also known as urinary calculi. In this condition, mineral crystals in the urine block the underdeveloped urethra

Get some books on goat health. These are great references. They will scare you because they'll list everything that can go wrong. But [after] raising goats for thirty or forty years, you won't see even half of those things, and, even then, you'll still be learning things about goats.

—Rikke D. Giles

Before you get a goat, read all you can about goats and talk to people who have them. Start with an older doe or wether and *then* get a kid. And don't buy goats at sale barns. Animals are usually sold at auction for a reason.

—Pat Smith

I want healthy goats. I look for clear eyes, moist noses, shining coats, and strong, straight backs with level toplines. I also look for strong, straight legs that don't have spun hocks or knobby knees. Seeing an animal run helps assure me that it has healthy legs. I'm also looking for lumps, abscess, crooked jaws, herniated navels, or cleft palates.

—Bobbie Milsom

I have a milk cow and milk goats. You need only two goats for them to be happy and content, and you can keep seven head of goats per one cow. Plus, goats are more intelligent, friendlier, and safer.

—Samantha Kennedy

and cause the bladder to burst; death occurs within a few days. Castration of pet and recreational goats is best postponed until the animal is at least one month old (later is better).

Whichever sex you're considering, be aware of one of the peculiarities of goat breeding: breeding polled goats to one another sometimes results in hermaphroditic offspring (displaying both male and female sexual organs). It pays to check, keeping in mind that male goats always have teats, so you don't end up with one these unusual goats.

The Sale

You've done your homework, and you're ready to buy. Based on your research, contact sellers who produce the sort of goats you want. Make appointments to visit and view their animals. It never hurts to ask for a seller's references in advance, especially when buying expensive goats. Be sure to check them out before your visit.

When you arrive, look around. Goats should be kept in clean, safe, comfortable surroundings. Do the goats appear healthy? Are they friendly? Are their hooves neatly trimmed? Is their drinking water clean and are feeders free of droppings? Evaluate the seller: does he or she seem knowledgeable, honest, and sincere?

Ask to see health, worming, and breeding production records (and milk production records for dairy goats) on any goats that you're considering. Virtually all responsible goat breeders and dairy operators keep meticulous records, so if the seller can't produce them, be suspicious.

Carefully inspect paperwork when buying registered goats. A seller can't legally transfer them into your ownership unless he or she is the certified owner of record. Does the description on the papers match the goat? Check ID numbers tattooed inside ears (and sometimes on the underside of tails) against numbers printed on registration papers or embossed on ear tags. Remember, without an

up-to-date registration certificate in your hand, you'll pay registered price for a goat that may be grade. Judgments based on intuition aren't always accurate, but if you feel uncomfortable with any part of a seller's presentation, seek elsewhere.

After the Sale

Sellers will often deliver your goats for a modest fee; it's the easiest way to get your purchases home. You can, of course, fetch them yourself if you prefer or if the seller doesn't deliver. Diminutive goats, such as kids and adults of some miniature breeds, are easily transported in high-impact plastic airline-style dog crates stowed in the bed of a truck (secured directly behind the cab to block wind), in a van, or in an SUV. Horse trailers, stock trailers, and topper-clad truck beds all suffice. Whatever you use, supply deep bedding for the animals' comfort and use tarps to keep goats out of direct wind and drafts.

Goats mustn't be stressed in transit; stress equates with serious, sometimes fatal, digestive upsets. Keep everything low-key. Avoid crowding. Provide hay to nibble en route, stop frequently to offer clean drinking water, and dose your goats with a rumen-friendly probiotic paste or gel such as Probios or Fast Track before departure and after you reach your destination.

Have facilities ready to receive your goats, and feed them the same sort of hay and concentrates to which they're accustomed. Many sellers will provide a few days' feed for departing goats if you ask. Begin mixing the old feed with the new feed to help the goats gradually make the change. You won't want to further stress newcomers by immediately switching feeds.

Isolate newcomers from established goats or sheep (goats and sheep share many diseases) for at least three weeks. Deworm them on arrival, and if their vaccination history is uncertain, revaccinate as soon as you can.

Build your goats a showplace barn, and they'll love it. Or hammer together a three-sided shanty built of recycled lumber and secondhand corrugated roofing, and they'll love it. Given a cozy, dry place to sleep in a draft-free shelter, goats are content. They're the essence of simplicity to house. Feeding is easy, too, once you've learned the basic rules.

The Right Housing

Goats hate being wet. Trees and hedges can provide sufficient shade from light showers, but goats in rainy and snowy climates need access to weather-resistant man-made structures, too. In most climates, a three-sided structure (sometimes called a *loafing shed* or a *field shelter*), with its open side facing away from prevailing winds, makes an ideal inexpensive goat shelter. Other basic shelters include movable plywood A-frames, commercial calf hutches, hoop structures designed for hogs, straw buildings, and even large prefabricated doghouses. Bucks are hard on housing; they bash, butt, climb, and scratch their surroundings. Build buck shelters, pens, and fences out of stout, sturdy materials.

If you breed goats, you'll need enclosed housing. Close-to-term does, does with newborn kids, and delicate bottle kids require dry, draft-free housing, especially during the harsh winter months. Dairy goat owners also need covered, weather-resistant areas in which to set up their milking stands. If need (or preference) dictates keeping your goats in confinement housing, you'll probably want to house them in a barn.

Goats like to see one another. Consider making interior pens out of pipe or heavy-duty welded wire panels in lieu of solid walls. Goats also love to climb. Elevated sleeping platforms make for happy campers, as do playgrounds built of recycled telephone cable spools, slanted walk-upon climbing planks, and elevated perches. Provide getaways where kids or low-ranking herd members can escape aggressors; airline-style dog crates are effective hideouts for your little ones, and pens with narrow openings will provide refuge from big, bad bullies.

No matter what type of housing you construct, you have to consider the basics necessary to ensure health and comfort: space, drainage, ventilation, flooring, and bedding. You also need to ensure that your herd has the right feeding and drinking accommodations. Last but not least, you must determine the best means for containing and protecting your charges.

The Structural Basics

Whether fashioning quarters to house a single 4-H goat or a vast herd of meat goats, allow at least 15 square feet of bedded floor space per goat. Make certain that drainage is adequate, and slope the roof away from the shelter's open side so rain and snow will cascade off the rear of the structure.

When building field shelters for small numbers of animals, keep the roof height as low as you can: 5–6 feet in front and 3–4 feet in back is just about right. Low-slung roofs hold body heat at dozing-goat level, essential in colder northern climates. The disadvantage is that squat buildings are harder for you to clean.

No matter a structure's size, goat housing must be adequately ventilated. Goats housed in damp, poorly vented barns are prone to serious respiratory ailments, as are goats (especially kids) exposed to drafts.

In most climates, packed dirt or clay floors are better than cold, hard concrete. In arid climates, wooden floors work well, too, but eventually they'll rot, necessitating replacement. Whatever the flooring, bed the structure with 4–6 inches of absorbent material, such as

Goats enjoy a place to bask in the sun as well as shelter from the elements when needed.

straw, discarded hay, wood shavings or sawdust, peanut hulls, ground corncobs, or sand.

Although some goat owners clean and rebed indoor stalls daily, many prefer deep-litter bedding. With deep-litter bedding, you continually add just enough bedding to keep floors dry, cleaning everything out to floor level only periodically (a few times a year is usually adequate). The deep-litter system is comfortable and warm as well as extremely simple to maintain. Whatever material you use, you'll need to find a responsible way to dispose of it when it becomes soiled. Compost it, give it away, or sell it, but don't let it pile up.

If you keep your goats confined, they'll need a safely fenced communal exercise area that allows at least 30 square feet of space per goat. Mixing horned and hornless goats in a pasture setting can work, but not in close quarters, where considerably more jostling and sparring occurs. Whenever possible, horned and disbudded or polled goats should be housed and penned separately.

Troughs and Feeders

Goats require copious supplies of fresh, clean water, kept reasonably cool in the summertime and liquid when temperatures dip below freezing. Install running water and electricity in your barn or shelter or locate your goat's structure within easy garden-hose and extension-cord reach of existing utilities. If you do have electrical wiring, it must be protected with conduit and kept well out of the curious goats' reach. Any glass windows should be protected by screens.

Provide multiple watering troughs or buckets in lieu of a single big one. It's infinitely easier to dump, scrub, and disinfect several smaller containers than it is a full-size trough. If one water source becomes contaminated with poop, there will be others for your goats to choose from.

Goats fed on the ground are prone to disease and excessive internal parasite infestation, and they just plain waste a lot of feed by trampling on it and soiling it (goats won't touch soiled feed). Buy commercial goat feeders or build your own. Prime requisites are poop-proof hayracks and feeders, which must be installed higher than your tallest goat's tail or at least easily cleanable; it must also be difficult for kids to climb into and be designed so goats can reach their feed but not get their heads stuck while doing so. Because sheep feeders work well for polled and disbudded goats, search for DIY plans for either goat hay feeders or sheep hay feeders.

Don't store feed where goats can help themselves. Overeating, especially of grain or of rich legume hay, can quickly kill even the toughest goat. Store grain in goat-proof covered containers with snug lids (55-gallon food-grade plastic or metal drums and decommissioned freezers work well). Secure the feed room door with a goat-proof lock; opening hook-and-eye closures are child's play to a nimble-lipped goat. You may need to add a padlock or something similar.

Fences

The cardinal rule of goat keeping, especially if you plan to pasture your animals, is don't buy goats until you've erected stout, goat-resistant fences. Goats are curious, mischievous creatures. When the urge to wander strikes them, your goats will do their level best to escape. They love roaming the countryside (especially the roads), and they live to tap-dance on your car. Your neighbor's prized tea roses are as yummy as wild ones—and garden veggies taste good, too!

Goats can squeeze through incredibly small gaps, so standard plank-and-post fences won't faze them unless lined with good wire fencing. Woven wire, also called *field fence* or *field mesh*, makes terrific perimeter, paddock, and pen fencing and can be used to render wooded fences goat-tight.

Ultra-sturdy welded wire livestock panels can be used instead of woven wire, albeit at much greater cost. One caveat: most welded wire panels have jagged, sharp, snipped-wire edges, so use a heavy-duty rasp to smooth them.

Barbed wire, the traditional farm fencing, can be effective when built using eight to ten strands of evenly spaced, tightly stretched 15-gauge or better wire, preferably augmented by add-on twisted wire stays installed between posts. Installing barbed wire is not for the faint of heart, and it can cause catastrophic injuries to man and beast. (It should never be used where horses are pastured or ridden.) Yet it's relatively inexpensive and is often used to upgrade existing farm fences to goat-resistant status. It's marketed in galvanized, high-tensile, and polymer-coated versions available at farm-supply stores.

Smooth, electrified wire fencing makes inexpensive, effective goat fencing, although a goat-resistant fence charger will set you back some bucks. Farm stores stock electric wire of numerous types and gauges, but for the long haul, high-tensile versions work best. To make it work for goats, you'll need to install stout braces and end posts and use six to nine strands of wire. Train your goats to respect the electric fence by luring them to it with grain until

Woven wire fencing consists of smooth, horizontal wires held apart by vertical wires called *stays*. It's sold in regular galvanized, high-tensile, and colored polymer-coated high-tensile versions. Verticals are placed at 6- to 12-inch intervals (wider verticals prevent horned goats and sheep from getting their horns caught), and it comes in heights from 26–52 inches. Horizontal wire spacing generally increases as the fence gets taller. When buying woven wire, check the numbers: 8/32/9 fencing has 8 horizontal wires, is 32 inches tall, and has vertical stays every 9 inches.

Disadvantages of woven wire are its cost and the time and effort required to install it. The advantages are that it's safe, it looks good, and it requires very little upkeep once it's properly installed. Most goat producers use 32-inch woven wire and stretch several strands of barbed or electrified high-tensile smooth wire above it for extra height. An offset electric wire installed inside the fence at adult-goat shoulder height prevents rubbing by itchy goats.

BEFORE BUILDING ANYTHING!

Before building goat structures (or renovating existing facilities) and before installing goat fencing, scope out applicable zoning laws and touch base with your county extension agent. Your agent will understand your needs and can assist you in making plans according to your climate and location. You can also visit the housing and fencing pages at the University of Maryland Extension's Maryland Small Ruminant website (www.sheepandgoat. com). Anyone installing wire fences should also peruse the University of Missouri Extension's online bulletin, "Selecting Wire Fencing Materials" (http://extension. missouri.edu/p/G1191).

POISONOUS PLANTS

Although goats safely process most plants, there are some that they simply ought not to ingest. These plants range from being mildly toxic to causing death with a single mouthful. These are some to watch out for.

Amaryllis *
Apricot (wilted leaves)
Avocado (leaves, fruit)
Azalea *
Baneberry *
Bindweed *
Bitterweed *
Black henbane *
Black locust (bark, seeds, new growth)
Black snakeroot *
Black walnut *
Bleeding heart *
Bloodroot *
Bracken fern *
Buckeye *
Buttercup *
Caladium *
Calla lily *
Cherry (all varieties, wilted leaves)
Clematis *
Crocus *
Crow poison *
Daffodil *
Daphne *

Death camas *
Dieffenbachia *
Dogbane *
Dutchman's-breeches *
Elephant ear *
English ivy *
English laurel *
Flax *
Foxglove *
Horse nettle *
Horsetail *
Hyacinth *
Hydrangea *
Iris *
Jimsonweed *
Johnsongrass *
Jonquil *
Lantana *
Larkspur *
Lily-of-the-valley *
Lobelia *
Locoweed *
Lupine *
Mayapple *
Milkweed (leaves)

Mistletoe *
Monkshood *
Mountain laurel *
Narcissus *
Nightshade *
Oleander *
Peach (wilted leaves)
Plum (wilted leaves)
Poison hemlock *
Pokeweed (seeds)
Potato (leaves, stems)
Privet (berries)
Rattlebox *
Red maple (leaves)
Rhododendron *
Rhubarb (leaves)
Scotch broom *
Sneezeweed *
Tomato (leaves, stems)
Water hemlock *
White snakeroot *
Wisteria (seeds and pods)
Yellow jessamine *
Yew *

* All parts of these plants are toxic or poisonous

Symptoms of plant poisoning include dilated pupils, teeth grinding, vomiting, labored breathing, cries of pain, racing or weak pulse, bloating, scours, muscular weakness or tremors, staggering gait, hyperexcitability, and convulsions.

If you suspect plant poisoning, remove your goat's feed and make it comfortable, supply it with lots of clean drinking water, and get it to a vet as soon as possible, taking along samples of any suspected poisons.

Ask your county extension agent which poisonous plants grow in your locale or visit the "Poisonous Plants and Other Plant Toxins" page at the Maryland Small Ruminant website (http://sheepandgoat.com/poison.html) for links to bulletins covering your state or region.

As ruminants, goats must consume ample amounts of grass and forage.

digestive enzymes—it's essentially a fermentation vat that houses the vast horde of friendly microorganisms that convert cellulose into digestible proteins.

As author Suzanne W. Gasparotto of Onion Creek Ranch succinctly puts it, "You are not raising goats, you are raising rumens." When rumen microbes stop processing cellulose for their host, the goat will sicken and sometimes die. Thus, it's important to learn to assess rumen health. Check to see if a goat is chewing her cud. Use a stethoscope to listen at the goat's bulging left side. A healthy rumen "rumbles" every 45–60 seconds, depending on time of day and what the animal has eaten. Listen to goats in

thcy gct zapped. High tensile versions should not be used to fence pastures shared with horses.

Goats Don't Eat Tin Cans

Feeding and supplementing goats is a complex subject that varies depending on your goats' ages and breeds, whether they're pregnant or lactating, and the types of feedstuff available in your locale. However, certain truths apply no matter where you live or what types of goats you have to feed.

Ruminate on This

Goats are ruminants, as are sheep, cattle, and deer; their digestive systems are very unlike those of simple-stomached species such as horses and humans. In lieu of a single stomach, every goat has four compartments: the rumen, reticulum, omasum, and abomasum. Each compartment has a specialized job.

In newborn kids, only the abomasum is functional. When a kid raises its head to nurse, a band of tissue called the esophageal groove closes and shunts milk directly from its esophagus to its abomasum. That's why it's important for bottle kids to be fed at doe-teat height. As a kid suckles its dam's udder and begins nibbling on plants, dirt, and the rest of its environment, it ingests the microbes she needs to kick-start rumen function. At three to six weeks of age, the kid is functioning like a grown-up goat.

The rumen, located on the goat's left side, is the largest (and first) of the four stomach compartments. The rumen does not secrete

DID YOU KNOW?

The forage and feed that a dairy doe eats and inhales can flavor the milk she gives you. According to the University of California Cooperative Extension's bulletin, "Milk Quality and Flavor," 80 percent of the "off" flavors in goat's milk are feed-related.

The best way to avoid objectionable flavors is to eliminate moldy hay and grain, grub suspected plants from your pastures, and remove certain feedstuffs at least five hours prior to milking.

Plants known to flavor milk include bitterweed, buckthorn, buttercup, wild carrot, chamomile, cocklebur, cress, daisy, fennel, flax, wild garlic, horseradish, wild lettuce, marigold, mustard, wild onion, pepperwort, ragweed, sneezeweed, and yarrow. Feeds best put out at least five hours before milking include alfalfa, cabbage, clover, kale, rye, rape, soybean, and turnip.

Toxins from certain poisonous plants, when ingested by cattle or goats, end up in your milk supply and can be a threat to you and your family. Abraham Lincoln's mother died after drinking cow's milk tainted with white snakeroot. It can happen today, so be careful.

A salt lick pan keeps the salt clean.

every situation—goats who are healthy and those who are sick, goats who have recently eaten and who have not. Note how much their left sides bulge.

When a healthy goat eats, it quickly tanks up on whatever looks tasty and then retires to "chew its cud." The goat burps up partially macerated material from her rumen, rechews it, and then swallows it again. The goat continues the process until its ruminal microbes have digested the food enough for the food to pass into and through the reticulum (where certain nutrients are absorbed) and on to the omasum. The omasum decreases the size of food particles and removes excess fluid from the mix. Finally, the material moves to the abomasum, where body enzymes complete digestion.

Let Them Eat Forage

The millions of microbes (bacteria, protozoa, and other microorganisms) that populate your goats' rumens require mainly cellulose fiber, meaning forage (browse, hay, or grass), to survive.

Concentrates such as commercial goat feeds, corn, and other grains ferment more rapidly than does forage, producing excess acid that can readily kill both the beneficial microbes and your goat. The bulk of all goats' diets must be forage, supplemented by concentrates only when the goats truly need them.

The best dry forage is long-fiber grass hay. High-protein hays, such as alfalfa, clover, and lespedeza, cause the same serious problems as high-protein concentrates: urinary calculi, acidosis, bloat, founder,

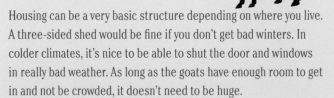

ADVICE FROM THE FARM

KEEPING THEM HAPPY, HEALTHY, AND WARM

I chuckle when I see people put up big, fancy goat houses [because] they always end up telling me, "Well, gosh, it was a waste of money. I should have just put up a lean-to."

A friend of mine just moved to Texas and bought a nice little ranch with an air-conditioned barn. She thought her goats would be in heaven, but they are terrified of the slight noise it makes and won't go into the barn. She has to shut off the air conditioner. Then they'll go in, but they don't spend much time [there] because it's enclosed, and without air conditioning, it gets so hot.

Just remember the number-one thing: goats hate getting wet. Make sure the building you put up is waterproof.

—Samantha Kennedy

Housing can be a very basic structure depending on where you live. A three-sided shed would be fine if you don't get bad winters. In colder climates, it's nice to be able to shut the door and windows in really bad weather. As long as the goats have enough room to get in and not be crowded, it doesn't need to be huge.

Don't store your feed where they can get into it. A separate building (or, if you have a larger barn, a separate room) is best. It's nice to be able to feed goats without having to go in with them. If you have a half wall or a walkway that they can stick their heads into, you can feed without being run over.

—Michelle Wilfong

The hay feeder should allow enough room for all of the goats to gather 'round.

wildlife won't contaminate the feed with disease-carrying droppings.

To supplement the diet, provide a high-quality mineral mix or lick formulated for your type of goats and your locale. Place licks and mixes where goats won't inadvertently poop on them. Goat products generally include copper in quantities that are toxic to sheep, so if you keep both animals, it's vitally important that you choose a dual-species (low-copper) mix or place goat minerals where your sheep can't reach them.

Goats are extremely selective eaters. Given the option, they'll nibble choice bits of hay but dump less savory morsels onto

milk fever, and ketosis (in pregnant does). Goats tolerate (and even savor) a weedier mix than many species do, but all hay must be green, sweet smelling, and absolutely dust- and mold-free.

Dairy does, late gestation and nursing does of all breeds, and most fast-growing young stock require grain. Choose clean, mold-free, commercial goat or horse mixes and cracked or whole cereal grains, and store them (and hay) where birds, cats, and

the floor or ground, where they'll eventually poop and pee on them. Most goats would starve before eating soiled hay, so plan on feeding from waste-resistant feeders and using discarded hay for bedding or for feeding to less picky species, such as cattle and horses.

Dietary Do's and Don'ts

Dietary changes must be made over a period of time—no exceptions. Abrupt changes trigger serious digestive upsets that will kill your goats. You must establish a routine and stick to it. Don't stress your goats by skipping or delaying their feedings.

Allow enough hayrack and feeder space for all goats in a group to eat at the same time; 12–16 inches of feeder space per goat is usually sufficient. Keep feeders clean. Goats won't (and shouldn't) eat or drink from fouled hayracks, feeders, or water sources. Make certain that each goat is eating. Goats that refuse their usual feed are probably ill.

Do not allow your goats to get fat. It's not healthy. Goats don't marble fat throughout their bodies the way most species do; it's deposited around their internal organs, where, in large quantities, it can inhibit vital function. Learn to assess body condition, keeping in mind that there will be individuals who are leaner or chunkier than the norm.

Pure, Clear Water

The cheapest, most essential nutrient of all is water. Goats won't thrive without 24/7 access to lots of sparkling-clean, good-tasting water. They need it to maintain digestive health. Lactating does require water to make milk, and without water, males form urinary calculi. Don't skimp. Keep those tanks and buckets filled and clean. Consider installing an automatic watering fixture. Your goats are sure to love you if you do.

DID YOU KNOW?

When humans hear the word *pasture*, they visualize meadows of lush, waving grasses. Not so for our friend, the goat. Like their cousin, the deer, goats prefer twigs, leaves, wildflowers, and weeds—with perhaps a spot of grass to round out the menu. They thrive on brush-studded rough and rocky land.

Furthermore, since they prefer different plants, goats can be pastured with livestock such as horses and cattle. Goats not only avoid the grasses that these other animals require but also grub out invasive weeds, briars, saplings, and brush, thus clearing land for grassland pasture.

Property owners appreciate goats' taste for such hard-to-rout nasties as purple knapweed, wild blackberry, leafy spurge, purple loosestrife, musk thistle, and multiflora rose. Some entrepreneurs rent goats to landowners specifically to grub out brush and weeds.

Goats are intelligent creatures. Exactly how intelligent is uncertain. Goats (like cats) spurn IQ tests devised by humans; in fact, a goat usually prefers to do things his own way. Consider the interesting similarity between the words *caprine* (relating to goats) and *caprice*. The *American Heritage Dictionary of the English Language* defines *capricious* as "characterized by or subject to whim; impulsive and unpredictable." That, in just a few words, describes goats.

GOAT

BEHAVIOR AND YOU

The same goat on different days and under different circumstances can seem as smart as a whiz kid or as dense as a box of bricks. It may lope down the driveway behind your truck, screaming in anguish because you're leaving, or crouch silently under the feed shed when it senses that you want to give it a shot.

Goats are at all times nimble and curious, a combination that means trouble unless you (and your neighbors) have goat-proof fences and finely tuned senses of humor. Goats do get out. Sashaying down the highway at 2 a.m. is quite the goatish thing, as is raiding your neighbor's lush garden.

Follow the Leaders

Instinct plays a major role in the lives of goats. Ten thousand years after they were domesticated, they still possess intrinsic knowledge of the ways of the wild goat herd.

Like most species, goats maintain a pecking order. The top goat eats or sleeps whenever and wherever he or she pleases. The second-ranking goat defers to the top goat but lords over the rest of the herdmates. And so it goes, down to the goat at the bottom of the heap, who eats last, gets the worst place to sleep, and has to jump when any other goat tells it to. In the wild, each flock is led by a herd king and a herd queen. On the farm, you will probably have only the latter.

The herd queen, a wise old doe, will lead until she dies or becomes too infirm to carry out her duties. She's respected by all; underlings rarely jostle for her position. At her passing, confusion reigns until a new herd queen (frequently a daughter of the old queen) takes her place. The herd queen shows the others what to eat and what not to eat. No one samples a plant before the queen. If she eats, they eat the same thing. If she tastes something nasty, she makes a grand production of spitting, sputtering, and wiping her mouth on the earth.

Fighting within the herd inevitably takes place, especially among newcomers and bachelor males. A fight is a one-on-one proposition; goats don't gang up on a single opponent, although they may figuratively stand in line for their chances to trounce a newcomer. Aggression occurs between both sexes, including wethers (castrated males), and consists of staring; fluffing of the coat, particularly along the spine, as the hackles rise; stomping of the front feet; pushing; rushing; threatening with the horns (chin down, with horns jutting forward); and outright warfare. Goats don't back up and then charge head-down as rams do. An aggressive, annoyed goat positions itself at a right angle to the opponent's body with its head facing the adversary. When ready, the attacker rears onto its hind legs and pivots, swooping forward and down, usually smashing into the opponent's flank, neck, or head.

Spend some time with your goats. You can easily tell which doe is your group's herd queen. You must kindly but firmly make it clear that she must defer to you so she'll accept you as her leader. As acting herd queen (no matter your sex), you can get your goats to follow wherever you lead them.

Get a Handle

When moving goats or handling them for procedures such as worming and inoculations, it's important to understand how your goats are likely to react and why. Being your goats' herd queen can be essential here because it's all but impossible to drive a herd of goats. Goats lack the keen flocking instinct that causes frightened sheep to mob together and move as a single mass. A startled group of goats scatters, although individuals

This Toggenburg cross wears a collar, allowing it to be led easily by its owner.

may whirl to face what spooked them. If still alarmed, they bolt, calling on their speed and agility to outmaneuver perceived danger. Goats are a herding dog's worst nightmare.

Goats dislike entering or crossing water or areas of deep shadow, they resist passing through narrow openings, and only with great reluctance will they attempt to navigate slippery surfaces. They readily move forward out of darkness toward light, from confinement toward open spaces, and into the wind rather than downwind, and they're more likely to go uphill than down.

Loud noises and sudden movements frighten goats. Startled or disgruntled individuals often lie down to avoid being driven or handled. They also aggressively butt and shove at the animals around them.

Most goats resist if you haul them around by their horns. It's better to lead or restrain a tame goat (even a horned one) by its collar or to temporarily immobilize it by cupping a hand under its chin and lifting its head. If you do handle a goat by the horns, avoid snapping off a tip (which will bleed like crazy); instead, grasp the horns down by the bases.

Harried goats are easily stressed. Since stress invariably leads to serious health problems, avoid stressing them as much as you can. Keep things low key when dealing with goats, and don't lose your temper. Give your goats a chance to understand what you're asking before you react; patience goes a long way in the goat yard.

Don't underestimate the power of a goat. When handling seldom-handled goats or when working with goats in close quarters, wear a long-sleeved shirt, long pants, and boots or steel-toed shoes. Keep small children out of the action altogether when working around fractious goats.

Goats are smart and have long memories. Depending on the actions you take today, things will either be better or a whole lot worse the next time you handle your goats.

Can You Read Me Now?

Goats communicate mainly through body language, but sometimes they vocalize as well. An alarmed goat stands rigid, poised to run but with legs firmly planted, tail curled over the back, head held high, and ears pricked forward at perceived danger. It may stamp one forefoot or snort to alert the herd. Its alarm snort resembles a loud, high-pitched sneeze. To assert their authority, dominant individuals

glare at, crowd, bite, and butt underlings. High-ranking, assertive individuals may try this with you, too. Nip aggression in the bud. You always need to be top goat in your herd.

Although goat vocalization traditionally has been called *bleating*, the sounds that goats make are now more commonly referred to as *calls*. Goats call in greeting to their human caretakers and to other goats, to demand food, and to locate their kids and other herd members. Does murmur tenderly to their newborn kids, and goats scream in terror and in pain. Some breeds call more than others do. Nubians are the noisiest of all.

The Birds and the Bees and Behavior Keys

Breeding season brings a new set of behaviors to the goat yard, some of them peculiar by anyone's standards.

It's a Guy Thing

As breeding time approaches, a buck goes "into rut," which means that he's in breeding mode and can think of nothing else. What better way to attract the ladies than by liberally dousing himself with "perfume"? Unless he's been descented, the glands on his forehead begin exuding a pungent, earthy musk. He adds to his allure by spritzing his face, beard, chest, and belly with thin jets of urine. He also grasps his penis in his mouth (yes, he is that agile) and sometimes urinates, whereupon he curls his lips in a grimacelike response (a behavior known as *flehmen*). And all of this before he even meets a doe in heat!

Once he's turned out with the ladies, he'll add new tricks to his repertoire. He'll trail a prospective girlfriend, sniffing her sides and under her tail, sometimes pawing her with a stick-rigid front leg, all the while flapping his tongue and making bizarre vocalizations called

You don't want your goats to go running when you need them to come to you.

Train Them Right

If you're a large-scale meat producer, you probably don't need to train your goats, except possibly the bucks. Because you'll probably keep them in separate quarters part of the year and handle them more often than you handle your does, it's always a good idea to educate bucks. If you call them while rattling a bucket, they'll stampede to your handling facilities. That may be enough training for your purposes. However, if you keep a few pet, 4-H, or recreational goats, you'll at least want to teach them to lead and tie (remain quietly attached to a hitching rail). Whatever your training goals, there are some points to keep in mind.

Goats work their hearts out for food, making them ideal candidates for clicker training. *Clicker training*, also known as *operant conditioning*, is widely used to train sea mammals, horses, and dogs. If you've never tried clicker training before, we recommend starting by reading one of the clicker training books listed in the Resources section. Although you probably won't want to teach your goat to fetch a soda from the fridge (then again, you might), most of the training routines common with other pets work exceptionally well with goats.

Reward-based training always works best, but when you need to thwart undesirable behavior immediately, reach for a high-powered water gun or a household pump sprayer with a long, strong jet. Goats despise water, especially when it's squirted in their faces. A loud "No!" coupled with a blast or two of water tends to grab the most errant goat's attention. Don't just yell and wave your arms and chase your goat away. To goats, chasing is play behavior, which means you're actually rewarding the goat for misbehaving.

Goats can be led using a halter or a collar; a halter tends to give you more control. Walk with your goat's shoulder at your right hip; until it understands, ask someone to follow and urge the goat along when it falls back or stops. A well-timed, brief shove on the rump works better than pushing or swatting. Reward the goat when it does well. It will learn much faster if it's having fun.

For safety's sake, never use choke-type collars or slip-style halters to tie up your goat. Use a slipknot in your rope so you can untie it quickly if it pulls back or somehow gets tangled. *Don't* go off and leave a semitrained goat tied up. You need to be ready to save it if it panics.

blubbering. When she urinates, he samples it and flehmens. When she stops running from him and stands to be bred, he mounts her. As he ejaculates, he flings back his head and then recuperates for a heartbeat before dismounting.

During rut, bucks become more pugnacious, even toward humans. Being rammed by a buck is more than a playful bunt. A big buck can knock you over and seriously hurt you; so can a little one if he happens to clip you behind both knees.

Bucks penned apart from the ladies practice their techniques among themselves. Some bucks, especially bottle-raised and pet bucks, may court their favorite humans. This isn't much of a problem when a 40-pound Nigerian Dwarf wants to rub his forehead glands on your leg, but it's a serious one when a 300-pound Boer wants to mount your twelve-year-old daughter. It's important to stay alert when working around bucks in rut. Children and vulnerable adults shouldn't handle them at all.

Does Just Want to Have Fun

Does have their own set of unique breeding behaviors, although some overlap with those of bucks. Females don't pee on themselves, but tongue flapping, blubbering, and mounting female herdmates and their human caretakers are not uncommon. This is referred to as "being bucky."

Other signs of estrus (heat) include allowing other does to mount them, frequent urination, decreased appetite, clear mucous discharge from the vulva, tail-wagging (flagging), a great deal of strident calling, and mood changes. The super-sweet doe may intentionally kick over the milk pail or attack her underling herdmates, while your old "picklepuss" wants to love you to death. A lactating doe's milk production takes a dramatic dive while she's in heat, too.

Goats are prone to a host of serious ailments, and that's a fact. However, it's just as true that properly managed goats rarely get sick.

We've said it before, but we'll say it again: don't buy trouble! Choose healthy foundation stock and take basic steps to watch over the health of your charges. Certain problems will still be inevitable, but it's easy to avoid major nasties such as foot rot; caprine arthritis encephalitis (CAE); caseous lymphadenitis (CLA); contagious ecthyma (CE), also called *sore mouth* and *orf*; and Johne's disease by buying from disease-free herds. At the very least, avoid purchases from livestock auctions or poorly managed herds.

GOAT

HEALTH, MALADIES,

AND HOOVES

Once you have purchased your goats, you should have relatively few health concerns to contend with if you feed and handle your goats properly, worm them when needed, vaccinate according to your vet's recommendations, monitor their health every day, and don't allow them contact with anyone else's goats or sheep.

Anytime you take your goats where other goats or sheep are present, you're exposing them to a host of communicable diseases. If the risk is necessary or unavoidable, at least hedge your bets against introducing sickness into your herd by quarantining incoming goats—whether newly purchased or home from an outing—for at least three weeks. To prevent accidental contamination, don't feed and handle quarantined goats until after you've tended to your main herd, and don't let visitors, pets, or poultry move freely between your quarantine area and your primary goat housing.

It's important to recognize illnesses and start treatment right away, so you'll have to learn to tell when a goat is feeling out of sorts. Stroll among your goats at least twice a day, watching for signs of illness, and remove suspect individuals to your quarantine area without delay. If only one goat is ailing, reduce the sick goat's stress by moving a second goat to the quarantine area and penning it a distance from, but within sight of, the sick goat.

Recognizing Illnesses

Although discussing all of the maladies that befall goats is beyond the scope of this book, you'll find the most important ones in the "Goat Diseases at a Glance" section in this chapter. Before identifying what ails your goat, you need to be able to recognize when there's something wrong. So you think you have a sick goat. You've isolated the animal, but you're not quite sure what's wrong. Call your vet immediately. Time is of the essence when treating coccidiosis, pneumonia, pregnancy problems, and toxemia, as well as a host of other serious caprine complaints.

Sometimes, however, your regular vet is temporarily unavailable, so you need to have other people and resources to consult. Start (before trouble arises) by finding a mentor—an experienced local goat owner who is willing to help you when the chips are down. Locate a mentor via goat clubs, directories, and e-mail groups, and establish a working relationship before you need help.

Assemble a caprine medical library and a well-stocked first aid kit. Print information from the Internet, arrange your printouts by subject matter, and file them in handy three-ring binders. Laminate important pieces, such as kidding diagrams, and keep them in your first aid or kidding kit. Still another option: download and store PDF files so you can access vital information with just a few quick clicks of the mouse.

It's important to keep a first aid kit handy not only for those inevitable emergencies but also for day-to-day treatment of cuts, scrapes, and dings. Keep the kit where you can readily find it, and replace each item as it's used. Three or four times a year, thoroughly inspect your first aid kit and discard any expired products.

Overvaccinating wastes money and stresses the goats. Always vaccinate your goats with C/D T combination vaccine, an over-the-counter combination product that protects them from *Clostridium perfringens* types C and D and tetanus. They need it no matter where you live. Discuss additional local and herd-specific needs with your vet or county extension agent before heading for the farm store's vaccine cooler.

Most goat keepers learn to inoculate their own goats. It's cost effective, and veterinarians are generally pleased to make fewer farm calls, but you must consult with your vet to formulate a vaccination program specific to your area and to your herd.

FIRST AID KIT

Meet with your vet and decide which emergencies you can face by yourself. If minor ones are all you feel prepared to tackle, a bare-bones kit is enough. However, if you're an experienced livestock keeper or live far from your vet, he'll probably suggest prescription drugs to keep on hand. This is what we keep in our kit.

In the Fridge

Keep these items together in a box so that they're easily accessible.
If you have children, use a lockable tackle box.

Antibiotics

· Biomycin-200 (as effective as LA-200, but LA-200 stings on injection)
· Penicillin
· Tylan-200

Vaccines

· C/D T toxoid
· C/D antitoxin
· Tetanus antitoxin
· Epinephrine for treating anaphylactic shock

Other Items

· Probios probiotic paste
· Goat Nutri-Drench
· Banamine
· Injectable vitamin B1
· Sterile water for flushing wounds

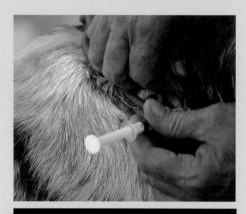

Vaccines are administered by subcutaneous injection.

In the Emergency Bucket

We store our first aid kits in a 5-gallon food-service bucket with a snug lid, clearly marked with permanent marker.
Group items together and store in plastic zipper-seal bags clearly marked with the contents.

· A lead rope and halters in several sizes
· Scissors, several disposable scalpels, and a sharp folding knife (store together)
· A powerful flashlight and extra batteries and bulbs (store together)
· Cotton-tipped swabs, stretch gauze, sterile pads, adhesive tape, two rolls of self-stick 4-inch bandage (we prefer Vetrap), and a digital thermometer (store together)
· Small containers of Kaopectate, milk of magnesia, Tagamet, and baby aspirin (store together)
· An assortment of 18- and 20-gauge, ¾-inch disposable needles and 1-cc, 3-cc, and 6-cc disposable syringes (store together)
· Latex gloves
· 4-inch-wide duct tape
· Betadine scrub to clean wounds
· Schriener's Herbal Solution and emu oil, our favorite topical wound treatments
· Nitrofurazone salve to protect summertime wounds from fly strike
· Blood-stop powder: never be without it!

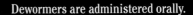

Worms

Goats are extremely susceptible to stomach- and intestinal-parasite infestation. Chronically wormy goats are scrawny, rough-coated, depressed, and anemic. They frequently suffer from diarrhea and usually die. Goats must be dewormed using the correct anathematics (worm-killing products) for the type or types of worms involved. Choosing dewormers at random simply won't work. Deworming is a complex subject to be discussed with your vet and your county extension agent, but in the meantime, keep the following points in mind.

The only way to know which vermicide to use is to take fresh manure samples for fecal analysis. When collecting samples, follow your goats with a sandwich baggie and collect "berries" as they fall (try to select nuggets that didn't come in direct contact with the ground). Take the samples to the vet, who will prepare smears and examine them under the microscope, searching for worm eggs. Your vet can then recommend a product based on your herd's exact needs.

Goats must be dosed according to weight. Underdosing is ineffective and leads to chemical resistance; overdosing can, depending on the product, kill your goats. Make sure you know your goat's weight. Though livestock scales work best, they are prohibitively expensive for most hobby farmers. With a little effort, you can use a standard bathroom scale to weigh small goats. First, weigh yourself. Next, pick up the goat and step back onto the scale; subtract your individual weight from the combined weight of you and the goat, and you have the animal's individual weight.

Parasites

Like all other warm-blooded creatures (and some that aren't), goats are plagued by external and internal parasites. Following are descriptions of both.

Flies, Lice, Mites

In addition to everyday, in-your-face stable and biting horseflies, two specialized types of flies plague our goats: keds and botflies. Keds primarily infect sheep, but they do prey on goats as well. They're wrinkly, brown, wingless flies that look like ticks and feed on blood. Botflies are fuzzy, yellowish-brown insects that resemble honeybees. They hover around a goat's nostrils, where they deposit newly hatched larvae. The larvae migrate up the goat's nasal passages, feeding on mucus, until they reach the goat's sinuses. This naturally annoys the goat and can trigger severe inflammation and bacterial infections. After a time, larvae work their way back down the nasal passages, drop to the ground, pupate, and emerge as adult flies. Two generations in a single summer are not uncommon.

Several sorts of lice also live on goats. Some feed on skin and hair; others suck blood. Louse infestations cause extreme itchiness, skin irritations, rough coats, and hair loss. Lice are species-specific: goats can't pick up lice from poultry or birds.

Mites burrow into the skin or feed on its surface, creating a fluid discharge and scaly, inflamed, denuded patches of skin called *mange*, *scabies*, or *scab*. Infestations are highly contagious and require aggressive treatment. One type, psoroptic mange, is a federal quarantine disease, so if you suspect that your goats are infected, contact your vet without delay!

WAIT AFTER WORMING

The milk from newly dewormed dairy goats must not be used for human consumption; withdrawal periods vary from product to product. Ask your vet.

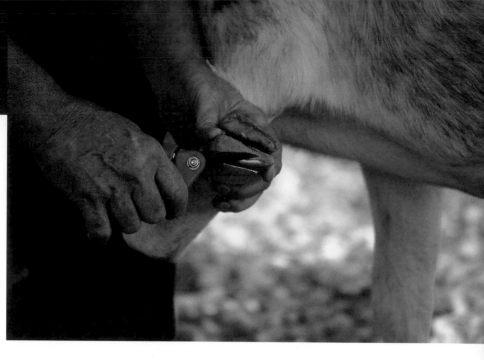

A standard pair of hoof clippers gets the job done.

Before using any dewormer, read the instructions. Not all are labeled for goats, so if your vet recommends an off-label dewormer, ask about cautions and restrictions. Follow instructions exactly.

Finally, if you choose homeopathic, herbal, or other organic dewormers, have fecal tests run on an ongoing basis. These products work well in some instances but fail miserably in others. Don't just assume that they're doing the job; for your goats' sake and yours, be certain.

Hooves

Soil moisture and type, time of year, and breed influence how fast hooves grow. Trimming protects the integrity of your goats' hooves. In general, plan on giving pedicures at least two or three times a year, timing them to coincide with other labor-intensive procedures, such as worming and vaccinating. Hooves are easier to trim when they're moist. Avoid trimming hooves during high-stress times such as extreme weather conditions, late pregnancy, or weaning time.

Trimming

You'll need proper tools for hoof trimming. Most folks use standard hoof shears, but trimming with horse hoof nippers, a hoof knife, and a rasp works well, too. It's a matter of experience and convenience.

Safely tie your goat to a secure object using a sturdy lead rope attached to its halter or collar. Squat beside the goat, perch on an overturned bucket, or stand and lean over to trim.

Start trimming at the heel and work forward. Trim the heel even with the frog (the soft, central portion of each toe) and then trim the walls level to match. If the frog is especially ragged, touch it up with a knife, taking paper-thin slices until you reach a hint of pink, but go no further than that; the frog is a sensitive structure. When you're finished, the hoof should be flat on the bottom and parallel to the coronary band (the area where hoof and hair intersect).

When trimming a goat with foot disease, trim the infected hoof/hooves last to prevent spreading disease to healthy tissue. When you're finished, disinfect your tools to prevent infecting your other goats.

Not every lame goat has foot rot. To evaluate a gimpy goat, watch it from afar. Which foot or feet is it favoring? How badly is it limping? Scan for foreign objects lodged in or between toes. Then carefully trim all four hooves. As you do, watch for signs of disease.

Dealing with Foot Rot

Foot scald and foot rot (or hoof rot) are closely related. In fact, they share a causative agent, the bacterium *Fusobacterium necrophorum*.

F. necrophorum is a common, hardy bacterium that dwells in soil and manure and is found on virtually every farm where livestock is kept. It causes thrush in horses and contributes to foot rot in cattle. It's an anaerobic organism (which means it can grow only in the absence of oxygen), so when animals are kept in dry, sanitary conditions, *F. necrophorum* poses no threat to them.

However, when hooves are continually immersed in warm mud and muck, bacteria invade the foot, often via a minor scratch or ding, causing foot scald, a moist, raw infection of the tissue between the sufferer's toes. Foot scald usually affects only one of the front feet. It's nasty and painful, and it frequently leads to full-blown foot rot.

Foot rot occurs when *F. necrophorum* is joined by *Bacteroides nodusus*, another anaerobic bacterium that thrives in the hooves of domestic goats. It gains access via foot-scald lesions and other injuries. When *F. necrophorum* is present, *B. nodusus* sets up house in the deeper layers of the skin, where it produces an enzyme that liquefies the tissue that surrounds it.

You can't miss foot rot: affected goats are very lame, and infected tissue is slimy and stinky. Infection beneath the walls and sole of the hoof causes the horny walls to partially detach. More than one hoof may be involved.

Foot rot is treatable, but it's a long, costly, time-intensive process and, in most herds, not an entirely successful one. The key to foot-rot control is not to bring it home in the first place. Goats can't get foot rot without coming into contact with *B. nodusus* bacteria.

To prevent its introduction, trim new goats' hooves on arrival and quarantine them well away from your main herd for at least three weeks. Do the same with returning 4-H and show goats, goats who

What a well-trimmed hoof looks like.

Vibrosis is caused by the bacterium *Campylobacter fetus*, subspecies *intestinalis*. When one or two does affected by vibriosis abort, they can trigger an "abortion storm." A vibriosis vaccine is available, often in combination with the EAE vaccine.

Toxoplasmosis, which is caused by the coccidium *Toxoplasma gondii*, is spread when a host cat contaminates goat feed and water with her droppings. There is no vaccination or treatment for toxoplasmosis.

When a doe aborts a kid, the fetus and tissues should be submitted to a laboratory for diagnosis; you can't treat the rest of the herd unless you positively know what's wrong. Your vet can tell you where to send the specimens. The material must be fresh, so store it in sturdy plastic bags, pack the bags in a Styrofoam box and surround them with chill packs, and then rush the package to the lab.

have boarded at your veterinarian's facility, or any other goat that leaves your farm and returns.

We can't say it enough: don't buy goats at livestock sales. Many producers knowingly dump infected stock at farm auctions. Even if the one you buy isn't infected, it's probably been exposed to infected goats.

Goat Diseases at a Glance
Abortion

Enzootic abortion (EAE) of does is a chlamydial disease transmitted from aborting goats and fetal tissues to other does. Infected does abort during the last month of pregnancy or give birth to stillborn or weak kids who soon die. An effective vaccine is available. (The second *E* in the abbreviation stands for "ewes," but this is a problem for goats as well.)

Bloat

Bloat is a buildup of frothy gas in the rumen, usually triggered when a goat tanks up on an unaccustomed abundance of grain, rich grass, or legume hay. Bloated goats can quickly die of the condition, so if you suspect that your goat has bloat, call your vet immediately.

VACCINATION TIPS

- Use disposable syringes and needles, and dispose of them in a responsible manner. Use a clean, new syringe for every session and a new needle for each animal. Sharp needles cause less pain and work better.
- Choose 16- or 18-gauge needles in 1/2-, 5/8-, or 3/4-inch lengths. Longer needles easily bend or break. Shorter ones are perfect for giving subcutaneous (injected under a pinch of skin) shots, which is how goat vaccines are administered. Always use a clean needle to withdraw vaccine from the bottle. A used needle contaminates the remaining contents.
- Give injections into clean, dry skin. Some vets recommend swabbing the area first with alcohol.
- To give a subcutaneous injection, pinch up a fold of skin and slide the needle under it, parallel to the animal's body. Slowly depress the plunger, withdraw the needle, and then rub the injection site to help distribute the vaccine. Shots can be given in the neck, over the ribs, or into the hairless area behind and below the armpit.
- Store leftover vaccines and antibiotics in your refrigerator, following the instructions on their labels. Discard leftovers after their expiration dates pass.

Youngsters are at a greater risk of contracting coccidiosis.

Caprine Arthritis Encephalitis (CAE)

CAE is an incurable viral infection caused by a retrovirus similar to the one that causes HIV in humans. CAE infects only goats. A relatively uncommon juvenile-onset neurological form of CAE causes encephalitic seizures and paralysis in kids, but CAE is primarily a wasting disease of adult goats. Early symptoms of infection include swollen knees, unexplained weight loss, and congested lungs. Sufferers eventually die of chronic progressive pneumonia.

Initially associated mostly with dairy goats, CAE has spread due to the practice of crossing meat-breed bucks with dairy and part-dairy percentage does to produce commercial meat goats. Because CAE is spread from infected does to kids via body fluids, colostrum, and milk, producers are breaking the cycle by removing kids from their dams at the moment of birth and then artificially rearing them either on pasteurized milk or on milk replacer. CAE testing of individual goats is possible, but because these tests aren't 100-percent accurate, it's best to buy from certified CAE-free herds.

Caseous Lymphadenitis (CLA)

CLA is a chronic, contagious disease of sheep and goats caused by the bacterium *Corynebacterium pseudotuberculosis*. The bacterium

---

ADVICE FROM THE FARM

THE BEST MEDICINE

Like lots of goat owners, we swear by a homemade blend of one part molasses, one part corn oil, and two parts Karo syrup. We use it to put final finish on an animal or provide quick weight gain to get ready for a show. We give a 50+-pound animal 50–60 cc every other day or so. It works, and we use it often as part of our weight gain routine.

We also use it on does in the early stages of toxemia, anemic goats, and goats who are sick and not eating as a support treatment.

—Robin L. Walters

A handful of German black sunflower seeds added to your goats' diet each day will prevent goat polio. These seeds are high in thiamine, the roughage is good for the rumen, and the oils in the seeds are good for the coat.

—Bobbie Milsom

Goats can break their horns, especially bucks who like to bash into things or aggressively head-butt their pen mates. If a horn is broken completely off and it isn't hemorrhaging or badly contaminated with dirt and debris, I've had good results from simply spraying the area with an aerosol antiseptic/sealer such as BluKote, keeping the animal isolated so that others will not bump the very sore head, and watching for any signs of infection.

—Melody Hale

When there's [an animal] here that is in dire straits, I spend every minute with that animal. When it opens its eyes, it sees me or feels my hand caressing it. It does matter.

—Donna Haas

Johne's disease can affect any ruminant.

breaches a goat's body through mucous membranes or cuts and abrasions. The animal's immune system valiantly tries to localize the infection by surrounding it in one or more cysts. If the ploy is unsuccessful, it will die.

CLA presents as lumps near the jaw, in front of the shoulder, and where a doe's udder attaches to the body. Some goats develop internal cysts, too.

Coccidiosis

Coccidiosis is a very common, potentially fatal yet easily prevented, easily treated disease of young kids. It's caused by uncontrolled proliferation of single-cell protozoal parasites called *Coccidia*, found in barnyard soil. Coccidiosis is species-specific, meaning that goats aren't bothered by poultry, canine, or sheep *Coccidia*.

Suspect coccidiosis when kids more than two weeks old experience severe abdominal pain (evinced by crying or reluctance to lie down) coupled with dark, watery, foul-smelling diarrhea streaked with mucus or blood. Take a stool sample to your vet for fecal diagnosis.

Although rehydration with electrolyte solutions and administration of antidiarrheal medications along with sulfa drugs, amprolium, or tetracycline usually effect a cure, it's easier to prevent coccidiosis than it is to cure it. Many producers choose to give their goats feeds laced with anticoccidial drugs called coccidiostats, whereas others add them to the goats' drinking water. Horse owners, take note: one such feed additive, Rumensin (monensin), is extremely toxic to horses.

Contagious Ecthyma (CE)

Commonly known as *sore mouth*, also known as *scabby mouth* or *orf*, CE is a contagious poxlike virus that causes the formation of blisters and pustules on the lips and inside the mouths of young kids as well as on the teats of the infected kids' mothers. The blisters pop, causing scabbing and pain so intense that occasionally a kid will starve rather than eat. Most kids recover in one to three weeks without treatment.

An effective live vaccine is available, but you mustn't use it unless you already have sore mouth on your property. Goats will shed their vaccination scabs, which will contaminate your property and almost certainly spread the disease to the rest of your herd. Because sore mouth is easily transmissible to humans, wear rubber gloves when handling stricken kids. Keep children away from all infected goats!

Enterotoxemia

There are two types of enterotoxemia in goats caused by *Clostridium perfringens*: Types C and D. Type C is a disease of young kids caused by an anaerobic bacterium found in manure and soil. It enters via newborn kids' mouths when they encounter dirty conditions while seeking their mothers' udders. Bacteria produce a toxin that causes rapid death. Treatment is usually ineffective; death usually occurs within two hours of the onset of symptoms, which include seizures and frothing at the mouth. However, kids from does vaccinated for enterotoxemia during late pregnancy develop immunity to the disease via their mothers' colostrum.

Type D is also present in the soil and manure. It attacks rapidly growing, slightly older kids who ingest the bacterium while investigating their environment. It, too, causes tremors, convulsions, and a host of strange neurological behaviors leading rapidly to death. A vaccine is available alone or in combination with type C or as a C/D and tetanus vaccine.

Floppy Kid Syndrome (FKS)

FKS affects kids between three and ten days old. Its precise cause is still uncertain. Kids autopsied as part of a study conducted by Texas A&M University had very distended abomasums full of acidic-smelling, coagulated milk. Scientists speculate that overconsumption of rich milk triggers an overgrowth of certain microorganisms in the digestive tract, resulting in systemic, often deadly, acidosis.

Afflicted kids show muscular weakness and depression, progressing to flaccid paralysis, and often death. In all cases, their abdomens are distended, and if gently shaken, they may "slosh."

Keeping your goats' living quarters clean goes a long way in promoting good health.

Johne's Disease

Johne's (YO-neez) is a deadly, contagious, slow-developing, antibiotic-resistant disease affecting the intestinal tracts of domestic and wild ruminants, including goats. The bacterium that causes Johne's, *Mycobacterium avium,* subspecies *paratuberculosis,* is closely related to the one that causes tuberculosis in humans. Infected goats are dull, depressed, and thin. Johne's disease, also known as *paratuberculosis,* is incurable.

Ketosis

Ketosis is a relatively common metabolic condition associated with pre- and postpartum does, especially overweight and underexercised does pregnant with more than one kid. Prepartum ketosis is also called *pregnancy toxemia*; it occurs within a month before birth. The rumen of a fat doe carrying multiple kids is scarcely able to hold enough nourishment to meet the nutritional needs of late-term kids, so her body begins burning its own fat reserves to provide energy.

Postpartum ketosis, also called *lactational ketosis*, occurs when the high energy demand on a doe nursing multiples (especially triplets, quads, and quints) causes excessive weight loss—she simply can't consume enough feed to meet their needs, so she dips into her own reserves.

In both scenarios, ketones produced by this process make the doe quite ill. Without intervention, she'll die. Early symptoms include listlessness, poor appetite, and possibly labored breathing, progressing to circling, stargazing, stumbling, and teeth grinding. She'll eventually collapse and, without aggressive treatment, lapse into a coma and die.

To prevent life-threatening ketosis, does should be fed a high-quality, balanced diet throughout their pregnancies and monitored closely the month before and the month after giving birth.

Listeriosis and Goat Polio

Listeriosis (also known as *circling disease*) and goat polio (also called *polioencephalomalacia* and *cerebrocortical necrosis*) are serious metabolic diseases with similar causes and symptoms. Both of these diseases occur mainly among confined goats who are fed relatively high-concentrate, low-fiber diets, and both of these diseases can be triggered by abrupt changes in feed and by moldy grain or moldy forage (especially silage).

Listeriosis is caused by the common bacterium *Listeria monocytogenes.* One type of listeriosis causes abortions; the other type, which is more common, causes encephalitis. Both types are usually seen in adult goats. Encephalitic listeriosis triggers inflammation of the brain stem and death (necrosis) of brain tissue, resulting in one-sided facial paralysis, drooling, stargazing, and stumbling; lack of appetite, depression, and fever are other common symptoms. Listeriosis can be passed to humans via the milk of sick or carrier goats. Without aggressive antibiotic treatment, afflicted goats die.

Goat polio is a deficiency of vitamin B1/thiamine, most commonly encountered in weanling and yearling goats. Ingesting excess amounts of grain, stress, prolonged or excessive use of antibiotics, or changes in feed drastically lower rumen pH, and beneficial microorganisms die off. This, in turn, decreases thiamine production. Thiamine is necessary to metabolize glucose. Without glucose to feed them, brain cells die, and neurological symptoms such as hyperexcitability, staggering or weaving, circling, blindness, and tremors appear. Untreated goats develop convulsions and usually die in 1–3 days, but when promptly injected with thiamine (best given intravenously by a vet), goats usually recover.

Mastitis

Mastitis is a serious infection of the mammary system. Substandard milking hygiene, delayed milking, and injuries are common causes. Symptoms include decreased milk production; clumps, strings, or

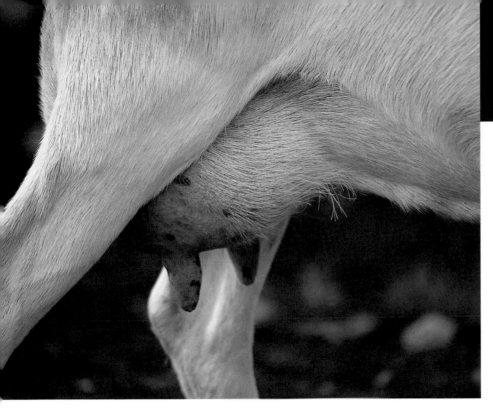

Minimize the risk of mastitis by keeping udders clean and milking areas sanitary.

scourge: only Australia and New Zealand are scrapie-free.

Scrapie is a slow, progressive disease that systematically destroys the central nervous system. It is far more prevalent in sheep than it is in goats. Symptoms typically appear two to five years after contraction and include weight loss, hypersensitivity, tremors, stumbling, blindness, excess salivation, lip smacking, and intense itchiness. Between one and six months after symptoms appear, infected animals die.

Goats residing on properties where sheep are also present must be identified through the USDA's mandatory scrapie eradication program. Contact your state Animal and Plant Health Inspection Service (APHIS) representative for up-to-date information.

blood in the milk; and pain, inflammation, and swelling in the udder. Usually only one side is affected.

Does must be milked in sanitary surroundings with clean hands and udders before, during, and after milking. Use home-based mastitis test kits to check for mastitis at weekly intervals.

If mastitis is suspected, seek professional treatment. Untreated mastitis swiftly leads to permanent udder damage, and one type even leads to gangrene and death.

Urinary Calculi

Urinary calculi are tiny stones or crystals that form in the urinary tracts of sheep and goats. Does get stones, but they pass through the larger female urethra (the tube that empties urine from the bladder) without difficulty. A buck or wether with a blocked urethra is in trouble, however, because his bladder is likely to rupture, and he'll probably die.

When bucklings are castrated, penis growth stops, so wethers castrated at an early age are especially troubled by calculi; their much tinier penises and urethras are easily blocked. A workable solution: don't castrate male kids younger than 4–6 weeks old.

A calcium-to-phosphorus ratio of 2:1 in the diet helps prevent calculi formation, as do small, measured amounts of ammonium chloride added to feed. Male goats should drink lots of water. Make it more appealing by keeping water sources readily available, full, and sparkling clean.

Pneumonia

Pneumonia is caused when one of a wide variety of opportunistic bacteria and viruses mix with stressed goats. Typical symptoms include depression, fever, coughing, and labored breathing. Because so many bacteria and viruses may be involved, accurate identification of the infectious agent is an essential part of successful treatment.

Scrapie

Scrapie is a transmissible spongiform encephalopathy (TSE) of sheep and goats similar to bovine spongiform encephalopathy (BSE, or "mad cow" disease) and to chronic wasting disease (CWD, which affects deer and elk). No human has ever contracted scrapie (or either of the human equivalents, kuru and Creutzfeldt-Jakob disease) from sheep or goats.

Scrapie appears to be caused by an infectious agent, but genetics also play a part. The disease was recognized in Britain and western Europe at least two hundred years ago, and it came to the United States in 1947 with British goats. Scrapie is a global

White Muscle Disease

White muscle disease is caused by selenium deficiency. Does grazing on selenium-poor land or those eating hay that was raised in depleted conditions require selenium/vitamin D supplementation during the last two months of pregnancy. Otherwise, their affected kids will have problems standing and walking; some will even become paralyzed. Prevention is the key to eliminating white muscle disease.

Whether you breed to get your dairy does in milk, you want to bottle-raise a pet or packing wether, or you raise kids for the meat market, learn all you can about the breeding process before you begin.

BRINGING
GOATS
INTO THE WORLD

Choosing Breeding Stock

If you buy a scrub goat at a farm sale and breed her to your neighbor's mixed-breed buck, you'll get a cute scrub kid. If your aim is to clear your woods of brambles and saplings, that may be enough. But if you want to raise quality animals, start with the best foundation stock you can afford.

Do your homework: know what types and breeds produce the kind of kids you want to raise. Don't limit yourself. By breeding dairy or Angora does to high quality meat breed bucks, many breeders produce marketable meat kids while pursuing their primary goals of producing mohair fiber or dairy products. Buy breeding stock from producers who keep detailed health, pedigree, production, and, in the case of dairy breeds, milking records. Ask to see verification when buying from purportedly certified disease-free herds.

Choose mature animals that have already sired or produced quality kids. When choosing dairy goats, even bucks, try to see a prospective purchase's dam and, if possible, its sire's dam, too. Udder quality is highly hereditary—as is a tendency toward multiple birthing. Choose goats (especially bucks) from twin, triplet, or quadruplet births to maximize their chances of also having multiple births.

Top-quality older goats can be best buys for entry-level and hobby-farm goat producers. Large-scale breeders often cull at six to eight years of age, but with a little extra attention, goats in this age group have years of productivity ahead of them. They already know the ropes, especially at kidding time, a definite boon when their newbie owners don't.

Sex in the Goat Herd

Goats reach puberty early: five to six months is the norm for most full-size breeds, but two-and-a-half-month-old bucklings have successfully impregnated their dams and sisters. To prevent unplanned pregnancies, castrate males not destined for breeding, and separate bucklings from the rest of the herd by the time they're twelve weeks old.

Most responsible producers breed doelings at eight to twelve months old; dairy breeders often say to breed them when they reach 80 pounds. In either case, doelings should be well grown and healthy, and they should be bred to bucks who (at maturity) are of the same size or smaller.

Most dairy goats are seasonal breeders. Their breeding season is triggered by decreasing daylight and runs from roughly late August through January. Some breeds, especially those originating in warmer climates, such as the West African Pygmy, South African Boer, and New Zealand Kiko, breed year-round.

Does cycle (come in heat) every eighteen to twenty-one days and remain in heat from eight hours to three days, ovulation occurring near the end of that period. Each doe's heat cycles differ from those of her herdmates, but her cycle generally follows a pattern. If she comes in heat every nineteen to twenty days and stays in heat for about forty-eight hours, unless illness or stress throws off her biological rhythms, you can count on her following this pattern most of her life. Goats don't experience menopause, which means that a doe will continue cycling until she dies. Many does kid into their mid-teens, but geriatric does experience more pregnancy-related problems than do younger does, so it's best to retire them at ten to twelve years of age.

In meat breeds like the Boer, bucks and does often live together, and pasture breeding occurs.

Should your buck live in the herd with your does? If you milk your does and your buck is in rut, definitely not. Does kept with bucks tend to give strong, off-flavored milk. It's hard to market "bucky-tasting" products. In addition, if you're establishing a CAE- or Johne's-free herd, you *must* be present to remove newborn kids before they can nurse their dams. Many other goat owners (myself included) simply want to be there when kids are born to assist if needed. If you don't know when each doe was bred, you could miss the big event, which is another reason to separate the buck from the does.

However, when a buck lives with his ladies, he has time and opportunity to properly court them. He's more likely to breed them as they ovulate, so conception occurs. Does are more likely to conceive with 24/7 exposure to a buck, which is why pasture breeding is the norm in meat-goat production.

If your buck doesn't run in the herd, you'll "hand breed" him, meaning that you'll lead him to the doe or vice versa and leave them in a pen together until the deed is done (several times, in fact). You'll repeat this performance every day until she rejects his advances.

If you don't own a buck, you can breed to someone else's by paying board and a stud fee. Or consider artificial insemination (AI). In addition to avoiding the cost and hassle of maintaining your own males, with AI you can choose top-quality bucks that complement each of your does. Goats can be inseminated using fresh cooled or frozen semen, generally resulting in a 60–65-percent conception rate.

Once does settle (become pregnant), they'll stop coming in heat. A vet can confirm pregnancy via ultrasound, or you can simply assume noncycling does have conceived.

The Waiting Game

Approximately 145–155 days after their last breeding date, depending on their breed, age, and previous production record, your pregnant does usually birth one to five kids.

It's wise to dry off lactating dairy does (take them out of production and allow them to stop producing milk) two to three months before kidding. This gives them time to rest and recuperate before their kids arrive and a new milking cycle begins.

Five or six weeks before kidding, boost pregnant does' C/D T vaccinations, trim their hooves, and worm them (read the labels;

some wormers trigger abortion in pregnant sheep and goats). At the same time, begin supplementing the does' diets with concentrates based on type, breed, and body condition. Consult your county extension agent or local mentor for specific advice.

If your property (or the land where your feed and hay are grown) is selenium deficient, give each doe a Bo-Se (selenium/vitamin E) injection at four or five weeks predelivery. If you don't know, ask your county extension agent or your vet.

If your does succumb to ketosis, it will happen during the month before or the month after kidding. Monitor does' weight, make certain they exercise, and keep treatment supplies at hand.

At least ten days before the first doe's due date, assemble a kidding kit or update the one you already have. If you use individual mothering pens instead of allowing your does to kid out on pasture, clean and disinfect existing pens or set up new ones in a well-ventilated, draft-free area in a shed or barn. Allow 25–35 square feet for each doe and her kids. Bed with dust-free material (sawdust can trigger respiratory problems in newborns) and fit each pen with an elevated waterer and feeder. Don't use 5-gallon food service buckets or other large water containers in which tiny kids can drown.

A week before a doe's due date, clip her udder, escutcheon (the area between her rear udder attachment and her privates), vulva, and tail, especially if you keep dairy, fiber, or other longhaired goats.

Throughout kidding time, keep your fingernails clipped short and filed in case you need to internally reposition a kid. Review kidding information and know in advance how to recognize problems and correct them. Have your vet's phone number programmed and ready to dial.

As the due date draws near, watch for signs that your doe is ready to give birth.

Delivery Day

A week or so before the first expected delivery day, start monitoring those does! Does tend to exhibit the same set of prekidding signals from year to year, but no two follow exactly the same routine.

Most first-time moms begin building (developing) an udder four to six weeks prior to kidding. The average veteran doe bags up (her udder begins filling with milk) beginning ten days before and continuing up to the very day she delivers. When delivery is imminent, does' udders are full and feel tight.

A doe's tail ligaments—the ones stretching from just above the spot where her tail joins her spine to her pin bones (those bony protrusions on either side of her butt)—become more elastic as delivery approaches. Check them twice a day. When they feel soft and mushy, she's roughly twelve hours from delivery.

TRAINING A BOTTLE KID

Training a bottle kid is fun, yet frustrating. We use this method and feeding schedule developed by veteran California Red sheep breeder Lyn Brown. It works just as well for kids.

Sit on the floor with your legs crossed. Place the kid in your lap, facing away from you, sitting on its butt with its legs straight out in front. Cup its jaw with your left hand, open its mouth and insert the nipple, and then steady the nipple using the fingers of your left hand. This keeps the nipple aligned with its jaw and its head in a natural nursing position, essential to keep milk from spilling into the rumen. It will do its best to avoid the nipple, but persevere.

Elevate the bottle just enough to keep the nipple filled with milk; as the bottle empties, you'll add more tilt. With your left hand on its throat, you can easily tell if its swallowing. This is important—you want milk in the tummy, not in the lungs.

Make certain you don't overfeed. "Most people kill their first bottle baby with kindness," Brown explains. "They overfeed it because the baby cries and they think it must be hungry. I know I did. Now I follow this feeding schedule, no exceptions. If our lambs [kids, in this case] cry between feeds, we give them Pedialyte or Gatorade. That won't hurt them as far as enterotoxemia goes while filling the void for them." The following amounts are calculated for full-size sheep or medium-size goats.

- Days 1–2: 2–3 ounces, six times per day (colostrum or formula with colostrum replacer powder)
- Days 3–4: 3–5 ounces, six times per day (gradually changing over to species-specific milk replacer)
- Days 5–14: 4–6 ounces, four times per day
- Days 15–21: 6–8 ounces, four times per day
- Days 22–35: gradually work up to 16 ounces, three times per day

"At about six weeks," Brown continues, "I begin slowly decreasing the morning and evening feedings and leave the middle feeding at 16 ounces until I eliminate the morning and evening bottle entirely (remember, they are eating their share of hay or pasture by now). I continue with the one 16-ounce bottle for about two weeks and then eliminate the bottle feedings entirely."

The kid's upturned head position helps milk flow directly into the abomasum.

A day or two before kidding, many does drift away from the main group, sometimes taking along a grown daughter or a friend. A doe may seem introspective or preoccupied and wander around as though looking for something. During the same period, a long clear string of mucus may trail from the doe's vulva.

From a day to just minutes before giving birth, does usually begin nesting—pawing the ground, turning around, and lying down and getting up again, over and over and over. A doe may stretch a lot or yawn or even murmur to her in utero babies in a soft, subdued voice.

When a doe gets down to business and starts pushing hard, stay calm. Don't help unless she needs it, but be ready to act quickly and definitively if she does. Does can deliver standing or lying down.

The first thing to appear at her vulva is a translucent bubble: the amnion, filled with amniotic fluid (this bubble can safely rupture at any time). As it emerges, you'll see first one and then another hoof, and eventually a little nose will appear. Once the shoulders are delivered, the kid usually plops right out. Strip excess fluid from its nose by running your fingers from below its eyes to its nostrils, or use a bulb syringe to suck it out. Place the kid in front of its dam so she can begin cleaning it. This is important because this is when the doe bonds with her kids.

If more babies are imminent, she'll repeat the process until they're all delivered. Never leave until you're sure the last has arrived! Move each one to the side immediately so it doesn't get stepped on. This is a good time to trim the kid's umbilical cord if it's more than 2 inches long. Hold a shot glass or similar container full of 7-percent iodine to the kid's navel area for several seconds. Make certain the cord is totally saturated, and use fresh iodine for each kid. *Don't omit this step.*

If the cord doesn't stop bleeding, apply a commercial navel clamp about an inch below the kid's belly or tie the cord's end off with clean dental floss.

After her ordeal, your doe will be tired and thirsty. Bring her a bucket of lukewarm water (perhaps adding a dollop of molasses as a pick-me-up), and give her a nice feed of hay. It's important to leave the new mom alone with her kids so that they can bond, but you still have a few more tasks to do.

Caring for the New Kids

Make sure that your kids get the right start in life—that means ensuring that they get the proper nutrients, stay warm, and are protected from disease. Bottle-fed kids have their own special needs. You must also make a decision about whether to castrate and disbud and when.

BOTTLE-FEEDING EQUIPMENT

You'll need proper equipment to feed your kids. We prefer the Pritchard teat: an oddly shaped, soft, red nipple with a yellow plastic base incorporating a flutter valve to regulate airflow. Pritchard teats can be screwed onto calibrated lamb and kid feeding bottles or onto any type of household bottle with a 28-millimeter neck (20-ounce plastic soda bottles are ideal). The first time we use a soda bottle, we measure out an individual feeding, pour it into the bottle, and mark the fluid level with a felt-tip permanent marker. We use a bottle for a few days, toss it into the recyclables, and replace it with a new one. Easy-flow human infant bottles and nipples also work well, as do the other soft nipples sold by farm-supply outlets.

You'll also need a measuring cup and, if you are feeding milk replacer, a mixing bowl and wire whisk. Keep all feeding supplies squeaky clean! Wash and rinse them after each use, and briefly soak everything except the nipples in a weak bleach solution (one part bleach, ten parts water) once daily.

Once a kid has gotten the black meconium (first manure) out of its system and has had enough milk, then the poop will appear yellow. If it makes a blob that sticks, then it has to be removed or the kid won't be able to poop, [which] will cause severe infection and death.

Just slip on a pair of those disposable gloves and carefully pull it off. You may have to wet it down to soften it enough to release. Once it's off, you can apply Vaseline or something that will keep it from sticking again.

—Rikke D. Giles

We have our barn arranged so that every stall has a baby access door to the center runway. That gives kids an escape and gets them socialized with the other kids they will be penning with later. We also have baby-height feeders in the runway so they can eat without having to fight to get to the food. It also gets them used to seeing and being around me. A good 90 percent of my kids are friendly enough to pet all the time; the other 10 percent of them are usually the ones who prefer to stay close to mama.

—Rikke D. Giles

Food, Shelter, Health

When the babies have arrived, milk a stream of fluid from each of the doe's teats to clear any wax plugs. This first milk, a thick yellowish fluid called colostrum, is packed with nutrients and antibodies essential to the kids' survival. The antibodies in colostrum are present for only about twenty-four hours after kidding. A newborn should ingest its first meal of colostrum within two hours; every kid should nurse before you leave. If kids don't nurse on their own, milk the doe and bottle- or tube-feed them the first meal. Once they've tasted this elixir of life, most kids will eagerly seek it for themselves.

Young kids must be kept reasonably warm. Some people install heat lamps above bottle kids' pens, but because these lamps often cause barn fires, their use is risky. A solid-sided pen in a draft-free section of the barn, when deeply bedded with long-stem hay or straw, is warm enough in all but the coldest climates. A comfy, well-bedded doghouse or airline-style dog crate with an old blanket or two draped over the top makes a dandy addition to bottle babies' quarters. Or install a bottle-kid pen in your home. Lift-top wire dog crates and puppy exercise pens with tarps spread beneath them make fine kid housing. Bedded with old, frequently laundered blankets, indoor kids produce little odor. They can even don human diapers. Our boys spent their infancy in our living room!

Kids are susceptible to conditions as diverse as constipation and scours, pneumonia, acidosis, enterotoxemia, floppy kid syndrome, coccidiosis, tetanus, goat polio, and white muscle disease. Learn all you can about these problems before kidding time. Weak kids must be tube-fed until they're strong enough to stand and suckle. This sounds scarier than it is. Ask your vet or mentor to show you how to pass a stomach tube and have one ready in case you need it.

Bottle Kids

Unless they are fostered on another willing doe, orphan and rejected kids must be bottle-fed—and what's more fun than raising a bottle kid or two? To bottle kids, you're Mom, herd queen, and best friend all rolled into one. They will carry that attitude into adulthood. In fact, recreational goat owners routinely bottle-feed for precisely this reason.

Many producers don't have time to bottle-feed, so they give away or sell the kids cheaply to those who do. If you'd like one, put out the word. Contact local breeders and post to regional goat-oriented e-mail lists. Tell local vets and your county agent. You're bound to find a kid or two.

If you have room, two kids are better than one. They'll entertain one another when you're gone, and, if you feed milk replacer, you'll save by buying in larger volume. Kids can be fed individually or in groups, using a rack-type bottle holder or multiple-nipple feeder.

The kids you accept should have fed on colostrum for at least the first twenty-four hours of their lives. If you're called to come pick up newborns and they haven't gotten any colostrum, ask if you can buy some from the breeder. If fresh or frozen goat colostrum isn't available, cow or sheep colostrum will do. Other alternatives include CL-Nanny Replacer Colostrum or CL-Ewe Replacer Colostrum or Goat Serum Concentrate fed with Goat Colostrum Replacer. Don't rely on colostrum "boosters" of any sort; they simply aren't enough to do the trick. If your kids don't ingest real colostrum or a viable alternative during those critical first twenty-four hours (forty-eight hours are better), they'll lack vital immunity. Many such kids do survive, but you must be especially vigilant and get them to a vet at the first sign of illness because they won't have the ability to fight it on their own.

What do you feed bottle kids after colostrum? Goat's milk always works best, but these mixtures using store-bought cow's milk work well, too:

- 1 part dairy half-and-half and 5 parts whole milk
- 1 gallon whole milk, 1 cup buttermilk, and 12 ounces of evaporated milk

If you use commercial powdered milk replacers, *always* buy high-quality milk-based products designed specifically for kids. Soy-based replacers, products designed for the young of other

A baby bottle made from a small plastic soda bottle fitted with a Pritchard teat is an ideal feeding tool.

species, and one-type-fits-all-species milk replacers absolutely *will not* do. If you use a milk replacer, read the label carefully and measure ingredients every time. Haphazard mixing leads to potentially serious upsets such as bloat, diarrhea, and enterotoxemia. Mix only enough replacer for a day. Keep it in the refrigerator, and don't return unused portions to the jug. Buy enough of the same brand to last your kids through weaning because switching products leads to gastric upsets. If you must switch, do it gradually over the course of at least ten days.

Castrating and Disbudding
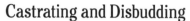

If you castrate male kids, the easiest and least expensive way to do it is to band them before they're two weeks old. A pliers-like tool called an *elastrator* is used to stretch thick, strong, rubber bands wide enough to slip over a kid's testicles. Lack of circulation causes the scrotum to wither and slough off in four to eight weeks. If you use this method, give each kid a shot of tetanus antitoxin (not tetanus toxoid) when you band him; *never omit this step!* However, since penis development ceases when a kid is castrated, authorities believe early-banded wethers are more likely to develop urinary

calculi than are later castrates and bucks. Many recreational and pet owners delay the procedure until kids are at least three months old and then have a veterinarian castrate them under sedation and local anesthesia. Meat producers selling 40–60-pound kids to ethnic markets needn't castrate at all. In fact, some communities pay premium prices for bucklings.

If you don't want horned goats, kids should be disbudded when they're three to fourteen days old. Disbudding is accomplished by destroying emerging horn buds with a red-hot iron; it's not a job for the squeamish or uninformed. Ask a vet or your goat mentor to show you how it's done and what equipment you need, or have your vet disbud kids using a local anesthetic.

ADVICE FROM THE FARM

BOTTLE BRIGADE

Make sure you stick to a schedule when bottle feeding. An overly hungry kid is going to eat too fast. In general, bottle kids, given a chance, drink too much and too fast, and this can lead to bloat. In order to help prevent bloat and other digestive problems, I give my bottle kids C/D antitoxin and Poly-Serum every three weeks; it works.
—Robin L. Walters

Whatever kind of milk you are feeding, be sure that it's warm enough, but not too hot. Body temperature. Use a soft nipple because kids don't like hard ones.

—Pat Smith

Kids don't have to be bottle-raised to be friendly. I have a 300-pound Boer buck that was dam-raised and pasture-bred. When I got him at eighteen months, the only time anyone had put a hand on him was to worm or vaccinate him. He is a big baby and as sweet as can be. He'll even come up to me and duck his head for me to scratch him.

—Robin L. Walters

BUILD A BETTER KIDDING KIT

Usually, kidding goes without a hitch, but glitches can occur, so the goat keeper should assemble a kidding kit to field possible emergencies. We pack our kidding supplies in two containers. The one we take to the barn is a hard plastic step stool with a storage compartment inside. It is sturdy and tip resistant and holds a lot of gear; on cold, wet nights, it sure beats sitting on the ground. It contains:

- **Sharp scissors** to trim the umbilical cord to an inch or so in length. We disinfect them after each birthing and slip them into a plastic zipper bag to keep them clean.
- A **hemostat** to temporarily clamp the umbilical cord if it continues bleeding (disinfected and kept with the scissors).
- **7-percent iodine** to dip the cord into after trimming. Some folks squirt iodine on the navel while the lamb or kid is lying down, but it's much cleaner to dip the navel into a shot glass while the baby is standing.
- **Dental floss** to tie off bleeding umbilical cords if needed.
- **Two flashlights**—we pack a backup in case the first light malfunctions. We tuck them in another plastic zipper bag to keep them dry.
- **Lots of lubricant** for repositioning babies. We like SuperLube and keep two squeeze containers of it in our kit.
- **Betadine scrub** to swab a doe or ewe's vulva before repositioning babies.
- **Shoulder-length OB gloves**—sterile and individually packaged. They're harder to find than nonsterile gloves, but they're worth the search.
- A **sharp pocketknife** so we don't have to use our umbilical-cord scissors for routine cutting chores.
- A **digital thermometer** that beeps.
- A **bulb syringe** designed for human infants. It can't be beat for sucking mucus out of tiny nostrils.
- An **adjustable, rubber lamb/kid puller**. This, the thermometer, the knife, the gloves, and other small items are stowed together in a single plastic zipper bag.
- **Nutri-Drench** (one labeled for goats) for weak newborns and exhausted moms, including a catheter-tip syringe with which to give the stuff.
- An **adjustable halter and lead**—it's easier to move most moms with one than without one.
- A **lamb and kid sling**—a back saver!
- **Towels**—soft, old towels for cleanup are stuffed into any remaining space.

Having once scrubbed spilled lube out of a kidding kit, we make certain the iodine, the lube, and the Nutri-Drench are individually double-bagged.

Our other container is a lidded 5-gallon pail. It stays in the house, and its function is to keep our other kidding supplies centrally located. It houses the following items:

- **Milk replacers** (specifically for kids)—we repackage them in plastic zipper bags and store 3–4 pounds in the container, the rest in tightly sealed tins.
- A plastic calibrated nursing bottle with a Pritchard teat and several spare teats.
- A flexible plastic feeding tube, a felt-tip marker, and a 60-cc syringe for tube-feeding weak newborns.
- A **16-ounce measuring cup**; it's also great to store the spare Pritchard teats in. It's a big one, so it can double as a milking receptacle.
- A **small whisk** for mixing milk replacer. All feeding supplies, including the measuring cup, are stored in a plastic zipper bag.
- Syringes and needles go into another bag, along with an **elastrator** and the **rings**.

Other folks add different items to their kits. What you need depends on your goat's breed and on where you live.

There are three avenues for making money with your goats: meat, dairy, and fiber. Each has advantages and disadvantages, so carefully weigh all factors before deciding which avenue will work best for you.

GOATS

FOR MEAT, MILK, AND FIBER

Meat Goats: The Mortgage Lifter

Farmers and ranchers across the United States and Canada call meat goats "the mortgage lifters" of the millennium. Easy-to-handle, easy-to-raise meat goats are selling at all-time high prices, yet more goats and more goat producers are needed to supply North America's burgeoning goat-meat market with market goats and the breeding stock needed to produce them.

In the 1970s, few Americans ate goat meat. Most North American goats were either dairy goats, Angora goats sheared for their mohair fleece, or Spanish scrub goats used for brush control. The scrub goats were sometimes used for meat, most of which was sold on the hoof and shipped to Mexico. More recent US Department of Agriculture (USDA) statistics tell a completely different story. According to the USDA National Agricultural Statistics Service's first annual goat survey, on January 1, 2005, the US goat inventory totaled 2.5 million: breeding goats totaled 2.1 million and market goats 0.4 million. Milk goats numbered 283,500 head, Angora goats 274,000, meat and all other goats 1,970,000.

Consider this: the USDA estimates that roughly 50–60 percent of America's meat goats are processed at USDA-inspected abattoirs. The first year that goat-slaughter statistics were kept was 1977, when 35,000 goats were tallied. By 1985, the figure had grown to 124,000. Only eight years later, 320,000 goats were accounted for, amounting to more than a 900-percent growth in only sixteen years. In 2000, federally inspected slaughterhouses processed 549,000 goats.

Why the Demand?

The reason for this industry growth is simple: 65–70 percent of all red meat consumed globally is goat meat, and America's expanding ethnic population is willing to pay premium prices to buy it. Families of Mediterranean, southern European, Middle Eastern, African, Southeast Asian, South American, Central American, and West Indian backgrounds all favor goat.

Cabrito (the flesh of 10–15-pound milk-fed kids) and *chevon* (the meat of older kids and mature goats) are favorite fare in Hispanic households. The US Census Bureau projected that between 1995 and 2050, Hispanics will account for 57 percent of the immigration into the United States and that Hispanics will account for 25 percent of the US population by 2050.

Kid goat is the traditional mainstay of Muslim feasts served before Ramadan, at 'Id al-Fitr, and at 'Id al-Adha. The Muslim population in the United States is a significant and growing segment. Muslim converts of non-Middle Eastern origin compose a substantial fraction of the total, and they are said to be particularly observant of traditional Muslim dietary preferences.

Another segment of America's population is clamoring for goat meat: health-conscious individuals turn to naturally lean goat meat for its nutritious qualities. Compared with beef, pork, and lamb, it's lower in calories and fat and equal or higher in protein. Although chicken is slightly lower in calories than goat, it is also lower in protein and higher in fat. People are also discovering just how great goat meat tastes.

Why Are Farmers Flocking to Goats?

At present, goat producers cannot supply enough market goats to meet North America's skyrocketing demand for goat meat. The fact that America's demand for goat meat far exceeds the domestic supply means a ready, established market for new goat producers.

Dairy goats are more angular than meat goats. This Saanen doe exhibits excellent dairy traits.

Other appealing aspects: goat producers can choose from a number of viable business options, depending on what best fits their interests and circumstances. Start-up costs are relatively low, business can be expanded rapidly, and land is not usually a problem.

Finally, goats are intelligent, friendly, and just plain fun to have around. Except for the occasional ornery buck, they're nonaggressive and easy to handle. Almost without exception, the goat producers I've talked to say that the best part of farming goats is the goats.

Business Options

Most commercial producers maintain large herds of unregistered and crossbred goats. Their objective is to produce fast-maturing, low-cost kids for slaughter. Commercial producers market live meat goats (usually by the pound) directly from their farms, through livestock auctions, or to buyers and brokers. Slaughter-goat prices vary depending on age, grade, and availability.

Registered show-goat breeders maintain fewer, but far more costly, registered animals. Their goal is to use popular genetics to produce goats capable of winning in stiff competition at major goat shows. Goats in this group can fetch five-figure prices.

Some breeding-stock producers market high-quality, fast-maturing registered goats of popular bloodlines to other breeders whose goal is herd improvement rather than show-ring victories. Such goats can sell for around $1,000 and up, with quality bucks selling higher than does.

Other breeders produce quality percentage stock of the popular breeds—usually Boer or Kiko—by breeding top-flight bucks to lower percentage does. They market does and bucks to other breeders, high-quality wether kids as 4-H/Future Farmers of America (FFA) show stock, and low-quality and excess bucks for slaughter.

Many goat dairies large and small are entering the commercial meat market by breeding their milking does to purebred or high-percentage Boer or Kiko bucks. Meat kids are raised on excess goat's milk or on milk replacer while their mothers continue working the milk line.

Costs, Expansion, Land

Compared with other livestock ventures, start-up costs for raising goats are unusually low, especially for entry-level commercial meat-goat producers. Good-quality commercial breeding stock is inexpensive and readily available. Although new goat fencing can be costly, in most cases, existing fencing, housing, and handling facilities are easily and inexpensively converted for goats.

It's possible to start small and expand rapidly by retaining doelings and marketing only male kids for slaughter. Routine multiple births (two kids are the norm, but up to four not uncommon) equate with rapid opportunities for herd expansion.

It doesn't take a lot of land to raise goats. Many registered breeding-stock producers—who don't need a lot of goats to show a handsome profit—operate from small farms.

Pastured goats, cattle, and horses prefer different plant species and can be pastured together or in rotational grazing programs. Measured in amount of lean product per unit of input, goats maintained on lush pasture or in a feedlot scenario can't compete with cattle, sheep, or hogs. However, pastured on brushy, weedy, rocky browse, goats top the others hands down. A further advantage: seven to eight goats flourish on the dry feed that a single beef cow would consume.

Goats prefer rough browse. With minimal supplementation, they produce marketable kids on land that would starve beef cattle. They produce marketable meat while improving woodlots and destroying noxious weeds.

With goats, location barely matters. Meat goats can be marketed live from the farm, at livestock auctions, or to brokers who truck large numbers of live goats to goat-slaughtering facilities. Although the majority of federally inspected goat-slaughtering facilities are located in Pennsylvania, Missouri, Texas, Delaware, Maryland, and Illinois, producers have formed marketing organizations in many other areas to cooperatively ship their own goats to slaughter, thus eliminating the middleman and earning additional profits for themselves.

 HOBBY FARM ANIMALS

ORGANIC GOAT MEAT

Health-conscious buyers are willing to pay a pretty penny for premium, "natural" meat. Organic chevon generally fetches the highest prices, but in most locales it's next to impossible to produce.

Why? In a word: parasites. Goat worms are legion, and goat producers must rely on chemical dewormers to keep them in line. But to market organic chevon, a producer must be enrolled in the USDA's Certified Organic Program, in which chemical dewormers, along with antibiotics and most chemical-based medications, are strictly verboten. It's hard to raise goats organically, so thoroughly investigate certification before you commit—and research hardy breeds such as the Kiko, developed specifically for enhanced parasite resistance.

A better approach is marketing "natural" or "grass-fed" chevon. Such kids are raised without growth hormones or stimulants. They can consume browse, grass, and hay but not concentrates. Goats are naturals for this type of meat production.

Keep in mind, however, that unless you're licensed to sell processed meat, you must sell organic, natural, and grass-fed goats direct (alive) to your customers or delivered to a USDA-approved slaughterhouse.

Starting Right with Meat Goats

Don't rush blindly into raising meat goats. Settle on a niche and then do your homework. You *must* have a viable market for your product, no matter what type of meat-goat enterprise you choose.

Talk with your county extension agent. Ask about local marketing opportunities. Contact others in your region who are successfully engaged in the business you choose. Tell them what you have in mind and ask for their feedback. Ask lots of questions. People already involved in goat-raising are your number-one source for local goat-business information.

Subscribe to meat-goat periodicals. Spend time online perusing sites and services; a tremendous amount of meat goat-material is available for free on the Internet. Attend meat-goat seminars sponsored by goat organizations and universities or private seminars such as Suzanne Gasparotto's Goat Camp.

Meat goats are hot, hot, hot. If you can raise them economically and find a steady market, you're almost sure to show a profit. And it'll be a long time before supply exceeds demand.

Dairy Goats: Got Milk—or Cheese?

Goat's milk and goat's milk products are in huge demand. If you have what it takes to succeed in the business, it's likely you can turn a profit milking dairy goats—but you have know what you're doing from the get-go.

Dairy businesses are labor intensive. It may be difficult to find reliable help at wages you can afford. If you can't, can you carry the operation by yourself? With your family's help? Is your family willing to help? Goats must be milked twice a day, at precisely the same hours, seven days a week—they never take holidays or weekends off. And milking is only part of the dairy worker's day. You must feed and clean up after your goats and doctor them when they are sick or injured. You must buy and store feed and bedding. You must sanitize the milk room and milking equipment twice a day. One person shouldering that load can burn out fast.

Don't assume that you can simply sell milk from your own back door. Most states enforce strict regulations governing the sale of fluid milk. Before setting up a business, be it milking ten goats or two hundred, you *must* contact the agency responsible for dairy regulation in your state and procure a license. Agencies vary from state to state. If you don't know where to start, ask your county extension agent or log on to the American Dairy Goat Association's website (www.adga.org), where you'll find up-to-date contact information for every state and several foreign countries.

You'll discover that in most states, to sell dairy products (especially raw milk) for human consumption, you'll have to set up a Class A dairy. For this, you will need a milking parlor, a separate milk room, regulation equipment, and an on-site waste system. The milking parlor must have a concrete floor (or one made of another impervious material); smooth, painted, or finished walls and dust-tight ceilings; approved lighting and ventilation; and metal or other nonwooden milking stands. The floor must slope away from the milk room.

There must be a separate milk room to house your bulk tank and cleanup area, and it must have a tight-fitting, self-closing door leading to the milking parlor. The milk room must incorporate the requisite ventilation, lighting, floors, and walls. A regulation hose port must be installed in one wall to transfer milk from your bulk tank to the milk transporter's tank, and your bulk tank must be installed according to strict agency specifications. A two-compartment wash sink with hot water under pressure is a must.

All milking equipment and the bulk tank must meet strict 3-A manufacturing standards. An approved on-site toilet and a dairy waste management system are also mandatory. For an in-depth look at typical equipment and operating requirements, visit "Grade A Dairy Goat Requirements" at Langston University's E (Kika) de la Garza Institute for Goat Research (www.luresext.edu/goats).

If you plan to sell fluid milk in bulk, investigate prospective buyers before you commit. In many locales, no such buyers exist. If you find

DID YOU KNOW?

Texas is the long-time heart of the American mohair industry and home of the Mohair Council of America.

A Pygmy doe is a good choice if you want milk for your home.

one, contact a representative and ask if he or she needs additional suppliers. If so, how much milk will the buyer purchase from you? What is the payment? How much is charged for hauling? Are you expected to supply milk year-round? Request the names of existing suppliers and contact all of them. Ask a lot of questions. Make sure the buyer is reliable before you buy goats (and expensive equipment!).

Before you sell milk or other dairy products direct from your farm, make absolutely certain it's legal. In most states, it's OK to sell to individuals who use the milk for animal food but not to individuals who use the milk for human consumption. Be careful. Fines for peddling illegal milk are often very steep.

Another legal way to market your good goat's milk is through livestock. Pigs, calves, and meat-goat kids thrive on goat's milk, and all are readily marketable as meat. Or market your milk as value-added products such as delicious, high-quality goat cheese.

When most people think of goat cheese, they visualize chèvre, the tasty, tart, earthy goat cheese from France. But goat cheese runs the gamut from spoonable, silky fromage blanc to mold-coated, Brie-like chevrita to cheddar-like firm, ripened cheeses.

You won't learn to make quality artisan goat cheeses overnight; you need to apprentice with an established cheesemaker, attend seminars, or take courses. By doing so, you can learn to produce an in-demand product marketable at farmer's markets, to restaurants, through retailers, or from home.

Before undertaking any goat-related dairy enterprise, contact Appropriate Technology Transfer for Rural Areas (ATTRA; http://attra.ncat.org). You can read and download many of their publications, such as *Diary Goats: Sustainable Production* and *Value-Added Dairy Options*, online, or you can call and speak with an ATTRA advisor who can compile a free information packet tailored specifically to your needs. Don't overlook this resource! You'll also want to visit Sustainable Agriculture Research and Education's website (www.sare.org). Simply click on "Learning Center," where you'll find access to books, fact sheets, newsletters, and more valuable resources dealing with the processing and marketing of dairy products.

Milk must be handled properly to eliminate "off" flavors. You'll spend time quick cooling, filtering, and probably pasteurizing your goat's milk; to do it, you'll need the right equipment. It's not difficult, but it does take some time and effort. Is it worth it? We think so, but educate yourself and then decide.

Fiber Goats: The Chosen Ones

Two types of fiber are harvested from goats: mohair and cashmere. Angora goats produce mohair, as do some pint-size Pygoras, and almost any breed or type can yield cashmere. Let's take a peek at both kinds.

Mohair

It stands to reason that Angora fiber would be clipped from Angora goats—but it isn't. Angora is the hair of Angora rabbits. Angora goats produce *mohair*, a word derived from the Arabic word *mukhaya*, meaning "the chosen." Sumerian tablets and biblical references set the breed's development sometime between the fifteenth and twelfth centuries BC. The name "*Angora*" is derived from *Ankara*, a Turkish city where the Middle Eastern mohair industry evolved. Mohair became so important to the Turkish economy that none of

WHY MOHAIR? WHY CASHMERE?

- The chemical composition of mohair and cashmere is similar to that of wool, but they have much smoother surfaces, so they lack the felting properties of wool.
- White cashmere and white mohair accept dye exceptionally well.
- Soft, strong, lustrous, and elastic mohair yarn is woven into garments of all sorts. The world's finest teddy bears and similar stuffed toys are crafted of mohair fabric, whereas dolls' hair is often made of natural, flowing locks of mohair.

its precious goats was sold abroad until the sixteenth century AD, when Charles V of Spain brought the first pair of Angoras to Europe. America's first Angoras arrived in 1849, a gift from the sultan of Turkey to a Dr. James B. Davis of South Carolina. Since then, Angora numbers have skyrocketed.

Angoras' manageable size and calm dispositions make them easy to handle. Adult does weigh 70–110 pounds; bucks average 120–185 pounds. Both sexes grow slowly, rarely maturing until they're three to five years old. Angoras breed seasonally. They produce longer than other breeds do (nine to fourteen years is not unusual), and well-cared-for Angora goats can live to eighteen years of age. Compared with other breeds, Angoras have lower birthing rates. Most does who carry a pregnancy to term produce a single kid, although twins aren't uncommon.

Conscientious hobby farmers can earn a considerable income by producing Angora goat fleeces for the handspinners' market, which has taken the country by storm. Because mohair becomes coarser as goats age, highest prices are for kid fleeces. The trick to greater profits is keeping handspinner-quality fleeces free of muck, manure, dirt, and bits of organic matter such as hay chaff, twigs, and burrs; it's almost impossible in a huge flock setting but manageable for the hobby farmer with around thirty or fewer goats. Fleece can be damaged by external parasites, especially lice. And, like most breeds, Angoras are highly susceptible to internal parasites. Angoras must be fed a nutritious diet to produce quality fleece; they are excellent browsers and brushers, but they definitely require supplemental feeding.

Angoras come in pure white and colored versions. Colored Angora fiber is very popular with handspinners, but commercial fiber-buying operations buy white fiber only. Colored breeding stock costs more than white stock does, but all things equal, colored fiber sells at a higher price per pound.

Angoras aren't as winter-hardy as other goat breeds are; they do best in semiarid regions. Angoras can be raised as far north as Minnesota, Wisconsin, upper Michigan, and New England, but suitable winter shelters are an absolute must, especially at kidding time and after shearing.

Angora goats *must* be clipped. Scissors and manual sheep shears can be used to clip small flocks, but larger operators use electric sheep shears fitted with a twenty-tooth goat comb. Angoras are shorn twice a year, generally in March (prior to kidding) and again in the fall, yielding 5–10 pounds of 4–6-inch wavy locks per clip. A well-managed Angora annually yields 20–25 percent of its body weight in mohair, making it the most efficient fiber animal in the world. Angoras are also popular as pets and as 4-H/FFA project goats.

DID YOU KNOW?

Because of their cushy softness, elegant drape, and hefty price tags, cashmere garments have always appealed to affluent buyers. Louis Bonaparte presented the first Indian pashmina cashmere shawl in France to Madame Bourienne in 1799. He started a trend. Emperor Napoleon Bonaparte later gifted Empress Eugenie with a total of seventeen pashmina shawls. Queen Victoria loved these shawls, too; her wearing them made pashmina shawls popular throughout Great Britain. Early nineteenth-century English dandy Beau Brummel was famous for his white cashmere waistcoats. The "original sweater girl," Lana Turner, wore a tight cashmere sweater in a 1937 film called *They Won't Forget*; due to her influence, cashmere sweaters remained popular through the 1950s.

Pygora Fiber

Pygoras are cute, scaled-down fiber goats developed by Oregon goat breeder Katherine Johnson, who bred Pygmy goats to full-size Angoras. Pygora does weigh 65–75 pounds and stand at least 19 inches tall; bucks and wethers tip the scales at 75–95 pounds and are 23 inches or taller. They come in all Pygmy goat colors and their dilutions, plus white. Each Pygora produces fleece of one of three types:

- Type A (Angora type)—lustrous, curly fiber up to 6 inches long. These animals must be shorn.
- Type B (blend type)—a blend of Pygmy goat undercoat (cashmere) and 3–6 inches of Angora mohair. These goats shed and can be shorn, plucked, or combed.
- Type C (cashmere type)—very fine fiber, 1–3 inches long, with no luster. These goats shed and may be shorn or combed.

Cashmere

The "Cashmere goat" is not a breed. Cashmere is the winter undercoat (down) produced by nearly every breed of goat. Cashmere goats are simply goats that produce a bountiful supply of undercoat, especially undercoat in the acceptable micron range.

Fiber diameter is measured in microns; a micron is one-millionth of a meter. Mohair measures 20–50 microns; Merino sheep wool, the finest wool of all, 18–25 microns; alpaca fiber, 15–35 microns; and cashmere, just 12–19 microns, making cashmere one of the cushiest animal fibers on earth.

Undercoat growth begins about midsummer and stops around the winter solstice. Cashmere-producing goats can be sheared prior to their natural shedding time (the undercoat is separated from guard hair by a commercial dehairing machine) or combed as the animals shed. A productive goat yields 4 ounces of guard-hair-free cashmere fiber.

Most cashmere fiber is produced in China, but America's cashmere industry is emerging as breeders realize that Spanish goats, dairy and meat breeds, and even some Boers produce cashmere worth shearing, making cashmere a money-making secondary cash crop.

PART FIVE

PIGS

BY ARIE B. McFARLEN, PhD

WHY PIGS?

Pigs are one of the oldest domesticated animals and one of the most valuable to humans. Today's pigs are descendants of wild boars first domesticated in Asia and Europe several thousand years ago, when human societies shifted from being nomadic hunters and gatherers to being based on settlements and agriculture. Traditionally, the pig served as a primary food source in civilizations around the world, and no part was wasted.

PIGS:
GETTING STARTED

Today, pigs are still an important commodity. Modern husbandry produces leaner specialized swine breeds for cured products such as ham, sausage, and bacon and for fresh cuts such as chops and spareribs. Pigs have important medical uses as well: pig insulin and heart valves have successfully been used to treat human diseases for decades. And, in some places, small breeds have become popular pets.

Hobby farmers interested in raising pigs for profit can do so easily while also producing meat for their own freezers. Many buyers—including gourmet home cooks, professional chefs, and specialty markets—are interested in knowing where their food comes from, the conditions in which the hogs were raised, and the nature and quality of the pigs' diet. People appreciate the ability to purchase directly from the farm.

Pigs can also be utilized to improve the land or complement other livestock production. A farmer can take advantage of a pig's natural habit of rooting to clear brushy, weedy, or rough areas of a property, thus enabling and preparing the area to be reseeded or planted with a valuable crop. Pigs are extremely efficient utilizers of feedstuff and can fatten quickly on the wasted morsels of other animals.

Understanding the basics of pig evolution, biology, and behavior can provide valuable insight into selecting the right breed for your farm and caring for your new pigs.

The Pig's Place in History

Prehistoric drawings of wild boars can be found in Spain's famed cave of Altamira, a dwelling of Cro-Magnon humans some 30,000 years ago. The artwork of ancient civilizations that followed depicted pigs in all sorts of settings, even in scenes with royalty and deities, suggesting that pigs have been familiar, useful animals throughout human history.

The domestication of wild pigs may have occurred first in central and eastern Asia. According to zooarchaeologist Richard Redding at the University of Michigan, 11,500-year-old pig bones have been recovered at Hallan Cemi in southeastern Turkey. Further research indicates that these pigs were domesticated, predating the cultivation of cereal grains. Anthropologists also believe that the Neolithic people of the Peiligang culture in China (7000–5000 BC) raised millet and pigs as their primary food sources. Evidence of this cultivation has been recovered from an excavation site located at Jiahu, led by archeologist Shu Shi. And the earliest known book on raising pigs was recorded in 3468 BC by Emperor Fo Hi of China.

Additionally, the pig has been an important food source in Europe for thousands of years, where it was both independently domesticated and introduced through trade and migration from the Far East. Pork products and lard were used to sustain Roman armies, and owning pig herds was a sign of wealth in Anglo-Saxon Europe.

Despite cultural attitudes that lowered their status, pigs were raised in large numbers by Europeans, who eventually took their pigs to the New World lands they explored and conquered during the Renaissance. There, a pig's adaptability to varying foliage worked against domestication. Many pigs turned loose in the New World quickly adapted to new habitats and became feral hogs, widely roaming the American colonies by the 1600s. Over the centuries, immigrants from all lands brought their native pig breeds with them, and the pig population of North America became a hodgepodge of mixed breeds.

The wild pig of Africa is known as the warthog.

Pig populations spread across the United States as settlers moved westward. By the 1840s, a growing proportion of American pigs were raised on the fertile soil and plentiful corn of the Midwest and Great Plains. With the advent of the refrigerated railcar, pork production in the United States became solidly concentrated in the Midwest.

Through industrialization and the growth of cities during the 1800s and 1900s, the backyard pig and small-scale butchers became a relic of bygone days. Beginning in the early 1900s, the general population gradually gave up producing meat in favor of supermarket convenience and the cost benefits of mass-produced products. The United States Department of Agriculture (USDA) stated that, in 1980, more than 600,000 family and commercial farms were producing pork in the United States. As of 2000, fewer than 100,000 pork-producing facilities remained. This means that fewer but larger farms are producing the bulk of the market hogs, creating niche-market opportunities for the hobby farmer or small producer.

Pigs: The Breeds

There truly is a hog breed for every person and husbandry method. Eight major hog breeds are raised in the United States, plus several Heritage, pet, and minor breeds. All breeds differ in growth rate, litter size, mature body size, time required to reach market weight, and grazing ability.

All major pig breeds and most minor breeds in the United States have their own official breed registries, which maintain pedigrees and statistical data. Registries are the best place to start when researching pig breeds. Registries also provide breeders' lists and contact information, further facilitating your.

In addition to color, size, and personality, pigs also can be selected based on statistical data regarding their expected performance. The Swine Testing and Genetic Evaluation System, or STAGES, was developed by the National Swine Registry to track the most economically significant traits of the breeds as well as to make predictions based on genetic potential. STAGES can track litter size and weights, days to market, back fat depth, and intramuscular fat. The information can be used to compare the potential of pigs from the same breed with the production goals of the farm and help swineherds make educated decisions about which pigs to buy.

Pig breeds can be grouped into one of four categories: commercial, endangered, Heritage, and pet. The breeds in each category have distinct characteristics and are suited to particular husbandry methods, climates, and production intentions.

Commercial Breeds

Commercial hog breeds are commonly seen in large-scale confinement facilities, which are animal feeding operations used primarily to raise pigs to market weight or to farrow sows. These confinement buildings house the pigs throughout their lives. Proponents of confinements claim that this housing method protects the pigs from weather, predators, and disease and allows for easier care, feeding, and management. Commercial hogs are selected for characteristics suited to mass production, such as rapid growth and physical uniformity. Most commercial breeds are hybrids or superhybrids (hybrids crossed with other hybrids) produced in professionally managed breeding programs with the primary purpose of improving growth and reproductive rates. These breeds include, but are not limited to, the Duroc, Landrace, Poland China, and Spotted or Spots. Although considered commercial, these breeds may do well on a pasture-based system.

Duroc

Durocs are red pigs with drooping ears. They are the second most recorded breed of swine in the United States. Developed from the

DID YOU KNOW?

A breed standard is defined as the ideal or required physical attributes an animal must possess. In pigs, breed standards define such things as ear size or shape, body type, color, mature weight, and sometimes production records. Breed standards for various breeds have been modified over the years as preferences and desirable traits have been identified.

A young Duroc sow. The Duroc is one of the United States' main commercial pig breeds.

are the fourth most recorded breed of swine in the United States. Descended from the Danish Landrace and Large White Hog, the American Landrace also includes Norwegian and Swedish stock. Landraces are known for their length of body, large ham and loin, and ideal amount of finish weight. Landrace sows are prolific; they farrow large piglets and produce an abundant milk supply. These traits have designated the Landrace breed as "America's Sowherd," and the breed is heavily promoted in crossbreeding.

Jersey Red of New Jersey and the Red Duroc of New York, Durocs can range from a very light golden to a dark mahogany. Popular for prolificacy (ability to produce large litters) and longevity, Durocs also produce a quality, lean carcass. Most hybrid breeds in the United States include the Duroc, which contributes improved eating quality and rate of gain. A high rate of gain means that the Duroc requires less feed to create a pound of muscle. Breed standard requirements include solid red color; medium-length, slightly dished face; and drooping ears.

Landrace

The Landrace is a white pig. Its ears droop and slant forward so that the top edges are nearly parallel to the bridge of a straight nose. Landraces, noted for their ability to farrow and raise large litters,

Poland China

The Poland China breed had its beginning in the Miami Valley, Butler, and Warren counties in Ohio, developed by crossing and recrossing many different breeds. Poland Chinas were originally bred for two important characteristics—size and ability to travel—because they were driven on foot to market and in some cases were required to walk nearly 100 miles. Today's Poland China hog is recognized as a big-framed, long-bodied, lean, muscular individual that leads the US pork industry in pounds of hog per sow per year. The Poland Chinas have very quiet dispositions with a rugged constitution.

Spotted or Spots

The present-day Spots descend from the Spotted hogs, which trace a part of their ancestry to the original Poland China. A later infusion

---

ADVICE FROM THE FARM

CHOOSE THE RIGHT BREED

Select your pigs based on the way in which you want to raise them. Genetics and previous environment determine a lot about how a pig will behave on your farm. Pick pigs from farms that are raising them in a similar way to what you will be doing.

—Al Hoefling

Pigs come in all shapes, sizes, and colors. Choosing a pig that matches not only your husbandry methods but also your

personal preference is important. Try to be objective with your requirements. Choose the breed that matches your personality, but also choose the best animals in that breed.

—Bret Kortie

When you purchase breeding stock, you should not be far from breeding, I feel. A lot changes in an animal from 50 to 250 pounds. The top animals should be kept for breeding, and that is hard to tell when they are really young.

—Josh Wendland

The rare Guinea hog's history is difficult to trace.

from two Gloucestershire Old Spots boosted the breed with new bloodlines. Spots are good feeders, mature early, and are very prolific, and they pass these characteristics to their offspring. Established in 1914, the National Spotted Swine registry has grown to one of the top-ranking purebred breed associations in the United States. Spots have continued to improve in feed efficiency, rate of gain, and carcass quality, making them popular with both small farmers and commercial swine producers.

Endangered Breeds

Endangered hog breeds are those classified as being in danger of extinction, either because of low numbers or insufficient genetic diversity to maintain the population. Although these endangered breeds are found in limited numbers, many dedicated conservationists and farmers work to maintained and even expand

their populations. The following breeds are included in The Livestock Conservancy's Conservation Priority List.

Choctaw (Critical)

Choctaw hogs are descendants of the pigs brought to the New World by the Spaniards and adopted by the Choctaw Indians of Mississippi. The Choctaw is a small to medium-sized pig, averaging 120 pounds. Physical characteristics typically include erect ears, wattles, and mulefooted (single) toes. Generally black, it may have white on the ears, feet, and wattles. The Choctaw is a long-legged pig, able to range widely for food.

Gloucestershire Old Spots (Critical)

Originating from the Berkeley Valley region in Gloucestershire, England, Gloucestershire Old Spots (GOS) were known there as Orchard pigs. Traditionally, GOS were used to clean up fruit orchards, nut tree stands, and crop residue. GOS are large white pigs with black spots. Their ears are large, are lopped, and cover the entire face to the snout. Mature GOS sows are known for large litter sizes and abundant milk. Full-size GOS will reach 400–600 pounds by the age of two years. GOS are well known as a gentle and sweet-tempered breed. Highly adaptable, they can be raised with a variety of management practices and in varied climates.

Guinea (Threatenend)

A small hog, weighing between 85 and 250 pounds, the Guinea is solid black and hairy rather than bristly. The Guinea is a true miniature pig, not a Pot-Bellied (dwarf) pig, and its body parts are proportionate for its size. Guinea hogs have a sketchy and disputed history but are now considered a unique American breed. Historically, Guineas were recorded as having come to the United

DID YOU KNOW?

The Livestock Conservancy groups breeds into the following categories according to population numbers.
- **Critical:** Fewer than 200 annual registrations in the United States and estimated global population of fewer than 2,000.
- **Threatened:** Fewer than 1,000 annual registrations in the United States and estimated global population of fewer than 5,000.
- **Watch:** Fewer than 2,500 annual registrations in the United States and estimated global population of fewer than 10,000. Included are breeds that present genetic or numerical concerns or have a limited geographic distribution.
- **Recovering:** Breeds that were once listed in another category and have exceeded Watch category numbers but are still in need of monitoring.
- **Study:** Breeds that are of genetic interest but that either lack definition or lack genetic or historical documentation.

A Saddleback mom feeds her hungry brood.

straight faces and snouts. Their large drooping (or lopped) ears nearly cover their faces. Large Blacks average 500–600 pounds and were originally raised in rough conditions, left to clean up residue and fallen fruit and nuts from fields, brush, crops, and hardwood forests. Mature sows average ten to thirteen piglets per litter and produce ample milk to feed them. Large Blacks stay in production for eight to nine years, a highly desirable trait for homesteads.

States via slave ships, or possibly with explorers from the Canary Islands, but the breed's history cannot be proved conclusively. Guineas are known for their friendly disposition and gregarious nature. Guinea hogs are highly adaptable and suitable for sustainable or low-input systems because they are able to forage and graze well, gaining nicely on grass and weeds. Guineas are good mothers, averaging six piglets per litter. Meat from the Guineas is fine flavored, though fattier than most other breeds, making it desirable for slow-roasting meats and adding rich flavor to other dishes.

Hereford (Watch)
Hereford enthusiasts are proud of their breed's appearance, with its colorful red coat and the white markings on the face, feet, and belly. The Hereford was created from a cross of Duroc-Jersey and Poland China hogs by John C. Schulte of Norway, Iowa, around 1920. Herefords have drooping ears; wide, slightly dished faces; and curly tails, with even bodies from shoulder to ham with a slight arch to the back. They can be raised on pasture or in semiconfined conditions. Their color and hardiness are well suited to outdoor production, but they need protection against sunburn. They grow well on a variety of feeds and like to root. Hereford boars are known for their aggressive breeding habits and are very prolific; litter sizes average eight or nine piglets. Full-size Herefords range from 600 to 800 pounds at two years of age.

Large Black (Critical)
The Large Black has intensely black pigment in the hair and skin to protect it from sunburn. Originating in the Cornwall and Devon areas of England in the late 1800s, Large Blacks today can be found in very small numbers in the United Kingdom, Australia, Ireland, Canada, and the United States. Large Blacks have long deep bodies and long

Mulefoot (Critical)
The American Mulefoot is a distinct breed recognized officially since 1908 and recorded since the American Civil War. The Mulefoot is a medium-sized black hog with a soft hairy coat and hooflike, instead of cloven, feet. Mulefoots produce succulent, flavorful, highly marbled meat. Mulefoots tolerate both heat and cold very well and can be raised in nearly any climate; they are excellent foragers and grazers. Adult Mulefoots weigh between 400 and 600 pounds by the age of two years.

Ossabaw (Critical)
The hogs of Ossabaw Island, one of the Sea Islands off the coast of Georgia, are descendants of hogs left by the Spaniards nearly 400 years ago. They have remained a distinct, genetically isolated, feral population ever since. Ossabaws are primarily black with a brown tinge, often with white splotches throughout the body. They have heavy coats with thick hackles, similar to a razorback's. Ossabaws are able to put on large amounts of fat during times of ample feed to sustain themselves through periods when food is unavailable. Ossabaws will hunt small mammals, birds, and reptiles for food, making them very self-sufficient. Their meat is very tasty, with a firm but not tough texture. Ossabaws are relatively large in the shoulder area, yielding more roasts and chops, and they can weigh from 100 to 250 pounds fully grown.

Red Wattle (Threatened)
The Red Wattle hog is believed to have originated in New Caledonia and come to North America via New Orleans with French immigrants in the eighteenth century. Although generally red in color, the Red Wattle may have black markings. The essential identifying feature is the breed's wattles—fleshy appendages of cartilage that hang like tassels from the lower jaw at the neckline. Each wattle is thumb-size

This baby Hampshire pig has the breed's distinctive white "belt."

in diameter and grows from 1 to 5 inches in length. Red Wattles produce a fine, lean meat that is said to have a unique taste, between that of beef and pork. Red Wattles are particularly good in bio-friendly systems in which the hogs are used to turn compost or root up marginal ground. A Red Wattle sow will typically farrow nine to ten large piglets and produce ample milk to fatten them. Red Wattles can be expected to reach a full size of 1,000–1,200 pounds by the age of three when fed a balanced diet.

Tamworth (Threatened)

The first Tamworths were imported to the United States from England in 1882, but the recorded history of the breed dates back to the late 1700s. Considered the oldest "unimproved" breed in England, Tamworths have remained relatively unchanged for the past 200 years. The Tamworth is a red-gold colored pig with a straight, fine, abundant coat. Primarily bred as bacon hogs, Tamworths produce lean and highly flavorful meat. They are heavy rooters, which is handy if a farmer wishes to till rough ground, rustle behind cattle, salvage crops, or raise hogs on marginal ground or in tree stands. Litter size averages about ten piglets, with an exceptionally high weaning rate. Full-size Tams weigh 600–800 pounds.

Saddleback (Study)

The Saddleback is an amalgamation of the Wessex and Essex breeds and is found mostly in England, raised as a fresh pork hog. Saddlebacks have white belts similar to the Hampshire's, with large flop ears and slightly dished faces. The Saddleback also shares several of the Hampshire's production qualities. Saddlebacks are medium-large pigs that produce large litters that grow quickly. Suited for foraging as well as confinement, Saddlebacks can produce well on a variety of feeds.

Heritage Breeds

Heritage pig breeds in the United States are those with a history of production, conformation, documented registries, and breed standards. They have large enough populations to not be considered endangered, and they include the following seven breeds.

Berkshire

Originating in Berkshire, England, the Berkshire is a large black pig with white feet, nose, and tail tip. The Berkshire was imported to New Jersey in 1823, making it the first purebred swine in the United

States. Berkshires were heavily crossed with other breeds to bring about improved rate of gain and hardiness. In 1875, the American Berkshire Association was formed to preserve the purebred stock. Berkshire meat is said to be richly flavored, dark red, and well marbled. Berkshire boars average 500–750 pounds, whereas sows average 450–650 pounds.

Chester White

The Chester White breed originated in Chester County, Pennsylvania, through a combination of Yorkshire, Lincolnshire, and Cumberland pigs from England; the latter two are now extinct. Three registries were combined to form the Chester White Swine Record Association, with registrations dating back to 1884. The breed standard requires a completely white pig with a slightly dished face, medium floppy ears, and a full, thick coat. Chester Whites are known for their superior mothering ability, durability, and soundness. They are preferred by producers and packers for their muscle quality and white skin, which dresses out to a light pink when processed.

Hampshire

This black pig has a distinct white belt known as "the mark of a meat hog" and produces lean meat with minimal amounts of back fat. Originating in southern Scotland and northern England, this breed has been highly developed and utilized in the United States. Admired for its prolificacy, vigor, foraging ability, and outstanding carcass qualities, the Hampshire has grown in popularity and demand. Hampshire females have gained a reputation among many commercial hogmen as great mothers and have extra longevity, remaining in production for up to six years. Hampshires are the third most recorded breed of pig in the United States, indicating that they are popular on hobby farms as well as in commercial productions.

There's no need to wonder how the Pot-Bellied pig got its name.

tail, four white feet, and a white blaze are desirable; ears must be lop or semilop. The breed is renowned for its quality of pork and bacon. Sandy and Blacks are touted as one of the best pig breeds for a first-time pig keeper because of the breed's docile personality, mothering ability, and ease in handling.

Large White

The Large White owes its origins to the old Yorkshire breed of England. Large Whites are distinguished by their erect ears and slightly dished faces. They are long-bodied with excellent bacon and hams and fine white hair. True to the name, full-size Large Whites average 600 pounds. The Large White is a rugged and hardy breed that can withstand variations in climate and other environmental factors. Although developed as an active outdoor breed, it does very well in intensive production systems. Its ability to cross with and improve other breeds has given it a leading role in commercial pig production systems around the world. Sows produce large litters and plenty of milk.

Middle White

Middle White swine originated in the Yorkshire area of England at about the same time and from the same general stock as did the Large White and Small White breeds. The main breed characteristic is the snubbed snout. The white-colored swine are well balanced and meaty. Early maturing, Middle Whites are valued when the object is to produce lightweight marketable pork in a relatively short time. Despite their smaller size, the sows have been found to rear an average of eight piglets per litter. They are good mothers and are known for their docile behavior. The Middle White can make a contribution to crossbreeding programs to improve eating quality.

Sandy and Black

The Oxford Sandy and Black is a British breed once thought to be extinct. The breed has been reestablished in its native England and was recently recognized as an independent breed by the British Pig Association. Although wide variations in color occur, the pigs must be sandy (blonde to red) with black blotches (not spots). A white-tipped

Yorkshire

The Yorkshire (or "Large White" in England, its country of origin) is white in color, with erect ears and short snouts. This is the most recorded breed of swine in the United States and Canada. Yorks are very muscular, with a high proportion of lean meat and low back fat. American Yorkshire breeders have led the industry in utilizing the STAGES genetic evaluation program, amassing the largest database of performance records in the world. In addition to being very sound and durable, Yorks have excellent mothering ability and large litters, and they display more length, scale, and frame than do most commercial breeds.

Pet Breeds

Some people enjoy keeping pet pigs around. Two favored breeds are the KuneKune and the Pot-Bellied.

KuneKune

The friendly KuneKune comes from New Zealand, where the breed has been domesticated since the mid-1800s. *Kune* means "fat and round" in Maori. The KuneKune's body is round and sturdy, with short legs, an upturned snout, and two tassels hanging from its lower jaw. Kunes are smaller than commercial breeds of pig, usually no heavier than 260 pounds, and they have a very good temperament. Kunes fatten readily on grass and are considered a grazing pig rather than a forager. Kunes produce a quality carcass although fattier than that of a commercial pig breed.

Pot-Bellied

Developed from the "Í" breed of Vietnam in the 1950s, the Pot-Bellied pig is a dwarf breed. It is usually black, with loose folded skin and thin hair. The breed has a deeply wrinkled face, a short snout, a low-to-the-ground hanging abdomen, and disproportionately short limbs. Most people who purchase Pot-Bellieds keep them as pets, although—like any pig—the Pot-Bellied can be eaten. In

A Tamworth/Berkshire-cross piglet has elements of both breeds' coloration.

comparison with other breeds, it is a much fattier pig, with an average dress-out of 36 percent meat and 54 percent fat (the remainder being waste). The average adult Pot-Bellied pig weighs between 100 and 250 pounds.

How to Buy

Don't rush into buying pigs. Although keeping pigs is relatively easy, it will serve you well to learn all you can about the care, feeding, and handling of pigs before you add them to your farm.

Spend some time at the library or bookstore, and study reliable online sources. Try to find a mentor or local swine breeding groups, and talk with your local veterinarian or county extension agent. If possible, visit other farms that raise pigs and ask questions. Most people love to talk about their animals and enjoy sharing their stories. Be considerate of their time and express your gratitude for their help.

First, you must decide if you are going to raise a couple of pigs for meat in your freezer or if you want to breed and raise pigs for market. If your primary goal is pork for personal use, buying two weanling barrows (castrated males) to feed and butcher is a good way to start and will be a good learning experience: you will gain invaluable knowledge about handling, feeding, and caring for pigs with little investment or risk. Starting with spring-born weanling pigs will make things even easier because you will have the pigs in the freezer before winter, which will eliminate the need for winter housing and other husbandry issues.

If your goal is to breed, raise, and market your own pigs, you should start with the absolute best breeding stock you can get. If the initial cost of the pigs is a deciding factor, start with fewer pigs of higher quality as opposed to more pigs of lesser breeding. Pigs that conform to their breed standard and have proven production records are the best choice for both short-term and long-term success.

Choosing the Breed

Within certain limitations of your farm and production purposes, you will pick your pig breed based on your personal preference. With that in mind, following are some recommendations about choosing the best breed for your farm.

In addition to being grouped by their official breed names and categories, many pig breeds are further classified as one of three

---

ADVICE FROM THE FARM

CHOOSE THE RIGHT STOCK

When choosing pigs for your farm, look for pigs that show an interest in you, come up to see who you are, or are friendly. Do not select pigs that are intensely frightened or do everything they can to get away from you.

—Al Hoefling

Don't buy the runt or a sickly pig, thinking that you can rehabilitate it or save it. Runts and sick pigs almost never catch up to the others, and many of them don't survive. Start with the best pig from the litter.

—Bret Kortie

Whenever you look to buy an animal, watch it walk: it should walk fluidly, not stiffly, and stand on its legs properly. The legs are important. Front legs should be straight and strong; back legs should have good bone.

—Josh Wendland

Visiting the breeder allows you to see the little ones and how they are cared for.

types—lard, bacon, or pork—depending on their primary use. Lard pigs have a higher percentage of fat overall and a distinctive shape, known as *chuffy*. Chuffy pigs look like sausages, with four short legs and rounded or arched backs. Lard, used as a dietary source of fat, was once the main reason for raising pigs, and meat was the by-product. Nearly all lard pigs fall in the Heritage or endangered categories.

Bacon hogs have longer legs, longer bodies, and greater depth of rib. This depth of rib is the area that bacon comes from, and pigs with larger sides produce more bacon. Some Heritage and endangered breeds are the bacon type, but most bacon hogs are commercial breeds.

Pork pigs tend to be leaner and grow faster than other breeds; they are used primarily for fresh cuts of pork. Most pork pigs fall in the commercial category, although excellent fresh pork can be had from Heritage and endangered breeds.

If you have a clear preference for fresh cuts or cured meats, select a pork or bacon type, respectively. If you have no clear preference, consider a multipurpose breed: a pig that can be used for lard and for fresh and cured meats. Indeed, many breeds, as well as many hybrids, are suitable for all purposes, although lard will not be prevalent in the modern commercial breeds.

You must also decide whether you want to raise purebred or crossbred stock. As with the breed choice, this decision depends on what you ultimately want to do with your pigs. People select purebreds to breed pigs, raise piglets, and sell breeding stock. Many farmers are also concerned with maintaining and preserving their chosen breeds. By raising purebred pigs, you are keeping your chosen breed in production and reducing the likelihood of extinction. Purebred pigs can always be sold as meat, but crossbreds cannot necessarily be sold as breeding animals.

If your goal is pork for your own use, consider that different purebred varieties have distinctly flavored meat, so you should select based on your taste preference. When evaluating breeds for home consumption, purchase some meat from the breeds you are considering to evaluate and compare the flavors.

Purebred pigs typically are registered or eligible to be registered. Registrations from an appropriate breed association can tell you a lot about your pig: its birthdate, parentage, and owners, as well as any championship titles in its pedigree. Contact these associations and visit their websites for breeders' lists and detailed information about the breed you are considering.

Crossbred pigs are of mixed-breed parentage and typically cannot be registered. Crossbred pigs certainly have their place on farms across the land. They are generally raised as market hogs, capitalizing on their rapid growth. Crossbreds of two purebred parents have what is called *hybrid vigor*. This means that the offspring of parents with dissimilar genes have the virtues of both parents to counteract genetic weaknesses, resulting in large, usually fast-growing pigs. If your porker is to be used for meat, a crossbreed is a less expensive and perhaps more efficient purchase. If you are raising a pig or two each year for freezer meat and do not intend to go into farrowing and raising piglets yourself, crossbred pigs will satisfy your needs.

Where to Buy

Pigs can be purchased from a variety of sources including breeders, small farms, and livestock and exotic animal sales. Ask other farmers if they can make a referral. Many feed stores have bulletin boards listing animals for sale, or you could place your own "pigs wanted" ad. Breed-specific or livestock newspapers and newsletters also frequently post ads for animals for sale.

Buying from Breeders

The best way to buy pigs is directly from the breeder. Careful screening is important, however, so contact a number of breeders and ask questions before you agree to any purchase. Reputable breeders should be willing to describe their operations and affiliations. Ask about their breeding program, how long they've been raising pigs, how the pigs are raised,

Piglets often have playful personalities.

and what illnesses and diseases they vaccinate for. Ask how the pigs are being fed to help you determine whether a particular breeder's stock will do well under your production method. Ask for references and get feedback from other customers to be sure you are dealing with an experienced, knowledgeable breeder.

A breeder truly interested in the welfare of the pigs will screen you, too! Be prepared to tell the breeder what you will be doing with your pigs and how you intend to raise them. If the breeder sees a potential mismatch, he or she should be willing to steer you toward a more appropriate breed.

An experienced breeder should also be willing to help you get started by offering specific guidance in raising your pigs. If you are purchasing purebred stock, the breeder should explain which association your pigs will be registered with. If you are starting with piglets, the breeder should offer information about weaning, vaccinations, and future breeding.

Visiting a breeder gives you the opportunity to view the herd, the breeder's management system, and—most important—his or her style of handling the pigs. You get to see firsthand how the pigs behave under their normal living conditions. Are the animals well cared for and the facilities reasonably well maintained? Are the droppings of the animals healthy, or are they pasty or watery? Do the animals appear to be well nourished?

It will be harder to find a local breeder of Heritage breeds, meaning that you will have to either travel to pick up your pigs or have them shipped sight unseen. If you buy pigs from a breeder whom you have not met, you have to rely even more heavily on his or her reputation. Again, get references.

Stockyard and Livestock Sales

Buying pigs at stockyard and third-party livestock sales is not recommended, especially for first-time buyers. Animals sold through such sales are typically culls from a herd, defective or sick. If they were not sick when brought to the sale, they can become sick from exposure to other animals at the sale. Usually, such sales do not afford the purchaser the opportunity to talk with the previous owner or get any information on the pigs' history or health.

The Selection

Once you have determined your breed and contacted some breeders, be prepared to evaluate the specific pigs offered for sale before you make your selection. Pigs should be evaluated on conformation, type, disposition, and health. If you want to purchase proven breeding stock, look at their production records.

Conformation

Pigs should be true to their type. Although it is difficult to exactly predict a pig's adult size by examining a piglet, some attributes can be evaluated. Piglets should be active, alert, and curious. They should have distinct muscle definition by weaning age, which is approximately eight weeks old, and be well rounded but not fat. Avoid runts or skinny or lethargic piglets.

Breeding-age pigs should be well filled out without being fat. Proven sows should have good teat placement (not pendulous or malformed), with at least ten teats on Heritage breeds or twelve teats on commercial breeds. Adult boars should have obvious testicles. If a boar has tusks, consider having him detusked before bringing him home.

All pigs should have a healthy coat and no bald patches or signs of skin conditions or parasites. Check feet for deformities, injuries, or excessively overgrown toes. Tails can be curly or straight.

Health

Never knowingly buy a sick pig. In many cases, careful visual inspection for symptoms will help you rule out a potential purchase. For example, pigs should be free from runny noses and eyes. Avoid pigs that cough, wheeze, or have diarrhea, as well as those with abscesses, lumps, noticeable swelling, or joint problems.

Your new pig will enjoy a chance to explore his new home.

In addition to doing your own inspection, make arrangements for a veterinarian to check out your potential purchase before you buy. The veterinarian can suggest blood tests that will confirm or deny the existence of diseases in the pig, check for parasites through feces and on the skin, and evaluate the animal's teeth, bone structure, foot health, and general quality. Furthermore, most states require some sort health certificates by a licensed veterinarian for pigs transported across state lines. Some states require blood tests to verify that the pigs are free from certain diseases that could be transmitted, such as pseudo-rabies, tuberculosis, or porcine reproductive and respiratory syndrome (PRRS).

The Sale

When you pick up your pigs, be sure to get a bill of sale identifying the animals you have purchased, their registration status if they are purebreds, the purchase price, and any guarantees the seller has claimed. If your pigs are registered, ask for the certificates of registry. If it is your responsibility to register your pigs, get a completed and signed application for registry.

Getting Your Pigs Home

Consider carefully how you want to have your pigs transported to your farm or if you want to pick them up yourself. Although many livestock haulers will deliver pigs for a fee, hiring a carrier that regularly moves pigs can expose your pigs to diseases during transport.

If you use a carrier, try to find one whose usual cargo is horses. Horse carriers typically are not exposed to any diseases that can be transmitted to pigs. Furthermore, most horse carriers are experienced handlers of livestock and may be more gentle in

handling your pigs than are carriers who typically haul pigs to the stockyard or slaughterhouse.

Don't just shop for price; remember to check references. If at all possible, look at the vehicle and trailer that will be used to haul your pigs to be certain that they are in good and safe condition.

If you are purchasing your piglets from a long distance away, arrange for the breeder to ship them via airline in approved dog crates. This is a relatively simple process and a very good way to reduce the risk of exposing the animals to contaminants or illness. Furthermore, airline rates for long-distance shipping are often more economical than are overland shipping rates. However, airlines permit only piglets or very small pigs to be shipped in regular pet carriers. Each airline is governed by the USDA standards for animal safety and may have additional specific requirements for shipping.

Large pigs can be transported by airline in custom-built crates, but they are shipped through the cargo department, not at the passenger counter the way a pet is shipped. This is not an economical way of getting only a few pigs to their new home.

When shipping pigs of any size, be prepared to pay for health certificates, any health tests required by your home state, the shipping crate, and the freight charge. Only use a known person, reputable carrier, or major airline for transport.

You can, of course, make the trip yourself to pick up your pigs. If you are transporting the pigs yourself, have proper bedding and water available for the trip. If you are traveling a long distance, bring feed, too.

Welcoming Your Pigs

Have your facilities ready to receive the pigs upon arrival. Isolate your new pigs from any pigs already on the farm. Worm and vaccinate them right away if this has not been done just prior to transport.

Provide water immediately. Allow the pigs to acclimate to their new surroundings for a short time before feeding them. When moved into a new place, pigs tend to jump around, spin, and trample over any obstacles they are not familiar with.

Feed your new pigs the same type of feed they have been eating. Be sure to inquire about this from the seller. You may be able to purchase a few days' worth of feed from the seller to start your pigs off at home. If you change the feed, begin mixing the old feed with new feed on day two, and gradually increase the proportion of new feed to old feed to help the pigs adjust to the change in diet.

It's the big day, and you are bringing your pigs home. Having a well-prepared pig house and fences will save you many hours and headaches once your pigs arrive, so planning ahead is a must.

Hundreds of years ago, areas were fenced to keep pigs out, not in. Pigs were left to roam the property and were prohibited from entering gardens or living areas that they could root and destroy. Today, most farmers need to fence in or otherwise confine their pigs to prevent them from trespassing on others' property. Pigs are crafty creatures, though, especially if they perceive food within reach. Careful site planning and materials selection will keep your pigs happy and safe

SHELTER AND FOOD FOR YOUR

PIGS

Selecting a Site

Avoid locating hog pens in any area where the ground becomes excessively muddy during rainy seasons. A slight slope will help drain away rainwater and urine and keep your pigs more comfortable. You, too, will appreciate a dry, slightly sloped site at chore time.

A herd of 200-pound pigs should be allotted at least 20 square feet per pig, or approximately ten pigs per acre. If you build an indoor/outdoor pen, design it so that 40 percent is indoors and 60 percent is outdoors; that is, each pig should be allotted at least 8 square feet indoors and 12 square feet outdoors. A sow with a litter needs at least 35 square feet, and a herd of sows should be limited to four sows per acre (if open grazing). Remember that these guidelines are absolute minimums. Pigs grow quickly, and crowding seemingly happens overnight. With crowding, you invite illnesses, fighting, boredom, and unpleasant habits. Give your pigs as much room as possible.

Fencing

If you raise your pigs outdoors, fencing is a primary concern. Hog fences should be at least 32 inches high for smaller pigs and up to 60 inches high for larger pigs. Fencing can be constructed of many materials, including wood, hog or cattle panels, pipe, field fence, electric wire, and barbed wire. Each has advantages and disadvantages, including cost, ease of construction, and longevity.

Before you purchase fencing materials, prepare a map of the area you wish to fence, including all measurements and post locations. Also carefully consider your gate placement and the direction in which your gate(s) will swing. Once you have your measurements, you can compare the costs of different types of fencing and the anticipated longevity of each. Constructing a fence that costs a bit more to build but will last many years longer is a sensible investment.

Wood Railings

Traditional pigpens or sties were made of wood rails, and if you keep your pigs on a concrete pad, then wood railings are quite suitable. Old concrete pads, such as those used for grain bins or previous barn sites, can be recycled effectively for pigpens. Wood used for pigpens should be at least 2 inches thick for strength and mounted on the inside of the posts. Lumber should be untreated because pigs tend to chew their pen materials and may be poisoned by wood treatment. Some manufacturers now offer nontoxic treated lumber, as is used in children's swing sets.

Wood typically does not last as long as does pipe or field fence, depending on the activity level of your pigs. If you have a ready source for wood railing, it may be the most cost-effective material for your pigpens. For areas that are wood-poor (such as the Plains states), wooden pens may be more expensive to build.

Hog Panels

An existing concrete pad also lends itself well to being fenced with hog panels. Hog panels are welded-wire fence panels, typically 33–54 inches high and 16 feet long. These panels have graduated openings, with small openings at the bottom and larger ones at the top. This helps keep piglets in and adults from climbing the fence.

KEEPING PIGS OUTDOORS

White pigs need to be gradually introduced to outdoor living, as they will sunburn easily. The sunburn causes the body to release prostaglandins, which will cause abortions [in sows].

—Robert Rassmussen

The most beneficial thing from raising pigs outdoors is that the fecal matter is spread out over a greater area, [which] reduces the manure and parasite load.

—Josh Wendland

Rotationally grazing pigs will give them the benefit of fresh air and sunshine and will virtually eliminate odors. Grazed pigs are happier pigs because they can behave as they would in the wild, moving about or lounging around at will.

—Bret Kortie

Hog or cattle panels are relatively inexpensive and sturdy. Farm supply and even home-improvement stores should have them available. A 16 × 16-foot pen, constructed of four panels, can be installed in less than one hour.

Hogs raised on dirt can also be penned with hog panels. Panels should be attached to wood or metal posts sunk into the ground at least 3 feet. Panels should be set at ground level or below: dig a trench, set the panel into the trench, and bury the bottom 6 inches of the panel. If you bury your panels, purchase the 54-inch height. Even very small piglets can get over shorter fences if they want to. Alternatively, mound up and pack dirt about 1 foot high on the outside of the fence.

A disadvantage of hog panels is that they do not stand up to excessive climbing. The welds will eventually break, creating a hazard for the pigs as well as an unattractive fence.

Pipe

Pipe is another effective fencing method. It will last for many years, and it cannot be chewed. Pipe needs to be welded to sturdy wooden or pipe posts. In areas with widely fluctuating temperatures, mount horizontal pipes with floating brackets to allow the pipe to expand and contract within the bracket without breaking the welds. If you have pipe available, you can make a very stout pen using recycled pieces. Used pipe is available through metal salvage yards. Depending on the cost of steel, pipe may be the least expensive option for some farmers. Be sure to select pipe that is sufficiently large in diameter to withstand the pressure and weight of your mature pigs.

Field Fence

Field fence, or woven wire, makes a very sturdy fence that will last for many years, but it also may be the most expensive type. A roll of field fence is 330 feet long and can fence an approximately 80 × 80-foot pen. Fencing must be secured on posts set about 8 feet apart, with wooden cross-braces set at the ends. Field fence comes in different gauges (weight of the metal wire used) and is priced accordingly.

Field fence can be challenging to install because it is very heavy and flexible until secured to the posts. If you are fencing a large area or a pasture for your pigs and feel certain that the fence will remain in place for many years, then field fence may be your most economical choice.

Electrified Fence

Pigs train well to a hot-wire or electric fence. If you intend to pasture your pigs and move them to fresh pasture regularly, an electric fence is a very effective method of keeping your pigs contained. Two hot wires set at 6 inches and 18 inches off the ground can contain pigs of most sizes. An additional wire at 30 inches high will keep in larger pigs. Each wire should be energized by a separate charger.

A perimeter fence is still recommended in anticipation of fence failure, power outage, or adventurous pigs. You should check your fence every time you feed or look in on the pigs to be certain that the charger and wire are fully functioning. Pigs can quickly discover a nonfunctioning wire and take advantage of the situation.

A hot wire is also effective in keeping pigs from rooting under other types of fencing. A single wire set 6–8 inches off the ground, 1 foot in from the existing fence, can prevent your pigs from digging out.

Keep in mind that pigs are very smart. Once they have been zapped by the wire, they will stay back at least a foot or two from the fence. This additional area should be considered when establishing pen size because it will make the utilized area of the pen much smaller. When using hot wire, provide a nonelectrified gate. This will make moving your pigs much easier, as they will not willingly cross an area that was previously barred with a hot wire.

Barbed Wire

The best use of barbed wire for hog fencing is only as an adjunct to existing fence. For example, placing a strand of barbed wire 6 inches off the ground on the inside of an existing fence may prevent digging. A strand across the top of hog panels or a wooden fence may keep the pigs from climbing.

Give your pigs a suitable mud hole within their designated area.

Barns, Huts, and Pens

You must provide adequate shelter for your pigs. A shady spot in summer and a shelter from the wind and the snow in winter are the minimum requirements. Shelter need not be elaborate or costly, however. Tree stands provide great shelter for hogs during all seasons. Simple A-frame structures, mini hog huts (Quonsets), or a barn with an open door are sufficient for shelters.

Pigs and Mud

In warm weather, pigs need access to clean mud to reduce their body temperature. If pigs become overheated, they will make every attempt to create a mud hole by tipping their water dish or digging deep enough to reach cooler ground. Neither of these habits is beneficial to pastures or paddocks.

Give your pigs and your property a helping hand by determining the spot where a mud hole will be situated and dumping water there for them. A sprinkler set up just above the area of your choice is also a good method. Be sure to place it over an area that can handle water or a mud hole you have selected where you don't mind water collecting. Running the sprinkler a couple of times a day will give the hogs a sufficient cool down.

A concrete wallow is a very shallow swimming pool intended for water, which may be more sanitary and manageable than an actual mud hole. To prevent drowning, the wallow should be no more than 1 foot deep for adult pigs and 3–6 inches deep for smaller pigs, with sloping sides to allow the pigs to get into and out of it easily. Equip the concrete mud hole with a drain for easy cleaning: prior to pouring the concrete, install a standard tub drain connected to PVC pipe extending beyond and draining away from the mud hole. Be certain that your drain releases the dirty water onto a slope away from the mud hole and onto gravel or well-drained soil to prevent creating a mudslide. Gravel around the mud hole will help maintain the area and provide drainage.

Give some thought to placement of the wallow for maximum utilization. Concrete mud holes are not practical for every pen because they can be expensive to build and will likely allow access to only a few groups of pigs at a time.

Plastic kiddie pools are great fun for pigs as well if they are not terribly destructive. As with the wallow, do not fill the pool more than 1 foot deep for adult pigs. A kiddie pool is not recommended in areas with small piglets because the straight, rigid sides will prevent piglets from getting out.

A-Frames

A-frames, or *arks*, are simple structures, usually made of wood, with steeply sloped sides that allow for drainage and snow shed. Another advantage of the A-frame's shape is that the slope provides cozy interior areas where piglets can lounge without being lain on by the sow. A-frames can have elaborate door structures or just an opening no more than half the size of one end. A burlap bag or similar cloth can be stapled across the top of the opening to cut down on drafts. Wooden floors are typical in A-frames, although a stout base can be left open, allowing for drainage under the hut. Wintering in A-frames is best done with a floored unit.

Quonset Huts

A Quonset hut is a corrugated steel structure shaped like an inverted U. One end is walled in, with an adjustable vent at the top. The other end is left open or partially walled to make a doorway. Quonset huts are fairly lightweight and easy to move; they need to be secured to the ground with rebar rods. Quonsets can also be outfitted with many extras, such as a guard rail around the bottom to provide a safe place for piglets, where mom can't get in and accidentally lie on them.

HOSE 'EM DOWN

A simple garden hose can be used to cool down your pigs in place of mud. Once they figure out that you are cooling them and not chasing them with the water, they will come running to get their showers. Many pigs will lie down as soon as the water hits them and turn their faces directly into the stream.

A small Quonset can provide shelter for several pigs at a time.

Existing Structures

Nearly any existing barn can be retrofitted to accommodate pigs. Old stables, lambing jugs, cattle pens, chicken barns, and the like can be modified to provide adequate or even optimal hog facilities. Horse stables usually require little more than adding a few extra boards along the bottom of the pen. Lower ceilings will provide more warmth in winter. Pigs do not like to traverse steps, so a ramp or opening directly into their pen is advisable.

Doors should be a consideration when building pens for pigs. A door can serve as a traffic control or stop when moving pigs around. In addition, the way a door opens can make cleaning chores easy or exasperating. Consider a swinging door that has removable hinge pins on each side, allowing the door to swing left or right. Secure locks, gate hinges, or hooks prevent the pigs from opening the gate by lifting, climbing, or rubbing on it.

Building Your Own Structure

If you are considering raising large numbers of hogs indoors, then building dedicated, customized pig housing is advised. Remember that pigs do best when they get fresh air and sunshine, so a barn with an outdoor exercise area is highly recommended. Evaluate designs for function, ease of feeding and moving the pigs, comfort of the pigs and stockperson, and cost. Many small-scale plans are available through county extension offices, USDA bulletins, and private companies that manufacture and build hog units.

Site location for such a unit is critical for several reasons. You will likely need to move pigs in and out, and you will need access to the building with trailers, trucks, or crates. If your feed is delivered, the delivery truck will need to be able to get close enough to the storage area to deliver feed. A well-drained site situated far enough away from a obstacles will make the building highly functional.

MOVABLE SHELTER

A-frames and Quonsets can be fitted with pens or small fenced areas to allow the pigs some exercise while containing them for various reasons. Either could be mounted on skids or wheels and moved about to allow access to fresh pasture as needed.

Bedding and Ventilation

It can't be said enough that pigs need dry bedding at all times, whether they are on pasture or in houses. Dry bedding makes the pigs more comfortable and keeps them warmer in winter months. Bedding also absorbs moisture and odors, making the whole environment more pleasant. If you are using portable huts, the bedding can be placed in the hut and then spread out over the pasture when the pigs are moved to a new area. Bedding dispersed over an area will provide compost and break down much more quickly than if left piled up.

A thick layer (about 4–6 inches) of wood chips or sawdust covered with a 2-inch layer of straw makes fine bedding. Other materials for bedding include old (not moldy) hay, wheat middlings, rice hulls, cedar chips, cornstalks, and commercial pellet bedding. Pigs like to nibble and chew on bedding, so a nontoxic material is a must. A decent layer of bedding will keep the pigs dry and free from odors and will also make stall cleanup easier.

Ventilation is likewise a vitally important issue in housing pigs. Proper ventilation moves air without causing a draft on the pigs and keeps moisture and ammonia odors from building up. If you choose to house your pigs, opening a door a few hours per day may be enough to solve your ventilation issues. Windows can provide light and ventilation if installed a foot or so above the height of the pigs. Fans installed in the peak of the ceiling will also help circulate air without causing drafts directly on the pigs.

Adequate water and a proper diet are two main factors in your pigs' overall health. Water and feed should be given in separate dishes, and the pigs should have access to fresh water at all times.

Water

Water is the most important staple in a healthy pig's diet. Under moderate weather conditions, a pig will consume ¼–½ gallon of water for every 1 pound of dry feed. During hot weather and while lactating, a pig's water needs will increase to ½–1 gallon of water per 1 pound of feed.

All pigs have individual personalities. Get to know them as individuals, and working with them will be much, much easier.

—Bret Kortie

By hand-feeding and watering your pigs twice per day, you give yourself the opportunity to look over your pigs and pens and recognize and correct any problems right away. This will save many pigs as well as equipment and money.

—Robert Rassmussen

Check out any pig that doesn't come up to eat. Any animal standing off by itself is not a good sign. Look for the source if any pigs are coughing, sneezing, or such. Dusty pens should be watered down to keep the dust down. If [a pig] can be treated and remain with the herd, that is best, as no animal wants to be isolated.

—Josh Wendland

There are several methods of watering pigs. The type of waterer or dish you choose will have a large impact on your chore load, so choose carefully based on your setup.

Water Dishes, Troughs, and Nipple Drinkers

The water dish is perhaps the simplest watering method. Made of rubber, metal, concrete, or durable plastic, water dishes can be moved easily or tied to a post or fence. Remember, however, that pigs love to turn things over with their noses, and a water dish is a favorite target. If you don't mind filling the water dish several times per day, this may be the right choice for you.

A trough is a more efficient choice if you are watering several pigs. A trough can be made by splitting an old hot-water tank in half lengthwise and either welding legs to it or anchoring it to some cinder blocks or 4 × 4-inch blocks of wood. Ready-made troughs are available from several manufacturers of livestock equipment and can be used as is and moved easily or bolted to concrete or otherwise permanently anchored.

A nipple drinker is another watering option. A piglet drinker is a 1-gallon tank with a nipple on the side. The nipple is a metal valve that releases water when pressed in with the pig's nose. This is a fairly sanitary way of watering small pigs because the animals are not able to step in or play in the water. You must be certain to anchor it securely to the side of the pen or a post because the constant pushing on the nipple will wiggle the unit loose. As the piglets grow, progressively raise the drinker to always keep the nipple at the same height as the piglet's back.

A nipple drinker can also be installed on a ready-made pipe fitting that is prethreaded to accept the drinker. You then hook the top end to a permanent water line or pressurized garden hose. These are really nice in the summertime or for heated barns but are not suited to outdoor facilities or unheated barns in the winter because the lines will freeze and break.

Watering Tanks and Automated Waterers

Hog watering tanks are sold in a range of sizes by several manufacturers. They typically hold between 35 and 250 gallons of water, and come equipped with a trough cut into the side. These automatically refill via a float valve. Tanks can be equipped with heaters for winter months.

Most watering tanks are challenging to keep clean. The drinker (trough) portion is set into the side of the tank, making it nearly impossible to clean out without emptying the whole tank and turning it upside down. In certain environments, such as on concrete or indoor barns that are not bedded, watering tanks are ideal. In dirt situations, the drinker portions tend to fill up with sediment because the pigs basically wash their faces in the drink well while getting to the water.

Automatic heated waterers are the most expensive but least labor-intensive method for watering pigs. Such a waterer is permanently installed on a small concrete pad complete with plumbing and electricity. The water flows automatically via a float valve. Most come with a flip lid that the pig lifts up while drinking. This lid prevents most contamination from feces or bedding.

Feeding

Many homestead hog raisers support feeding pigs on the ground. Scattering the feed around the pen encourages the pigs' natural behavior because they move about, looking for morsels, just as they would in the wild. If your pens are kept relatively clean, and you feed in the feeding area—not the bed or toilet areas—then this is a very practical method. Only large grains, cubes, or pellets work sufficiently, though. If you are feeding finely ground feed, you will need to provide troughs to avoid waste.

Feeders
Feed Dishes and Troughs

Feed dishes have the same drawbacks as water dishes—pigs love to overturn them, so heavier is always better. Metal troughs are durable, are relatively easy to clean, and can be secured to a fence or wall. Troughs must be divided into sections if used for more than one pig, however, to prevent the dominant pig from standing in the trough or pushing the other pigs out.

Some pigs have other uses in mind for the trough.

Wall Feeders and Bin Feeders

Until they are six weeks old, piglets should have wall feeders within a creep—a portion of the pen that is walled off from the sow. The creep typically houses the starter ration and water and may include an extra heat source for the piglets, allowing them to eat, drink, and rest without the threat of being accidentally crushed by the sow and without having to compete with the sow for food. Starter feed is typically finely ground or compressed into small pellets, and it needs to be placed in a wall feeder at a height suitable for small pigs. If you put feed in shallow pans, much of it will be wasted by trampling and spilling.

Wall feeders are also made for larger pigs, and they can be a clean and easy way to take care of small numbers of hogs. Wall-mounted hay feeders designed for horses are also functional for hogs. They are sturdy, can be mounted at any height, and can be filled easily with hay, clover, or alfalfa.

A bulk bin feeder is another option. This apparatus costs more than other feeders but offers significant labor savings. Also called a *self-feeder*, a bulk bin can hold from 50 pounds to several tons of feed at one time. The trough portion is a series of flap doors that the pigs can lift when they want to eat. An apparatus within the bulk bin can be adjusted to release the amount of feed needed, so that only the proper amount is accessible to the pigs at any given time. The pigs use their snouts to work the feed out from under a bar inside the trough. This style of feeder is especially efficient when raising market hogs, which should eat large amounts frequently; you can adjust the rations accordingly.

Highly technical and advanced knowledge of hog-feed formulation certainly contributes to better animal husbandry practices, but you do not need to be an expert in the subject or seek out specialized suppliers to find the right feed for your pigs. An advantage of raising pigs on a homestead or hobby farm is that you will be able to produce or supply a large amount of the feed necessary to raise healthy pigs. Developing your own feed ration is truly not a complicated process. Armed with knowledge of your pigs' basic dietary needs, you will find that feeding your pigs can be as simple as you want it to be. A proper ration will have the right balance of protein, fiber, vitamins, minerals, and carbohydrates.

Seasons and stages of life alter the nutritional needs of your pigs. When on pasture, pigs may need supplemental protein or nutrients. Excessively cold weather may increase your pig's need for food because it is using calories to maintain body temperature. Extremely hot weather can often decrease the feed intake of pigs because they simply do not feel like eating if they are too warm.

ROTTEN AND MOLDY

Rotten food and moldy grain can make pigs sick. Mold on grains can contain aflatoxins, poisonous substances that cause neurological impairment, abortions, internal bleeding, and death. It is far better to recognize and eliminate contaminants than to have to treat a sick animal or lose one later. Piglets are especially susceptible to toxins in mold, and piglets can suffer from bloat, liver damage, and death. Often, contaminated feed is not recognized until after it has caused sickness or death.

Molds and toxins in feed may not be obvious to the naked eye. Mycotoxins can be seen under the presence of a black light—they will look fluorescent green. Barring black lights or sick pigs, a lab is a sure way to identify contaminants in the feed. .

Purchasing your feed from reputable granaries or feed mills that routinely test their feed for contaminants is your best option. Do not knowingly buy grain from sources that store the feed outdoors, in leaking bins, or in damp or wet areas. Feed that is obviously moldy, has any peculiar odor, or looks dirty should be strictly avoided.

POISONOUS PLANTS IN THE PASTURE

Before turning your pigs loose on pasture, inspect the area for weeds that can be poisonous to your pigs. Some common poisonous plants are pigweed, Jimson weed, two-leaf cockleburs, lambsquarter, spotted water hemlock, pokeberry, black-eyed Susans, and nightshade. Pigs are not highly discriminatory when it comes to feeding, so it is up to you to safeguard your property. The ASPCA's Animal Poison Control Center maintains a current list of plants that may be poisonous to your pigs. If you cannot identify the plants on your property, contact your local county extension office, USDA office, or Natural Resources Conservation office for assistance.

Pasture-Raised

If you are able to pasture your hogs, you will be satisfied with the growth and health of your pigs. Pasture-raising adds many benefits to the pig's life as well as to the quality of the meat. By grazing, pigs are able to exercise and perform many natural behaviors, such as rooting, wallowing, and exploring. Your meat quality will be improved by the addition of healthy grasses and plants, healthy omega fatty acids from these plants, vitamin D from the sunshine, and fresh air. Pastured pigs have little odor because the wastes are not concentrated in any one area.

Well-managed rotational pastures can provide up to two thirds of a hog's food ration. Supplemental grain or milk is usually required for the remainder of the ration to assure proper nutrition and normal growth. Popular pasture mixes include orchard grass, alfalfa, and clover. Winter vetch, oats, and peas are also a desirable pasture mix for pigs.

Good Pasture Land

Good pasture land for pigs includes an area with a wide variety of plant sources. Legumes such as alfalfa, peas, and clover provide an ample amount of protein in the diet. These plants are also highly palatable to pigs, so the pigs will eat them voraciously.

Grass varieties provide nutrients, fiber, and variety in the hog diet, but various grass species contain differing nutrients. Stemmy, thin grasses typically have less nutritive value than do lush, thick grasses, although this is not always the case. A purely grass-based pasture will not provide all of the nutrients that your pigs need, so they will require supplemental feed. Grass also contains high amounts of water and fiber and needs to be balanced with other feeds. Nongrass forbs—flowering plants with nonwoody stems, such as wild radish—are also highly nutritious and palatable to hogs.

Aside from the nutrients your pigs derive from the pasture plants themselves, your pigs gain added nutrients from the soil. Soils vary greatly across the United States, but most contain trace minerals such as iron, copper, and zinc, as well as salts and vitamins. An added bonus is that the pigs will also acquire added protein as they root up bugs, worms, grubs, and small rodents.

Rotational Grazing

A rotational grazing system is one in which the pigs are grazed in one area and then moved to another to allow the first pasture to regenerate, to allow sufficient composting of manure, and to prevent your pigs from completely rooting up and destroying any given area.

Rotational grazing of your paddocks will help diminish internal parasites, allow for fresh grazing and air, and improve your soil. Pigs are curious creatures, and a change of scenery also gives them something new to explore. Any large area can be partitioned into consecutive paddocks to allow the pigs to move from one paddock to the next with ease.

If a pasture or paddock area becomes muddy (an exception being a mud hole), damp, or filled with feces, you must move the pigs to avoid infection and disease. Moreover, allowing your pasture to become overgrazed or overrooted will destroy the plant root system, and you will have to reseed it.

On the flip side, a pasture full of thistles can be reclaimed for better crops by running hogs on it until it is rooted up. Hogs will dig down along the sides of the tap root in a thistle, munch off the root, and leave the top to die. Letting your pigs do the work is a healthy alternative to herbicides and other chemical treatments. Thistles are also highly nutritious and healing plants.

Pastures and grass paddocks can be set up in rows with gates between them to make rotating an easy task. Another clever method is to create a circular set of paddocks divided into pie-shaped wedges, with the water and feed at the center. This makes moving the pigs a simple task. Chores are more manageable because the water and feed are at the center and do not have to be moved when the pigs are moved.

Once a pasture has been grazed by your pigs, seed the area with other varieties of vegetation, such as clovers and small grains. By allowing each paddock to rest for at least three weeks between grazings, each will have the opportunity to grow additional healthy vegetation.

Benefits to the Land

By allowing your pigs to graze openly, you will be benefiting your land. Pigs will be naturally fertilizing the area on which they graze,

GRADUAL TRANSITION

It is best to turn your pigs onto lush pastures for a few hours at a time to start. By gradually adjusting the amount of time the pigs are allowed to graze each day, you will also allow their digestion to adjust to the increase in water and fiber in the diet. A gradual introduction to succulent feed will prevent diarrhea in most hogs.

Grazing is an instinctual behavior for pigs.

returning valuable nutrients to the soil. By selectively grazing certain areas, you will be able to eliminate some noxious plant species. Moving your pigs as needed will allow the more favorable species to recover and regrow.

Pasturing: Alternative Feeds

In some areas, pasturing is practical only during certain seasons of the year or when there is ample rainfall. When pasturing is not feasible, you must provide alternative feeds by looking to the feedstuff available in your area. Alfalfa, clover, and grass hay are all beneficial feed sources for pigs. If your neighbors are growing wheat, barley, triticale, milo, oats, or corn, start with a ration from these grains. Buying local is always the best method; it will save you money and time, and you will be supporting some of your neighboring farms.

Many old texts relating to swine husbandry refer to the "trio" as the best feed for hogs. The trio consists of milk, alfalfa (fresh or hay), and a grain mixture. If you are able to feed your penned pigs with this ration, you will have healthy, well-nourished pigs.

Commercial Rations

When pasturing is not possible and penning is necessary, the most convenient way to feed your pigs is with a commercial balanced ration formulated for the age and weight of your pigs. Some commercial rations contain antibiotics or other unnatural ingredients, such as choice white grease (pig lard), chicken feathers, or other animal by-products, but organic feeds are becoming more readily available as the demand for these feeds grows. Organic feeds are certified to contain only organic ingredients, are not permitted to contain any animal by-products, such as bone or fish meal, and are free from antibiotics, pesticides, organophosphates, and chemical fertilizers.

You can have feed mixed by your feed mill, enabling you to determine the ingredients and amounts yourself. Moreover, custom mixes are usually a lot less expensive than premixed rations are. Buying custom mixes by the ton will also save you money. If you lack storage containers or cannot use a ton of feed within a two-to-three-month period, buy smaller amounts and have the mill bag the ration for you. Bags must be stored where they will not get damp or wet and cannot be chewed into by rodents.

Grains

Grains are the most common and easily acquired hog feed in the United States. Most commercial producers and many small farms feed grains as the only feed source to pigs. Even pastured hogs can benefit from a supplemental grain ration when fed at 40 percent of the total feed ration. Grains provide a variety of nutrients, fiber, carbohydrates, and bulk in the diet.

Corn

Corn is the most common ingredient found in hog feed in the United States. Of traditionally grown grains, corn is the second richest in fattening food nutrients (wheat is the first). Corn is an excellent energy feed and is an especially good finishing feed because it contains a high amount of digestible carbohydrates and is low in fiber. Corn is also used in the diet to add firmness to the fat on a pig.

In spite of all of the benefits and virtues of corn, it will not produce healthy hogs all on its own. Corn contains between 7 and 9 percent protein on average but is deficient in nearly all required amino acids. Corn is also deficient in calcium, minerals, and vitamins. Pigs fed on corn alone will eventually die from malnutrition and related diseases. Therefore, corn must be supplemented with vitamins, minerals, and a protein (amino-acid) source. When properly supplemented, corn is an excellent feed for all ages of swine.

Wheat

Wheat is considered to be equivalent to corn as a feed source. Wheat is slightly superior to corn in quality and quantity of protein, at 11–14 percent, and contains more lysine (an essential nutrient for growing and gestating pigs). Wheat can replace corn 1:1 in feed rations, and it will decrease the amount of soybean meal or other protein required to meet the desired protein level. Wheat should be rolled or coarsely ground because finely ground wheat may form a pasty mass in the mouth and become less palatable than corn to some pigs.

Commercial formulations are made especially for piglets...for eating, that is!

mixed within the rye seeds. If you are unsure about whether rye has ergot contamination, a laboratory can positively identify the toxin. Ergot-infested grain should never be fed to pigs.

Sorghum

Sorghum (milo) has about 92–95 percent of the feed value of corn. Milo should be ground and mixed into the ration because the small, hard kernels may not be chewed sufficiently. Milo has nearly the same benefits and deficiencies as corn and should be mixed with the appropriate supplements.

Oats

Oats are not considered a high-energy feed. Whole oats have about 30 percent fibrous material, mainly from the hull. Although fiber adds bulk to the diet, it is of no nutritive value to hogs. Oats fed to fattening and grower hogs should not constitute more than one third of the total ration. On the other hand, oats are an ideal ration when fed to sows in late gestation and early farrowing because the added bulk helps keep the sow regular and feeling full so she does not overeat. Hulled oats are an excellent feed for any size pig, although this extra processing step does increase the cost of the feed. This feed may be practical for getting weanling pigs off to a good start or when fitting a pig for show.

Barley

Barley can replace corn in a ration up to about 75 percent. Barley is less palatable to pigs and should therefore be mixed with the protein source and fed as a full ration to ensure that it is ingested. Comparable to corn in protein, barley also lacks the proper amino-acid balance. Like corn, barley must be mixed with the proper supplements if it is to supply adequate nutrition. Scabby barley (barley affected by the *Fusarium* fungus) possesses a vomitoxin, which, when ingested, causes vomiting and nausea. Additionally, *Fusarium* causes liver, intestinal, and reproductive disorders in pigs.

Rye

Rye is considerably less palatable to swine and should not make up more than 20 percent of the total feed ration. Rye should be ground and mixed with more palatable feeds. Rye is often infested with the fungus ergot (*Claviceps purpurea*). Ergot causes dementia and abortion and reduces feed consumption and growth rates. Infested rye can be identified by the black kernels on the seed heads or

Milk Products

Skim milk, buttermilk, dried milk, and other surplus milk products contain nearly all of the nutrients that grains lack. Milk products are also highly palatable, easily digested, and of great value to recently weaned and growing pigs. Homesteads and hobby farms producing their own milk will have a ready consumer in the pig. A gallon of whole milk weighs 8.6 pounds, and skim milk weighs about 8 pounds. Feeding 3½ gallons of skim milk is roughly equal to ½ bushel of corn, or 28 pounds. Adult pigs also benefit from the nutrients in milk and can be fed as much as 1½ gallons per pig per day in conjunction with a grain ration and fiber source.

Whey is the by-product of cheese production, the cloudy liquid that remains after the milk solids are rendered from the whole milk. Whey contains a high percentage of water and little protein. It is of little value in feeding growing or market pigs. It is best used for adult hogs as a supplement to grain and water. If you are lucky enough to be located near a dairy or cheese processor, whey is an inexpensive addition to a balanced ration and will greatly enhance the palatability of the grain ration.

Proteins

Proteins are the most significant category of nutrients in swine rations. As mentioned, grains contain too little protein and are deficient in too many vitamins and minerals to be sufficient as a complete feed. Proteins are broken down into various amino acids during the digestion process. These amino acids assist the pig during growth, reproduction, lactation, maintenance, and healing. Amino acids cannot be manufactured by the body, so they must be provided in the form of properly mixed feed.

Pigs will appreciate almost anything from the kitchen that you add to their rations.

Overfeeding protein is a waste on all counts. Protein-rich feeds suitable for pigs are generally more expensive than are grains (per pound), so a combination of grain and protein puts less strain on the pocketbook. Also, excess protein is excreted or adds weight to the pig's liver and pancreas without increasing muscle or back fat. Excreted protein has a very high concentration of nitrogen, a substance known to contaminate groundwater. Finally, pigs are inclined to overeat protein supplements if fed free-choice. Mixing the proper amount of protein into the grain ration is highly recommended.

Beans and Seed Meals

Soybean meal is probably the most economical source of quality protein for hog rations. It is highly palatable to pigs but should be mixed with the grain ration to prevent pigs from overeating the soybean and neglecting the grains. Soybeans are fairly well balanced in amino acids but deficient in minerals and vitamins, so a vitamin/mineral supplement is an absolute must. A combination of soybean meal and alfalfa meal mixed 3:1 will also balance the ration.

Raw soybeans contain high quantities of trypsin inhibitors, which block normal protein digestion in pigs; therefore, raw soybeans should not be fed to baby or weanling pigs. Raw soybeans and peanuts should not be fed to grower or finisher hogs (market hogs) because they create an undesirable carcass with poor meat, lard, and fat quality.

Linseed meal is well liked by pigs. It has a slight laxative effect and can be used efficiently in sow herds. Although it contains high amounts of calcium and adequate B vitamins, linseed meal is highly deficient in

lysine, so supplementation is necessary. Linseed meal should not make up more than 20 percent of the total protein supplement.

Cottonseed meal is not very palatable to pigs and also contains a toxic substance called *gossypol*. It is not recommended.

Tankage, Meat and Bone Meal, Fish Meal

Other products commonly added to pig feed as protein or fat supplements are tankage, meat and bone meal, and fish meal. Animal proteins are typically, but not always, more expensive than plant-based proteins.

Tankage is a product derived from rendering animal carcasses that is separated into protein and fat, respectively. Typical protein levels in tankage are about 60 percent, so you only need to add a little to obtain the desired protein level. It is generally liquid in form.

Meat and bone meal and fish meal are derived primarily from the by-products of butchering and processing. All are typically ground and dried and added to feeds in a meal form. Fish meal has been known to add a peculiar (fishy or weedy) taste to meat if used in large quantities.

All of these sources are deficient in tryptophan, and none should be used as a single source of protein. They are good sources of calcium and phosphorous, however. Although not as palatable as soybean meal to pigs, these meals can provide up to half of the protein supplements needed in a ration.

Alfalfa, Clover, Grass, and Succulents

Alfalfa meal is not generally fed as the only source of protein in a pig's diet. Dehydrated meal contains about 20 percent protein, vitamins A and B, and high fiber. Alfalfa meal is not highly palatable to pigs and is usually mixed with other protein sources and grain for a complete feed. Its higher fiber makes it a less desirable supplement for grower and finisher pigs.

TANKAGE PRECAUTIONS

Carefully check the source of tankage if you wish to use it in your ration. Tankage may be the product of euthanized or sick animals or roadside carcasses. Only those companies guaranteeing the source of the tankage should be used. Additionally, tankage, fish meal, and other animal by-products are not allowed in the diets of certified organic meat animals and in some naturally raised systems.

ASSESSING YOUR PIG'S NEEDS

Pay attention to your individual pigs to gauge the amount and type of feed that keeps them fit and healthy. By feeding two or three times per day, you develop a relationship with your hogs and can better determine their needs. Baby pigs will rarely overeat, but adult pigs may continue to stuff themselves if given the opportunity. An excessively hungry pig is likely lacking in nutrients and fiber. Try adding alfalfa or clover hay to the feed ration. Feed needs will change with age and sexual activity and during gestation. Carefully evaluating your hogs and adjusting the nutrient levels of their feed according to life stage will keep them in top shape.

Alfalfa pellets can be mixed into a ration for junior and adult pigs as part of the protein supplement. The pellets seem to be more palatable to the pigs. If alfalfa hay is not available in your area, pellets are an excellent alternative. Additionally, if you are feeding only a couple of pigs, pellet feeds can be very convenient to manage because you can purchase them in 50-pound bags.

Alfalfa and clover hay are usually well liked by pigs. Fed free-choice, the pigs utilize the hay for nutrients, fiber, and as a means to occupy their time. These hays are high in protein and can replace pasture in most cases.

Grass hay has little protein but is high in fiber. Older pigs that do not need the additional growth or calories do well on grass hay, but it is not recommended for very small or grower pigs as a primary feed source. Grass hay is also beneficial for helping a pig lose weight. Feeding free-choice and limiting the other feed available can help a pig trim down without too much anxiety.

Kitchen Scraps and Surplus Foods

Pigs appreciate variety in their diet just as much as humans do. Fresh kitchen scraps, vegetable peelings, and surplus vegetables all provide additional nutrients and interest to the pig. It is the benefit and even the duty of any homesteader to utilize this waste as pig food. Leafy vegetables also add nutrients and variety, but they will need to be accompanied by other more substantial feedstuff. Apples, pears, and other treats can be stored for winter months to add vitamins and variety to the diet. However, a diet high in fruit should be balanced with a protein to avoid diarrhea or weight loss.

Eggs are also a valuable protein source for pigs. Most farms have a surplus at some time during the year. Eggs can be fed, shell and all, mixed in with the ration. Pigs love eggs. In fact, beware of pasturing laying hens with pigs. Once the pigs learn where the eggs comes from, they will spend time following your chickens around and gobbling up any eggs they can find!

Day-old bread and other bakery waste can be fed to hogs with caution. Bakery products are high in carbohydrates and may or may not contain large amounts of fat. Protein content averages about 10 percent. As a carbohydrate, bakery waste can be substituted for corn for up to 50 percent of the ration. Protein, vitamin, and mineral supplementation should be the same as for grain rations. Excess carbohydrates in the diet will create more fat than meat, and prolonged use may produce weakness in the feet and bones. Moistening the bakery products with water, whey, or milk makes them much more appealing to hogs and will help balance the ration.

Fat

Fat is considered a high-energy feed source, but it is only a supplement, not a ration, and cannot be fed as a primary food source. There are times when additional fat may be required, such as when a sow has been depleted from nursing or when an infirm pig has lost a significant amount of weight. Most often, though, a proper diet supplemented with dairy products will be sufficient to restore a pig to prime condition.

Additional Feedstuff

Molasses is typically added to feed as a binder to prevent dustiness and to make unpalatable feeds more appealing, thus reducing waste. Molasses is also used as a laxative for sows just prior to farrowing. Molasses contains many vital nutrients, such as calcium, manganese, copper, iron, and potassium

Acorns and chestnuts are premium feed for hogs. In fact, in Spain and Italy, hogs finished on diets of acorns and chestnuts, respectively, command top dollar. You can fence an orchard or tree stand in the fall months, allowing the pigs to clean up anything that falls to the ground.

Pigs can also be fed grass clippings, sunflower seeds, weeds, and leaves as long as they have not been sprayed with pesticides.

Vitamin and Mineral Supplements

Vitamins and minerals are of great importance to the health and development of pigs. If feeding a ground ration, it is best to have a premixed supplement added to your feed. These premixes are formulated based on the age and needs of your pig. If you live in an area that is particularly deficient in some mineral, an additional supplement can be added. Check with your veterinarian or local feed mill for specific deficiencies that might be present in your area.

If your hogs are on pasture, they will meet a large portion of their mineral and vitamin needs through grazing. A pastured hog should be offered mineral and vitamin supplements free-choice to allow the pig to balance its diet naturally. Forcing pastured pigs to eat minerals and vitamins mixed in with their ration may cause toxicity. Block salt and minerals are not appropriate for hogs because they will not spend the time licking them as other livestock breeds may do, and some binders used to hold the blocks together can be toxic to pigs.

Pigs are social animals and should be raised in groups. Raising them without any companionship is cruel. If you must raise only one pig at a time, provide it with the companionship of another species. Of course, you can also be the pig's companion—if it is allowed to go wherever you go! Pigs raised in groups always eat better and are less destructive because they are not as prone to boredom. If possible, move groups of pigs together when taking them to new areas.

UNDERSTANDING

PIG

BEHAVIOR

Pigs' groups should remain stable throughout their lifetimes. Introducing a single pig into an established group will jeopardize the new pig's safety because the other pigs will fight with it in an effort to run it off. Reintroduction of pigs to their old family groups should be done through a fence to allow them to safely get reacquainted. Boars should never be moved into existing sow pens because sows are very territorial and will instigate fights with the boars if they feel as if they are being invaded. When breeding, always take the sow to the boar pen for mating.

Pigs are highly intelligent animals, and their intelligence can be used to your advantage. Pigs can be easily trained to come for food with a bell, your voice, or other signal. They can be taught to do tricks such as open doors, roll over, or sit on command. Pigs are a particular favorite at petting zoos because they want to interact with humans.

Pigs have very poor eyesight, which may explain why they become easily upset and frightened when they are being moved. If a pig is being moved into an unfamiliar space, allow it time to look around and evaluate the situation. Trying to move a pig into a dark space will cause anxiety, and it will not readily go.

Pigs typically have excellent hearing, and they communicate with each other through grunts and squeals. For example, when a sow is nursing her piglets, she communicates when it is time to eat by certain grunts, and she will sing a sort of lullaby to the piglets when they are nursing. An angry or upset pig will perform a series of barking noises to warn the antagonist to back off. When happy, a pig may make a sort of purring or gurgling sound.

PIG VITAL STATISTICS

Temperature: 101.5° Fahrenheit

Pulse: Seventy beats per minute

Respiration: Twenty-five to thirty-five breaths per minute

Expected Life Span: Three to fifteen years

Sexual Maturity: Females six to eight months; males four to six months. Although they are able to reproduce at this early age, females should not be put into breeding service before six months of age.

Heat Cycles: Female pigs will cycle about every twenty-one days, with the actual heat period (time in which she can become pregnant) lasting about one or two days at the end of the cycle.

Gestation: 3/3/3; that is, approximately three months, three weeks, and three days, or about 115 days

Color: Pigs can be nearly any color, including sandy, brown, black, white, red, and blue (gray), and any combination of colors

Hide: A variety of coats can be seen, from excessively hairy to nearly hairless, as well as curly, straight, smooth, and rough coats

Pigs get muddy, but they are not dirty animals.

A sense of smell is imperative to a healthy hog. Pigs identify each other and locate food through smell. The snout is used to gauge distance, to dig up food or make a hole to lie in, for courting mates, and for interacting with other pigs.

It is a popular misconception that pigs wallow in mud because they are dirty animals. This is simply not true. Pigs are not equipped with sweat glands, and they require adequate shade or other cooling measures to be comfortable and to survive. Mud not only cools the body but also creates a barrier against biting insects.

Pigs are, in fact, very clean animals and will designate certain areas of their living spaces for different functions. Pigs will choose separate areas for eliminating, sleeping, lounging, and eating. When setting up pens, be sure to always locate the feed away from toileting and sleeping areas. Once the pigs have chosen their "bathroom" area, place extra bedding there to absorb waste.

Pigs have a variety of personality traits. Getting to know your pigs will help you manage your herd. Most pigs will settle in once they get to know who you are. They may come running to see you but may stand back, away from the fence, when a stranger approaches. If they feel threatened, they will snort, bark, and paw at the ground. This is a sign of confrontation and aggression. Try to talk calmly to the pig and move around it in a slow, nonthreatening way. Do not try to touch an angry pig because it may try to bite you in an effort to move you away.

Friendly pigs are always fun to work with because you can scratch, pet, and move them much more easily. Do your best to identify your pigs' personalities and work with them according to their dispositions.

We know that pigs like to root, play, and explore their living areas. Bored pigs will resort to bad behaviors, such as chewing fences, digging, or destroying food and water dishes. To prevent such behaviors, provide your pig with an enriched environment, perhaps by placing a bowling ball in the pen or by providing fresh greens or grass clippings.

Much research has been done on pig behavior in an effort to reduce stress in handling for both the pig and the handler. Studies by Dr. Temple Grandin are very informative on the topics of stress-free animal handling and the psychology of animal behavior. Through her works, a person can learn to see the world through the eyes of their animals and learn how to anticipate their behaviors. Humane and stress-free handling will lead to a trusting and more manageable relationship with your pigs.

Safe Handling

Pigs are easy to live with if you spend some time learning about their basic needs and behaviors. Routine health care is minimal when a pig herd is well fed, well housed, and well appreciated. Most health issues can be avoided or successfully treated immediately, preserving your herd and its longevity.

An understanding of pig psychology will save you time and energy when handling pigs. Pigs can be extremely stubborn and will typically do the opposite of what you want them to do. By routinely handling your pigs, they will become accustomed to your care and movements and will be more likely to cooperate when you want to move them.

DID YOU KNOW?

Pigs can scream loudly. The scream of a jet engine taking off measures 113 decibels. The scream of a frightened pig can measure 115 decibels.

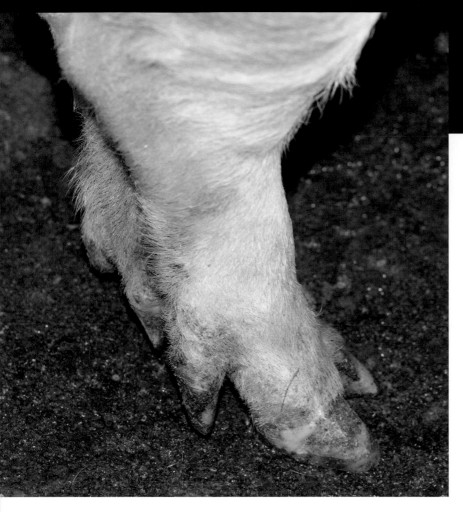

Pay attention to the health and condition of your pigs' hooves.

Catch a small pig by grabbing a back leg, swinging the pig clear of the ground, and then maneuvering it into a manageable position. Catch a medium-sized pig by a back leg and then place the pig into a sitting position between your legs while you hold on to the front feet. Hold the front feet in toward the head to maintain control of the pig and avoid getting bitten.

Large pigs will need to be snared. A snare is a device that, when placed over the pig's snout or top jaw, can be tightened or manipulated to hold the pig fast. A rigid snare is a metal rod with handles at both ends; you put the handle portion into the pig's mouth and then lift the rod into the air, wedging the snout in the handle. An overly aggressive pig can usually be restrained with a rigid snare. The disadvantage of a rigid snare is that you may take the flesh off of the pig's nose.

A cable snare is essentially a noose through a rod that is tightened around the snout. Cable snares have less potential to cause injury to the pig and can be released quickly in an emergency. If a snared pig breaks free from your grip, step back, away from the pig. It will usually flail its head about in an effort to release the snare. A flailing snare can easily break a bone on a human or cause serious injury. After a few tosses of the head, the snare will fall off the pig's snout. You can then pick it up and try snaring again.

Holding a pig will be an ear-piercing experience, so have a set of earplugs handy. Small pigs are best held with one hand under the

Catching and Holding Pigs

The shape of a pig's head does not lend itself well to being held, so different methods and equipment are used to move pigs. A pig board is a lightweight board with handles cut out along the top and one side. Typically about 30 inches long and 2–3 feet high, a pig board can be used to block a pig or direct it to wherever you want the pig to go.

You can also drive pigs into crates for handling. Premade crates are available, but a simple one made of stout boards will serve just as well.

A D V I C E   F R O M   T H E   F A R M

HANDLING YOUR PIGS

Get to know your pigs, and let them get to know you. Talk to your pigs and announce your presence when you are approaching their pens. They will learn to recognize your voice and will feel safe. Don't intentionally make loud noises or upset them, as they will shy away from you the next time you come around. When you are done handling them, give them a bit of food as a reward.

—Bret Kortie

Pigs always move better if it is their idea. It takes some practice to be a step ahead and know what the pig is going to do. Be calm and look at it from the pig's point of view. He is 2 feet off the ground, you are 5 feet off the ground. Look at his level and see that he has a clear path.

—Josh Wendland

What a pig lacks in eyesight, it makes up for with a keen nose and ears.

belly and chest and the other pressing on the back. Pigs do not like to be flipped on to their backs or picked up, and they should never be suspended by the nose or tail. Avoid carrying small piglets unless you are training them to be handled in this fashion.

Casting or Throwing Pigs

Casting, or throwing, pigs is a method used to lay a pig down and restrain it for treatment, such as administering vaccinations, detusking, or checking a hard-to-reach area. You can easily cast a small pig by picking it up and laying it on its back. Cast a larger pig by reaching under the pig, grabbing the two legs on the opposite side, and pulling the legs toward you. The pig should fall away from you. Hobbles (tethers for the legs) can be used to secure the hog while you are working on it.

Routine Care

Hoof trimming, detusking, and ringing noses are all practices that the small-scale hog producer should be familiar with. If these are practices you wish to maintain in your herd, expect to detusk at least once per year and trim hooves up to six times per year, depending on the surface on which your animals spend most of their time. Have a qualified veterinarian perform the procedures the first time and explain them to you step by step. Although most hobby farmers may never need to perform these tasks, all should know how.

Trimming Hooves

Some people trim hooves on show pigs to create a well-kept and even appearance; otherwise, it is usually not necessary to trim hooves. Most pigs raised outdoors get plenty of exercise and continually wear down their hooves. However, pigs raised on very soft bedding or in a confined situation will need their feet trimmed periodically. If the hooves of your pig grow too long or split, you will need to trim the feet to avoid abscesses, infections, and rot.

Splits typically happen on one toe, usually when the ground has become too dry or hard or when the pig is doing lot of stomping; for example, when flies are bothersome. The actual wall of the hoof will develop fissures or cracks that can lead to more serious problems. Split hooves may require the assistance of a veterinarian to properly evaluate the extent of the damage and the possible remedies.

Hire a veterinarian or experienced pig person to teach you how to trim hooves if you have never trimmed before. Serious injury can result from improperly trimmed hooves, so you need to understand how much to take off of the hooves during trimming. Smaller pigs can be held on their backs for hoof trimming, but mature animals will likely need to be cast in the manner mentioned previously.

Use a regular hoof knife or pocket knife for the initial trimming. Trim each side by cutting from the back of the hoof toward the tip, so that the inner side of the toe is slightly shorter than the outside. Trim with a light hand to avoid cutting into the quick (soft tissue), which will cause lameness and bleeding. If you accidentally cut into the quick, use blood-stop (styptic) powder or cornstarch to clot the blood and stop the bleeding. A splash of a copper solution will prevent any bacteria from entering the cut and will firm up the remaining hoof wall. After removing excess hoof with the knife, use a common rasp to even up the hooves. When the hooves are trimmed properly, the animal should stand straight up on its toes.

Cutting Tusks

A boar's tusks sometimes grow to dangerous lengths. Tusks can grow as much as 3 inches per year, and they should be trimmed annually. A boar constantly sharpens his tusks by scraping the top tusk against the bottom tusk, resulting in a sharp, fine point on the lower tusk. A boar that will be used for breeding should be detusked to avoid injury to the sow during the premating introductions and in case the sow rejects the boar. Show boars should also have their tusks trimmed.

Detusking a boar is a two-person job, and an experienced swine person or veterinarian should teach you how to perform this task. A

Ringing Noses

Pigs use their snouts for smelling, rooting, and digging, as well as to maneuver or nose their potential mates. Sometimes it is necessary to ring the nose of a pig to prevent it from excessively rooting good pasture or to prevent it from digging under fences.

There are several sizes and styles of rings, each requiring its own special pincers to set the ring. Choose a ring appropriate to the size of the pig; an overly large ring will quickly be torn out. Insert the ring either in the rim of the snout or through the partition (septum) between the nostrils. One ring is usually enough when set through the septum, but up to three rings may be needed for the rim of the snout.

Hold a small pig in a sitting position for ringing, but a larger hog needs to be restrained by a snare or snub. Set the ring quickly and release the pincers right away to avoid tearing the ring out if the pig jerks its head away. A newly ringed pig may stop eating for a few days while adjusting to the ring's presence.

Never ring a pig that is younger than three months old because this will retard its growth. Ringing a boar that will be used for breeding purposes is strongly discouraged. A boar uses his nose to stimulate the sow or gilt and to position her for mounting. A ring may cause the boar pain, thereby rendering him unwilling to breed.

boar in a snare can still flail his head around and injure a handler, so you will need a strong person to hold the pig.

Use a molar cutter (which was originally designed to trim a horse's teeth), a wire saw, or very sharp bolt cutters. Snub (tie) the boar to a post or snare the boar. Slip the molar cutter over the pig's tusk and cut off the tusk so that it is even with the surrounding teeth. You can use a wire saw in the same manner to remove the tusk, working the wire back and forth until the tusk comes off.

If you use a bolt cutter, be careful to cut the tusk even with or slightly above the height of the surrounding teeth, and make sure that the bolt cutter is sharp. A dull bolt cutter may actually crush the tusk instead of cutting it and may fracture the tusk down into the gum line, which could cause abscesses, infection, or other serious problems.

HANDLING TIPS FOR PIG OWNERS

Here are some helpful tips to use when handling and moving pigs:

- Don't try to make pigs do anything. Try to entice them or outsmart them because it is unlikely that you will be able to outmaneuver them.
- Don't frighten pigs by making loud noises, moving suddenly, or shouting. Move slowly and talk to them or offer a scratch or a pet.
- Driving and pulling pigs is difficult. It will be much easier on you if you set up panels or gates and crowd the pigs along, giving them only one way to go. You can also coax them with feed.
- Don't try to force pigs into a pen or chute that is dark on the opposite end. Place a light at the end of a tunnel, and they will be more likely to walk in.
- Provide yourself with an exit if necessary, either over a panel or by working outside the alley or tunnel. Do not allow yourself to be cornered by a pig. A frightened pig may become vicious.

Through proper management and feeding, most illnesses can be avoided. If your pig herd is subject to repeated outbreaks of a particular illness, a thorough consultation with your veterinarian is in order. A vaccination program may be your only defense against repeated illnesses.

Prevention is still the best medicine with pigs. For example, allowing pens and pastures to lie idle for months or, ideally, a year gives some infectious pathogens time to die off before pigs are reintroduced to the area. If it is practical under your management schedule, stagger farrowing, weaning, and other movements of pigs to allow each group to become well established. This will also make it easier for you to identify and isolate problems and provide the needed care and treatment.

Sudden Illnesses and Emergencies

Expertise grows with experience; until you have experience in raising pigs, contact your veterinarian for advice and treatment, especially for sudden illnesses or injuries.

Be prepared to describe all the symptoms your pig is displaying. For example, if your pig has diarrhea, is it watery or pasty? What is its color? Does it have a particularly foul smell? If the pig is displaying any unusual behavior, make note of that also. For example, is the pig paddling, spinning, or backing up? Describe any discharge around the eyes, ears, nose, or mouth. Has the pig recently been weaned, or is it an older pig?

If you suspect that your pig has eaten something poisonous, be certain to tell your veterinarian right away. Immediate action may save your pig.

Vaccinations

In some areas and under certain management practices, vaccines are necessary. Speak with your veterinarian on this point. If you purchase disease-free stock and make no additions, routine vaccinations may not be necessary. If your herd will be taking in new stock, or you will be rotating feeder pigs through your farm, then a vaccination program may be in order. Vaccines are available for many pig illnesses, but any illness should be properly diagnosed by your veterinarian before a vaccine is administered. Certain regions and breeds may be more susceptible to some diseases, while other areas rarely see an outbreak. Avoid vaccinating for illnesses that you are unlikely to encounter. Again, your veterinarian can inform you about the disease threats in your area or make a conclusion based on blood tests from your pigs.

Ask your vet to teach you how to give shots to your pigs. Being able to vaccinate or treat your own pig will make you a better manager of your herd. You will no longer need to wait for the vet to arrive, and you will be able to vaccinate as time permits.

Worming and Internal Parasites

Preventing and treating parasites can be a challenge at first, but with proper management and preventive worming practices, you can eliminate the bulk of parasites from your herd. You will never have a herd that is completely free of parasites. Reinfestation is common because pigs constantly have their noses on the ground. The objective is to keep the parasites at a minimal level.

Prevention

Your chosen management system is the determining factor in the parasite load carried by your pigs. It is a good practice to isolate and worm all incoming pigs to your herd. Feces from quarantined animals must be removed and disposed of in an area that your pigs do not have access to.

HEALTHY FEET

Healthy pigs should have a strong bone structure, which will help support a large amount of weight on small, short legs. The feet of pigs are typically cloven hooves except in the case of a mule-footed, or single-toed, hog. The feet must be sound to support the weight of the pig but also be of good structure to avoid cracked or split toes, abscesses, and arthritic conditions.

Rotational grazing is healthier for your pigs and your pasture.

available, and piglets younger than seven days should be treated orally. The best defense is a clean environment and highly absorbent bedding material.

Enteritis

Enteritis is a disease characterized by diarrhea and caused by one of several species of the bacterium *Clostridium*. Characterized by an inflammation in the small intestine, enteritis usually affects piglets within seventy-two hours of birth. Sudden death may occur before you have time to recognize the illness. There are vaccines for some types of *Clostridia*. Special attention to cleanliness, such as washing the sow before farrowing, may help control the incidence of enteritis.

Erysipelas

The bacteria *Erysipelas* can be found on most pig farms. It is always present in either the pig or in the environment because it is excreted via saliva, feces, and urine. It is also found in many other species, including birds and sheep. Infected feces are probably the main source of contamination because the bacteria can survive outside the body for several weeks or longer in light soils.

WASH YOUR HANDS

Although most parasites and diseases are particular to each species of mammal, some may be transmitted between animals and humans. These are called *zoonotic* diseases. Trichinosis, tapeworm, and the fungus ringworm, for example, can be transferred from pigs to people.

To reduce your risk of infection, wash your hands thoroughly after handling pigs. Be careful not to transfer pig feces into your living areas by tracking it in on your shoes or clothing. Biting your nails is also a way to expose yourself to worms, fungus, and bacteria. Children tend to put their hands in their mouths, so they must thoroughly wash their hands whenever they are exposed to anything soiled by pigs.

Symptoms include diamond-shaped skin lesions, high fever, abortions, mummified piglets, and variable litter size. A vaccine is available. If an outbreak occurs, immediate treatment with fast-acting penicillin is indicated. Moving the pigs to clean areas will help diminish the disease.

Flu

Swine flu is caused by a number of influenza A viruses. The onset is typically rapid and drastic, lasting from twenty-four hours to ten days. It is virtually impossible to keep swine flu from affecting your pigs at some point. It can be introduced by people, birds, carrier pigs, and other animals and is exacerbated by stress, fluctuating temperatures, wet bedding, and poor nutrition.

Symptoms include lethargy, coughing, fever, abortions; severely ill pigs appear to be near death. Most pigs recover without treatment, but secondary infections and pneumonia can result from the initial viral infection, and affected pigs should be treated with antibiotics.

Glässer's Disease

Known also as *Haemophilus parasuis* (Hps), this disease is found all over the world, even in herds with top health records. Hps attacks the smooth surfaces of the joints, coverings of the intestine, the lungs, the heart, and the brain, causing pneumonia, heart sac infection, peritonitis, and pleurisy.

Seen mainly in piglets, symptoms include a short, repeating cough; fever; depression; loss of appetite; lameness; and death. Hps is thought to be brought about by exposure to pneumonia, porcine reproductive and respiratory syndrome (PRRS), flu, drafts, poor environments, and stress. Treatment with penicillin or oxytetracycline is effective if administered immediately.

FIRST AID KIT

All pig owners should have a basic first aid kit for life's little emergencies. Stocking and periodically updating a portable tackle or tool box is a good way to organize first aid supplies and ensure that you have what you need on short notice. The following supplies are part of a basic kit:

- Antibiotic liquid to be given orally
- Antibiotic powder that can be added to the water supply
- Betadine solution, hydrogen peroxide, and triple antibiotic cream
- Coppertox (for treating foot injuries)
- Electrolytes or instant sport drink mix
- Eye dropper or syringe for orally dosing piglets
- Fly spray
- Hypodermic needles and syringes, assortment
- Marking crayon or paint
- Obstetric lubricant
- Pepto-Bismol

- Phone number for your veterinarian
- Prescription medications as recommended by your veterinarian
- Razor blade or scalpel
- Rectal thermometer used for babies
- Scours (diarrhea) remedy that can be added to the water supply
- Snare
- Spray bottle with Listerine
- Veterinary manual for pigs
- Wound spray

Leptospirosis

This disease is caused by bacteria that are carried by most mammals and is transmitted through the urine of animals, from pig to pig, and from other wildlife to pigs. Symptoms include loss of appetite, depression, abortions, stillbirths, and weak piglets that die shortly after birth.

If an outbreak occurs, pigs should be treated with the tetracycline family of antibiotics. Additional management practices include removal of urine, allowing wallows to dry out completely between uses, control of rodents and other livestock, and vaccinations.

Leptospira organisms have been found in cattle, pigs, horses, dogs, rodents, and wild animals. Humans become infected through contact with water, food, or soil containing urine from infected animals. The disease is not known to spread from person to person. Symptoms of leptospirosis in humans include high fever, severe headache, chills, muscle aches, and vomiting and may include jaundice, red eyes, abdominal pain, diarrhea, or a rash. If the disease is not treated, the patient can develop kidney damage, meningitis, liver failure, and respiratory distress.

Lice

Lice are external parasites that can be seen with the naked eye on many farm animals. Lice suck blood from their hosts and can transmit bloodborne illness and anemia. Lice can be transmitted to humans and other livestock.

Treatment is a simple pour-on delouser or anthelmintic. These treatments are ineffective against the eggs, so a second treatment, given ten days after the first, is required.

MMA Syndrome (Mastitis, Metritis, Agalactia)

MMA is a series of illnesses affecting the post-farrowing sow. Mastitis is an udder infection, characterized by swollen teats, fever, loss of appetite, and an unwillingness to suckle piglets. Metritis is a bacterial infection of the uterus, characterized by fever, loss of appetite, and a brown discharge from the vulva. Agalactia is a loss of milk production. These illnesses do not always follow each other or happen as a syndrome; each can appear on its own for various reasons.

The key to good milk production and the health of the sow during and after farrowing is cleanliness. Mastitis and metritis are treated with oxytocin (to promote milk let-down and uterine contractions), antibiotics, and vitamin injections. Adding fiber to the sow's diet and providing the opportunity for exercise also help stimulate milk production. Agalactia that has been brought on by mastitis or metritis can usually be resolved by these measures. Agalactia brought about by old age in the sow cannot be reversed.

It is imperative to monitor the piglets. If they are not receiving milk from the sow, then an immediate intervention must be made because piglets are highly susceptible to starvation during the first few days of life.

Care and cleanliness are essential for helping the sow take care of her babies.

Mange

Mange is an unsightly and uncomfortable parasitic skin disease caused by mites. It causes them to scratch to the point of scabs, scrapes, and punctures. In severe cases, pigs develop an allergic reaction to the mites and have an outbreak over the entire body. Spread from pig to pig, mange can be easily introduced to a herd via new animals that have not been treated and quarantined. It is rare for mange or pig mites to be transmitted to humans because they are host-specific parasites.

Treatment is through oral, injectable, or pour-on medications. Repeated treatment is necessary. Two doses, given ten days apart, should eliminate most cases.

Pneumonia

Characterized by a chronic dry cough, pneumonia is a common bacterial or viral infection in growing pigs and is occasionally seen in gilts with compromised immune systems or infections. Mortality in naive herds (those that have never been exposed to or vaccinated against the particular strain of pneumonia) is high, between 10 and 15 percent, if not treated immediately. Various strains of pneumonia exist, and blood tests can help determine the exact pathogen associated with an outbreak and thus determine the best treatment.

Other symptoms include fever, rapid breathing, loss of appetite, huddling, and loss of condition. Pneumonia can be compounded by the type and level of pathogens (bacteria and viruses) present in the living quarters, the immunity of the pig at the time of exposure, and the management practices of the herdsman.

Pasteurellosis

Caused by the *Pasteurella multocida* bacteria, pasteurellosis is commonly indicated by respiratory diseases between ten and eighteen weeks of age. Pasteurellosis is characterized by severe and sudden pneumonia with rapid breathing, high fever, and high mortality. Less severe cases may include mild pneumonia, coughing, nasal discharge, and wasting. Thought to be a secondary condition to PRRS, flu, or enzootic pneumonia, it should be treated the same way as pneumonia.

Porcine Reproductive and Respiratory Syndrome (PRRS)

PRRS infects all types of herds, irrespective of size and including those with high health standards. Severe economic loss can result from an outbreak of PRRS in a herd.

PRRS was first identified in the midwestern United States and dubbed the "mystery swine disease." Caused by an *arterivirus*, PRSS has no known cure.

Symptoms are varied depending on the age of the pig affected, but they include respiratory and reproductive disorders and can range from loss of appetite to abortions and stillbirths. Fertility issues are also implicated in the disease, with breeding and cycling being delayed. Symptoms can last up to three months and may recur. A vaccine used to control and prevent PRRS has had only limited success in the United States.

The best ways to avoid PRRS is to purchase only from herds tested to be PRRS free; to quarantine all incoming animals to avoid exposure to existing pigs; to maintain dry, clean facilities that are not in continual use by pigs; and to minimize outside visitors to the farm.

Pseudorabies

A herpes virus found in pigs, pseudorabies can remain dormant in the pig's nerves for long periods and then become reactivated. Once introduced to a herd, the virus remains and affects fertility to varying degrees. Periods of stress may activate the disease, which can be transmitted through the air and nose-to-nose contact. Other sources of contaminants can be feral animals, people, vehicles, semen, and contaminated carcasses.

Healthy piglets will be energetic and playful.

Scours

Bacteria, viruses, illnesses, and diseases may cause pigs to scour (have diarrhea). Piglets are particularly at risk when scours develops because they can quickly dehydrate and starve. An inexperienced hog person should contact a veterinarian immediately at the first sign of scours. Determining the cause will be your first step in curing the pig and preventing further episodes.

Adding electrolytes to the water source to bolster the pig's electrolyte balance during increased fluid loss will help stabilize the pig while you determine the cause and begin proper treatment. Electrolytes come in powdered form in small pouches and can be purchased at most farm stores. In a pinch, a bottle of sports drink will also provide the needed nutrients and replace fluids, and the sweet taste may encourage your pig to drink.

Prominent symptoms, such as coughing, sneezing, lack of coordination, reproductive failure, abortions, stillbirths and mummified piglets, and piglet death involve the nervous and respiratory systems. A vaccine is available but should be considered with extreme caution. Once a pig is vaccinated for pseudorabies, it will test positive for the virus, which may prevent you from selling animals as breeding stock in some states.

Pseudorabies can be transmitted to cattle, horses, dogs, and cats, causing rapid death. It is not known to cause disease in people.

Salmonellosis

Salmonellosis is caused by forms of the bacterium *Salmonella*—*S. choleraesuis, S. typhimurium,* and *S. derby* are the most common acute forms seen in pigs. If a pig is infected by a large number of *S. choleraesuis* organisms, it is likely to develop severe disease, starting with septicemia (blood infection) and followed by severe pneumonia and enteritis. Mild exposure to the disease may be asymptomatic because it can lie dormant within the intestines of the pig. Afterward, the bacteria may migrate into a variety of tissues, including the central nervous system and joints, resulting in meningitis and arthritis. *S. typhimurium* and *S. derby* may also cause septicemia and pneumonia, but enteritis is usually the only persistent manifestation. Subacute forms are characterized by foul-smelling diarrhea.

It is imperative that professional help be sought for the diagnosis and treatment of salmonellosis. Laboratory analysis of fecal samples can determine the type of bacteria and indicate the appropriate treatment. Maintain the highest cleanliness standards, minimize rodent infestation of feed, and do not mix groups of pigs.

It is important to remember that *Salmonella* organisms are transmissible to people and are one of the most common causes of food poisoning. Ensure that everyone working with the pigs adopts a high standard of personal hygiene to minimize the risk of infection.

Toxoplasmosis

Toxoplasmosis is a protozoan that affects humans and animals. Cats are the primary source of *Toxoplasma*, and they shed the disease in their feces. Pigs may become infected through eating infected cat droppings, rodents, or other pigs. The organisms form cysts within the pig's muscles and organs. When humans ingest the meat, the cysts can develop into mature parasites.

In humans, pregnant women are most at risk from toxoplasmosis because it may cause abortion or birth defects. Infection can be confirmed with a blood test; most humans never show clinical signs of infestation.

Clinical signs of toxoplasmosis in pigs are uncommon. Confirm suspicions with a veterinarian. Toxoplasmosis is typically treated with sulfonamides.

Transmissible Gastroenteritis (TGE)

TGE is a highly contagious disease caused by the coronavirus. Immediate treatment is necessary, and the older a pig, the better the chance for recovery. Piglets must be monitored and provided electrolytes, neomycin, extra bedding, and an extra heat source. There is a vaccine available.

Symptoms in small pigs include diarrhea, vomiting, and dehydration, which can cause death. TGE can be spread by the herdsman, and care should be taken to clean boots between pens, to keep dogs and birds away from your pigs, and to not visit other farms that raise pigs. All incoming stock should be tested for TGE.

Breeding pigs can be quite challenging, especially for the first-time pig breeder. It may be best to start with previously weaned piglets and raise them strictly for freezer meat. Alternately, purchasing an already proven sow may be a more manageable method for the first time. This will give you some experience in handling the animals, and you will become more familiar with the behavior of pigs.

If you are contemplating keeping a breeding herd, including boars, sows, and growing pigs, keep in mind that a sow eats approximately 2,000 pounds of feed per year and may not breed on schedule or perhaps not breed at all. Then consider the boar, which eats even more and complicates your management system by needing his own pen and other special concessions even when he is not in service.

PIG

BREEDING AND FARROWING

Then look at the averages in piglet rearing. Producers lose, on average, about 25 percent of all live-born piglets, most within the first four days of life. Piglets are susceptible to many diseases and health issues during their first eight weeks. Mothering ability also varies greatly among sows.

Breeding and raising your own pigs may sound like a dismal proposition, and, in fact, many things can go wrong. However, it is a very rewarding part of farming. The skills of swine breeding can be acquired over time, improving your chances of raising healthy piglets to adulthood.

If you decide to rear your own piglets, prepare yourself as much as possible by talking with other breeders and researching farrowing. Classes are sometimes offered at colleges, at extension offices, and through artificial-insemination companies. Ideally, a good sow will do all of the work, and you will have to do little more than look on in admiration.

Selection

Selecting breeding stock requires as much knowledge and expertise as possible. Learn as much as you can by talking to experienced pig people, reviewing the breed standard for the breed of your choice, and looking at pictures of quality pigs of your breed. You will need to evaluate each potential candidate's conformation and disposition as well as review any family progeny records available.

Sows and Gilts

Evaluate proven sows based on their age and productivity. Examine their progeny records for the number of pigs weaned in relation to the number of pigs born. Just because a sow has given birth to twelve piglets doesn't mean that she is a good mother. How many has she weaned, and are all weanlings healthy and of good weight for their age? A sow that births twelve and raises two is not nearly as valuable as one that births six and raises six.

Gilts are young female pigs that have not had a litter, so they cannot be evaluated on progeny records. Gilts should be chosen according to their conformation and size; both sows and gilts should have a refined, feminine appearance. Avoid overly masculine traits, such as coarse head, large shoulders, large hams, and burly overall appearance. Count the teats to be certain that the gilt has at least ten well-shaped, equally spaced teats. Avoid those with deformities such as inverted teats, needle teats, or teats that are randomly spaced.

If you intend to breed her right away, the gilt should be of sufficient age to breed, usually about eight months. If available, check the gilt's mother's records to find out the litter size she came from and what type of mother she had.

Boars

Boars determine the success or failure of your breeding program more than any other factor. Usually, a boar is used on all breeding females in a small herd, thereby making all of the offspring susceptible to any problems he may have. Evaluate a boar by looking at his reproductive organs. The testes should be large and well formed, and the penile sheath should be close to the body. A boar should also have at least ten well-formed teats because this trait will pass to his daughters. Boars should have a masculine appearance and be athletic and active, with thick, stout legs to support his weight during breeding. Large hams will help him support himself while mounting females.

DID YOU KNOW?

The world's largest litter of pigs was thirty-four piglets, born to a sow in Denmark in 1961. Labor took two days. Fortunately for mothers, most pigs have only eight to twelve babies in a litter.

The Breeding

Get your breeding area set up ahead of time. Have an easy method of moving the sow to the boar, such as an alleyway directly from one to the other or a pig carrier to move the sow. Get out of the way during mating, but be prepared to remove the sow quickly in the event of excessively aggressive behavior on the part of either pig.

Introductions

Introducing a young boar to a breeding group can be dangerous for the boar. Do not put your young boar in a pen of several females of various ages and sizes and expect him to fend for himself. Females may rough up the boar, causing injury not only to his body but also to his pride. A chastised boar may be hesitant to breed for several weeks or months, setting your program back.

It is best to start a young boar with one female at a time, allowing him to learn the ropes. There is something to be said for pairing an experienced sow with an inexperienced boar, but they should be fairly close in size to avoid any problems. The same applies to gilts being bred for the first time. Mating a very large boar with a small gilt may be disastrous. A small female may lie down and refuse to stand to be mounted. The weight of the boar may cause injury to the female's back.

An older, experienced boar won't take the time to find out if the female knows what to do. Some older boars are rough with their mates, so you should watch carefully to be sure that the boar does not tusk, bite, or otherwise injure the female.

Breeding pairs can be introduced through a strong fence to minimize fighting or to confirm whether your female is in heat. If the female backs up to the fence and allows the boar to nose her, it is likely that she is ready to mate. A female ready to mate also stands rigid, referred to as *standing heat*.

SIGNS OF HEAT

A female goes into heat every twenty-one days, typically displaying the following signs:
- Swollen or red vulva
- Sticky vaginal discharge
- Excessive grunting or squealing
- Mounting or being mounted by penmates
- Nervousness
- Standing heat

Hand versus Pasture Breeding

In hand breeding, you bring the sow to the boar for mating and then move her back to her own pen immediately after the boar is finished mounting. Mating takes place within a few minutes if the sow is ready. If she squeals, runs, and tries to get away from the boar, you may have misjudged her cycle. Put her back in her own pen. Once exposed to the boar, a sow may come into heat quickly as her hormones kick in, so bring the sow back twelve hours later to try again.

Pasture breeding is by far the easiest method for the handler. If you have an experienced boar and sows, you can safely put them together in groups on the pasture, and they will do what comes naturally. The only drawback to pasture breeding is being able to time your farrowing. If you do not see the actual mating, you may not know that it has taken place.

The Mating

Mature boars should be able to service two sows per day, up to forty per month. Junior boars can service one sow per day, up to twenty-five per month. When pen or group breeding, sows should be limited to six to eight per mature boar and four to six per junior boar.

Allow your boar to remain with the sows through only two heat cycles. If they do not successfully mate in that time, you will need to determine the problem. Remember that a heat cycle happens every twenty-one days and can last from two to five days per cycle. By limiting the mating opportunities to two cycles, you will better be able to predict farrowing dates. Any pigs that do not breed within the two-cycle time frame should be evaluated for health issues and culling.

If you are maintaining more than one boar and you are breeding piglets strictly for meat production or as terminal replacement animals, you may benefit from placing one boar with the sows in the morning and another boar with the sows in the evening. This may increase their fertility and litter size. This practice cannot be utilized for purebred herds, however, as all litters are recorded with the actual parents of the piglets. If mating with two boars, it is impossible to tell which sire is responsible without a DNA test.

Gestation and Birth

Pig pregnancy lasts for approximately three months, three weeks, and three days, an average of 115 days. Up until about eighty days, your sow will show no visible signs of pregnancy. Portable ultrasound can be used to determine whether the pig is pregnant.

During the first two-thirds of her pregnancy, a sow will do well on a maintenance ration with a protein content of about 14 percent. Gilts and very thin sows need to be fed additional feed or a feed with higher

CHOOSING THE RIGHT SOW

When choosing a sow, look for a good underline. The sow should have twelve to fourteen functional teats that are evenly spaced. The teat ridge should be under the middle of each side, and the teats should hang straight down. A sow with teats that are angled will not be able to nurse well.

—Jeremy Peters

Sows must have excellent mothering skills. They need to talk to their piglets and remain lying down when farrowing. Sows that get up and circle around are going to crush piglets. Don't keep a circling sow.

—Al Hoefling

In a small herd, a sow should give you two litters per year and wean at least six piglets each time. She needs to be able to raise them to weaning without additional supplement above what the rest of the herd gets.

—Josh Wendland

Note: Many Heritage hogs have only ten teats. This should not be seen as a defect.

fat content so that they continue to gain weight and adequately provide for the piglets they are carrying. The fetal piglets surge in growth during the last third of gestation (roughly the last five weeks).

The extra demands on the sow will continue throughout lactation. Increasing the feed by 25–50 percent is needed to maintain the sow's weight and provide adequate nutrients for the growing piglets.

Prefarrowing

About four weeks prior to farrowing, your sows and gilts should be treated for any internal or external parasites. Review your vaccination program to see whether any animals need booster shots.

Get all of your farrowing pens in order, but don't add bedding yet. Add bedding just prior to moving the sow in to prevent dust accumulation and contamination from other animals, such as barn cats. Make sure that your farrowing pens are clean, disinfected, and well lit. Have your feeders and water dishes ready.

Farrowing

Several signs can foretell farrowing, but not all sows display these signs. During the last three weeks of gestation, sows typically start to show a round protrusion just behind the rib cage. A teat ridge will develop, indicating that the udder is filling up with milk. The teats will start to stiffen and protrude. You may notice a reddening of the teats or that they feel warm to the touch.

Just before moving the sow into the farrowing area, place the bedding and water and feed pans in the pen. Bedding should not be more than a few inches thick: piglets will burrow into extra-thick bedding, making it easy for the sow to lie on or trample them. Use a mildly coarse material with small particle sizes. Wood chips are ideal for the first few days. When piglets become mobile (about three to four days) and have learned to get out of the sow's way, you may add straw to allow for extra drainage and some material to burrow in.

Move a sow or gilt into the farrowing quarters about one week to four days prior to their due date to give her time to settle in. Moving a sow too close to her farrowing time may result in a loss of piglets because she may be anxious over her new surroundings or afraid when separated from her usual penmates. An anxious or nervous sow will spend more time getting up and down during the farrowing and will be more likely to crush her piglets or step on them. If you are farrowing several sows that are normally housed together, try to place them near each other in the farrowing facility.

Allow the sow to move the bedding around as she sees fit; this behavior is a natural nesting instinct. Sows usually build nests two days to a few hours before farrowing, which is another indication that they are getting ready.

Approximately one week prior to farrowing, the sow's ration should be adjusted to one with a 14 percent protein level and an added laxative such as molasses. The laxative softens the stool without dehydrating the sow and reduces milk production for the first three days. Reduced milk production prevents the piglets from scouring by overeating, and it prevents udder edema and milk's caking on the udder.

The Sow's Behavior

The sow will become restless just prior to farrowing. She may get up and down several times, trying to position herself for the birth. Try not to disturb her while she is settling in or during the actual birth process.

A sow in labor appears to be in a meditative state. She will roll up on her side and extend all four legs outward. During contractions, she will pull the back legs forward and against the stomach.

TEMPERATURE

Boars are less likely to breed in very hot weather; excessive heat actually reduces sperm count. Limiting breeding to the cooler parts of the day or evening may increase fertility. High heat can also have adverse effects on sows, as heat can promote the loss of embryos.

A good sow will start to build her nest twenty-four to forty-eight hours before farrowing. She will build a nest that is a mound, not a hole. A mound will help prevent mashing the piglets. Good mothers talk to their piglets, and they do not get up during farrowing or shortly after.

—Al Hoefling

Gilts don't grow much after they've farrowed, so breeding a young gilt will result in a small sow. Her capacity will be impaired by her size, and she'll have small pigs.

—Josh Wendland

Heritage breeds tend to have excellent instinctual habits. Give them a clean, healthy environment, and they will do the rest. The less you interfere, the better the piglets will be.

—Bret Kortie

Once farrowing starts, the piglets should come along at about twenty-minute intervals. Some people believe that the smallest pigs are born first and last. Piglets can be born head first or feet first; either way is normal. Piglets are typically born with the umbilicus still attached to their bellies. The umbilicus is highly elastic and stretches to great lengths. There are also several weak spots along the cord that allow it to break away from the piglet's body, freeing it from the placenta. Unless the cord is excessively long and interferes with the piglet's movement, it does not need to be removed. Remove an extra-long umbilicus by grasping it at about an inch from the body and at an inch farther down and then pulling the two sections apart.

Sometimes, a herdsman will cut the cord to a uniform length and dip the umbilicus in a weak iodine solution. This is not as critical to the health of the piglet as it may be to other newborn livestock. If you are able to handle the piglets as they are being born, clipping and dipping is a good precautionary measure against infection. A clean, dry environment is the best way to avoid umbilicus contamination.

After all the piglets are born, the sow will pass the afterbirth, or placenta. A sow generally eats the afterbirth if it is not removed, so remove it. Some believe that eating the placenta will encourage cannibalism in sows, and most commercial facilities remove the afterbirth to prevent the spread of disease and for hygiene benefits to the piglets.

Aiding the Birth

Sows generally farrow without trouble. If the sow has labored for more than thirty to forty minutes without producing a piglet, she likely needs help. If you have never assisted in a birth before, contact your veterinarian immediately. Your sow may have a large piglet stuck in the birth canal or a stillbirth, or she may have just exhausted herself beyond the point of continuing.

If you are assisting the sow, clean your hand and arm thoroughly and lubricate it well with an obstetric lube from your first aid kit. A mild antibacterial soap will suffice in a pinch. Clump your fingers together in a bird-beak shape and slowly enter the vulva. Gently push forward until you feel an obstruction is felt. The birth canal

is very small, without much room for maneuvering. If you feel an obstruction, try to identify which end of the piglet you have.

If the piglet is presenting head first, gently grasp the head and pull toward you at a downward angle. If it is presenting back feet first, grasp both back feet and pull toward you. Do this very slowly; do not yank the piglet out.

Once you have removed the obstruction, the birth process should resume, and she should pass the placenta after the last piglet. A retained placenta can cause a severe infection. A shot of oxytocin, which causes uterine contractions, should help shed the placenta if the sow is not passing it naturally. Consult your veterinarian if you have not had experience with this drug or with removing placenta.

If your sow had serious birthing problems, talk to your veterinarian regarding the causes and remedies. For example, an overly fat sow will have more problems birthing, and you may have to initiate weight-management practices before breeding her again. If, however, the sow had problems of a genetic or dispositional nature, you should consider culling her from your herd.

Postfarrowing Watch List

Once your sow has farrowed, monitor not only the piglets but also the sow's demeanor. A lack of appetite, constipation, low milk production (hypogalactia), or other sicknesses are indications that your sow may not perform her mothering duties well. If the sow seems constipated in spite of her farrowing ration, give her an injectable laxative or add Epsom salts to her ration at the rate recommended on the package—typically administered by the weight of the patient. Constipation may cause straining, lack of appetite, and discomfort, which in turn will cause a sow to stop nursing. Loss of appetite can severely inhibit milk production, so encourage the sow to eat every time feed is offered. Adjusting the ration to a more palatable mix may help. If your sow has a favorite food, such as greens or bananas, add a small amount to her ration to encourage eating.

Failure to produce enough milk to feed her litter is a serious problem in a sow. If the temperatures are sufficiently high, a sow

Feed your sow properly during nursing to ensure that the piglets get a good start in life.

may be too uncomfortable or dehydrated to produce milk. Reduce the temperature in the farrowing area with a cooling pool of water, a sprinkler, or a fan.

The Newborn Piglets

A piglet is born wet, with a thin membrane covering its body. This membrane is designed to come off the piglet easily and should do so on its own. If the piglet has an excessive amount of the membrane on its face, it is best to clear it away to prevent suffocation.

Some breeders like to dry the piglets. Although drying is not necessary, it will not harm the little ones. A warm environment or a heat lamp will dry a piglet in short order.

Watch your newborn piglets carefully to see that all get a belly full of the first milk, or *colostrum*. Colostrum is vitally important to the health of the piglet because it is high in fat (energy) and antibodies that provide natural immunity from disease. A piglet will establish itself on a particular teat and will continue to go back to that teat for the duration of the nursing process. Because of this habit, it is important to keep a watchful eye on the sow's teats to be certain that some do not dry up or develop mastitis because the piglet will continue to suck the same teat even if it no longer produces milk. If this happens, the piglet will lose weight and eventually starve to death. If a piglet begins to lose weight, it is best to move it to another pen for special feeding or to graft the piglet onto another sow for nursing.

Although most sows do a fine job of raising piglets to weaning age, your husbandry methods affect many aspects of the piglets' lives. What you do to maintain the sow's comfort—adequate heating or cooling of the facility, cleanliness, and the proper feed supply—plays an important role in the health and well-being of the piglets.

Nutrients for Sows and Piglets

Sows need additional vitamins and minerals during gestation and nursing. By properly feeding the sow, you are properly feeding your piglets. The best diet for a sow does not provide all of the nutrients that your piglets need, however. Milk does not contain enough iron to create red blood cells in the piglet and must be supplemented through shots or access to iron-rich dirt. If the piglets are not being raised on dirt, a shovelful of dirt (about a gallon or so) can be brought to the pen for the piglets to nibble on. Remember that iron is a trace mineral, and only minute amounts are needed to maintain proper blood-cell formation. If given proper access, a piglet will eat what it needs for a balanced diet.

If you don't provide dirt for your piglets, you will need to supplement them with iron. Check with your veterinarian for recommendations and read the label of the iron supplement carefully to be sure you understand the proper dose.

The easiest method of administration is by injection, and each piglet should be treated within the first three days of life. Injecting a piglet in the ham is an easy, one-person job. Hold the piglet upside down by the rear leg that is to be injected. With the thumb of that hand, push the skin aside from the injection site in the fleshy part of the ham and then give the injection, remove the needle, and release the skin.

Watch for signs of anemia in your piglets. Symptoms include pale skin, rapid breathing, sloppy diarrhea, and weakness. If you see these symptoms, administer an iron shot immediately or contact your vet to administer the shot.

The goal in rations for a lactating sow is to provide enough nutrients for optimum milk outputs without depleting her vital nutrients and diminishing her physical condition. A lactating sow's daily nutrient load requirement is three times higher than that of a gestating sow. Feeding three times the normal ration, however, is neither practical nor healthy. Therefore, a specific ration must be provided for lactating sows.

ANTIBIOTICS AFTER BIRTH

If you had to assist the sow during delivery, you may have introduced any number of bacteria to the womb, and she will need an injection of antibiotics, such as LA-200 or penicillin.

CARE OF PIGLETS

When several litters are born at the same time, the piglets can be grafted onto other sows to even out the litter sizes and the size of the pigs in each group. This will allow the runts to get their fair share of milk and ensure that the sows can feed all they are nursing.

—Robert Rassmussen

The first week is most critical; piglets need to get enough milk to get on their way. Sows that get up and down a lot smoosh pigs: let them farrow in a clean pasture. Let the pigs behave naturally, and let the mother raise them. Castrate at one day old; it's the best time.

—Josh Wendland

Premixed mineral and vitamin supplements made especially for lactating sows can be added to feed for this purpose. Corn and soybean meal are also concentrated sources of energy and protein. High-fiber or high-moisture feeds such as alfalfa hay, beet pulp, oats, or wheat add bulk to the diet but may dilute the nutrient intake. When fed by volume or weight, these high-fiber feeds should be fed as free-choice supplements to the lactating sow, not as a replacement.

An old-fashioned method of feeding the lactating sow is to provide the traditional corn-and-soybean ration with the appropriate supplements and bulky ingredients in a separate, free-choice feeder. Increasing feedings to three times per day encourages the sow to ingest the proper amount of nutrients.

Lactating sows need plenty of fresh, clean water as well. If a sow starts to dehydrate, her milk production suffers immediately. If the dehydration continues, milk production will stop and, in most cases, cannot be revitalized. Be sure to provide water in a container that is piglet-proof or sufficiently high to keep piglets from getting in and drowning.

Identification

You will most likely want to establish a method of identifying each piglet, its parents, and its date of birth. There are a number of ways to do this, including ear notching, tagging, and tattooing.

Ear Notching

There are many systems of ear notching. If you raise purebred pigs, check with your pig's registry to see what system they require. In most systems, a particular right-ear notch indicates the litter number that the pig came from, and a particular left-ear notch indicates the pig's number in the litter—for example, 2 of 9. Ask your local 4-H club or county extension offices for notching instructions or have an experienced pig person instruct you in the proper way to notch piglets.

A notcher is a convenient tool for this task, although small scissors or a razor blade could work. Remember to notch only about ¼ inch because the notch will get larger as the pig grows. If you are clipping needle teeth, you should also notch ears at the same time. These procedures should be done within the first twenty-four hours of life.

Tagging

Tagging is another method of identifying pigs. If the piglets are not mixed with other litters, you can wait to tag the piglets until just before weaning or mixing with others. Piglets can be tagged early, but you will have to use a very small tag. This may be inconvenient later because you probably won't be able to read the small tag without catching the pig. However, large tags on small piglets will cause them to walk sideways or in circles until they get used to the tags. Tagging the piglets early may also upset the sow, and she may refuse to feed them or may even attack them.

Tattooing and Branding

Tattooing is usually reserved for large producers. People who already own a tattooer (used for goats, sheep, and cattle) may find that tattooing is the best method for them.

A freeze brand is a newer type of pig identification. A branding iron, usually made up of numbers or letters, is "frozen" with liquid nitrogen or dry ice and alcohol and then placed on the skin for approximately thirty seconds. The brand will swell, eventually scab over, and then grow back hair void of pigment.

Optional Procedures

Procedures such as clipping needle teeth, docking tails, and castrating are performed at the discretion of the pig owner. If you decide that these procedures are appropriate for your herd, seek the advice and instruction of an experienced swine person. You may have the opportunity to visit another farm and learn these practices firsthand; otherwise, a veterinarian can assist you on your own farm.

Cutting Needle Teeth

Newborn piglets have eight needle teeth: two on each side of the top and bottom. These teeth are not permanent but remain in the pig's mouth until about six months of age, when the permanent adult teeth grow in. Needle teeth (sometimes referred to as "wolf teeth") are very sharp and can cause serious injury to the sow's teats by slashing or cutting, resulting in her unwillingness to nurse the piglets. Needle teeth can also cause injury to the other piglets as they spar with each other and jockey for space to nurse. Runts may benefit from retaining

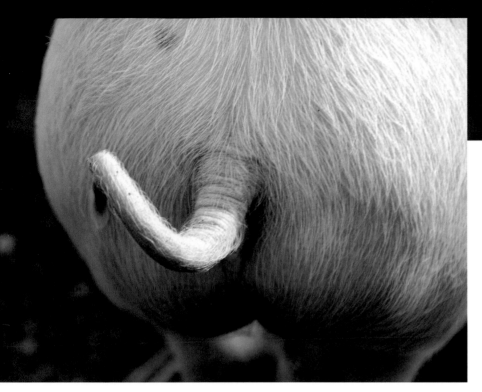

Castrating

Castration is a traumatic experience for boar piglets. It should be done while they are still suckling, preferably a week or so before weaning. If you wean your piglets early (before eight weeks), you may have to wait for the piglets' testicles to be sufficiently large to perform the procedure. Commercial producers typically castrate at three days of age, but these are experienced handlers who have castrated pigs many times. Ask a vet or other experienced person to show you how to perform this procedure.

Assemble your castration tools ahead of time. A new razor blade or scalpel is required, as is disinfectant for the wound. Castration requires either two people or the use of a sling or castration box. A sling is a V-shaped or curved cradle in which you lay a piglet on his back and then hold his back legs forward with a rod or brace to expose the genital area. A castration box is a unit in which the piglet is suspended from its back legs by placing the legs into two straps and tightening. The piglet relaxes once hanging, and the straps keep the legs far enough apart to allow you to perform the castration. Either unit is also handy for giving shots, clipping teeth, or any other chore that requires more than two hands.

If you are using two people to castrate, place the piglet on his back in the lap of the holder. Have the holder keep the back legs held forward (toward the holder). This secures the piglet and presses the testicles tightly against the scrotum.

their needle teeth because this may give them an added advantage over the larger piglets.

Many swineherds with smaller herds do not cut needle teeth and have no adverse effects. Many sows and gilts nurse their piglets with ease, regardless of the needle teeth. After all, pigs in the wild have dealt with needle teeth for thousands of years.

If you decide to cut needle teeth, it is easiest and most efficient to do it within the first twenty-four hours of the piglet's life. If you trim the teeth within six hours of birth, the piglets won't screech too much, and you are less likely to raise the sow's ire.

Nail clippers, small side cutters (pliers with an angled blade), or cuticle shears are the best tools for this task. Holding the piglet with its back against your chest, place your thumb to the far back of the piglet's mouth to pry it open. Snip the needle teeth even with the other teeth; do not cut the teeth off at the gum line. Take care not to cut the gums or tongue. Mark each piglet with a grease stick so that you know which ones have been clipped.

Docking

Docking (shortening the tail) is typically performed in facilities that raise large numbers of animals in indoor crowded spaces. Overcrowded pigs tend toward aggression, and tail biting is one of the ways in which pigs act out on each other. On a farm that has plenty of room for growing piglets, and where the space at the feed trough is sufficient to service all at the same time, docking may be an unnecessary procedure.

If you choose to dock, remove only the last third of the tail, the part thought to be less sensitive. Snip the tail with wire cutters or a similar tool and then treat the wound with a disinfectant. If you dock, do it within the first twenty-four hours of the piglet's life.

MANAGEMENT PRACTICES

According to the Farm Animal Welfare Council, "mutilation practices" such as tail docking, nose ringing, clipping needle teeth, and castration of boars are inhumane and undesirable. The council does, however, recognize the need for such practices under management conditions in which the practice may help avoid worse problems (such as aggression, tail biting, or teat slashing by piglets). These practices do serve a purpose and have been utilized for many years. Make informed decisions about your own practices by learning as much as you can on the subjects.

Young piglets stay warm under an infrared heat lamp.

Swab the area with a disinfectant. On one side, press the testicle against the skin between the thumb and forefinger. Make one cut, approximately ½–1 inch long, to expose the testicle. Make sure you cut low on the scrotum, which will actually appear high on the overturned pig, to ensure proper drainage of the wound.

As soon as you cut, gently pull the testicle out and away from the body. It will be attached to two cords: the sperm duct and a blood vessel. Continue pulling the testicle out until resistance causes the artery to snap off. This may sound cruel or disgusting, but the blood vessel automatically shrinks back into the body, stopping the blood flow.

Repeat the procedure on the other side. No stitches are needed. The wounds can be treated with disinfectant or fly spray if desired, but this treatment is not necessary if the piglets are kept in sanitary facilities. Do not dress the wounds in heavy creams or petroleum-based products because they will inhibit proper drainage and may induce infection.

Place the piglet in a clean, dry, freshly bedded pen. He will likely try to nurse or run to his littermates for consolation. As long as the castrated piglets remain active, alert, and interested in food, you can rest assured that they are not in any distress. Some swelling may appear, but it will go away within a few days. If excessive swelling, bleeding, or pus is evident, contact your veterinarian for proper treatment.

Food, Water, and Amenities for Piglets

Providing the best possible living quarters, feed, and water sources for your piglets are the key to weaning the maximum number of piglets per litter.

Creep Feed and Water

Piglets should be provided creep, or starter, feed that is formulated specifically for very small pigs, within the first ten days of life. Starter feed is a milk-based, as opposed to a soy-based, feed. Milk-based feed is highly palatable and easily digested by piglets; milk solids provide the needed protein (amino acids) for sufficient piglet growth and energy. Plant-based proteins are not easily digested before the piglets are about two months of age, as their intestinal flora has not developed sufficiently to utilize plant proteins.

Creep feed should be placed in an area that only the piglets have access to, such as under a rail or near the heat source. Most will not show interest in the feed right away. Provide only a small amount at first, and refresh or stir it every time you do chores. They will eventually start to nibble and eat.

Water should be provided specifically for the piglets in very shallow pans or in a nipple drinker. Piglets usually start to eat and drink at about two weeks, although some take much longer.

Your piglets must be eating well for a week or two before weaning to ensure that they will not lose weight once they stop nursing. Additional milk can be provided at weaning, but is usually not necessary unless a piglet has stopped eating altogether.

Bedding

When the piglets are nursing, it is better to add bedding on top of the existing bedding in their pen than attempt to replace old bedding. If the bedding is terribly wet and soiled, some removal might be in order, but proceed with caution. Disturbing the sow while she is still nursing may cause her to become agitated and aggressive toward you or the piglets, even to the point of savaging (also known as *cannibalizing*) her own piglets.

Heat Source

Piglets do best when they are maintained at a temperature of about 70°F. This is not practical for most people, especially with winter farrowing. Do your best to provide plenty of bedding, minimize drafts, and provide a steady heat source. Heat lamps are effective for even the coldest of barns. Situate the lamp low enough to provide heat for the piglets but high enough that the sow cannot chew on it. If at all possible, create an area off-limits to the sow, about 3–4 feet from her resting place. Set the heat lamp in the restricted area, and the heat and light will lure the piglets in. The piglets will rely on the

When the time for weaning eventually comes, it will be stressful for both mom and babies.

heat lamp for warmth instead of trying to lie tightly against the sow, which will help to minimize the risk of piglets getting crushed or stepped on.

Heat mats are also effective for keeping piglets warm. Heat mats are rigid plastic mats with an internal heating element. The piglets lie on the mats for warmth, and no additional bedding is needed. Heat mats can be expensive, so place them well out of the sow's reach to prevent her from destroying them.

Power cords from any heat sources, lights, fans, and so forth must be kept well away from both sows and piglets to prevent chewing, electrocution, and risk of fire.

Protection

You will be amazed by how quickly your piglets grow. Seemingly in no time, they will venture out to explore their surroundings. Piglets are master escape artists, but they never stray far from the milk and return frequently to make sure that the sow is still there. Do not allow your piglets to roam into other adult pigpens. Pigs are known cannibals, and a bite-size morsel may be too difficult for an adult pig to resist. You can use panels specifically designed for piglets to keep them with the sows. These panels have small openings at the bottom that piglets cannot climb through; typical cattle panels will not suffice. Boards along the bottom of the pen will also keep the piglets in. They can climb over partitions up to 18 inches high, so place your boards adequately to prevent the piglets from climbing over.

Weaning

When it is time to wean the piglets, it is best to move the sow to her new pen first, usually to be returned to the boar. You will have a much more difficult time removing piglets from a pen if the sow is still in it, and she may get aggressive toward you. If the piglets are to be moved to new pens also, move the sow out of earshot.

In addition to enduring the stress of being taken from the sow, making the transition from milk to a dry feed ration is very challenging to young pigs. Your pigs should be eating well from the creep feeder prior to weaning to make this transition easier. Additionally, the amount of liquid provided by sows' milk needs to be replaced with water. Piglets drink a large amount of water for their size. Ample sources of water, sufficient to allow all piglets access as needed, are an absolute must.

The main goal in feeding starter pigs is to provide ample nutrition for rapid growth. Fresh and dried milk products contain proteins that are highly palatable and digestible to young pigs. Soybean meal may cause allergy symptoms, such as diarrhea and weight loss, in very young pigs. Only a very small portion of the starter feed should contain soybean meal (less than 20 percent) initially; after about two weeks on the starter ration, the soybean meal portion may be increased.

While living with the sow, the piglets were likely already eating the sow's ration. The addition of milk products will help balance the piglet's nutritional needs. Unless you wean your piglets at a very early age (younger than twenty-one days old), you will be able to feed them the sow's ration with no ill effects. If you practice early weaning, it is best to buy a ration specifically designed for early-weaning pigs.

There may be a time when you must wean piglets early, such as when a sow is not producing milk or has cannibalized her piglets. You can properly provide for your piglets by feeding about one cup of whole milk per piglet in addition to the creep feed or mixed with the creep feed as a mash. Piglets relish a warm, juicy mash. Do not feed more than they can clean up between feedings, and do not allow the mash to become sour or moldy. If the mash is heavily covered in flies or has a sour smell, discard it immediately.

ALL IN THE TIMING

Try to wean your sow on a Thursday. The sow will usually come into heat on the following Monday. Placing the boar with the sow on Monday will make your due-date calculation much easier. Your sow should farrow midweek.

Delicious home-cured meats are the highlight of any hobby farm or homestead that produces its own pork. There are substantial secondary benefits as well: you can be confident that your meat is wholesome because you know how it was raised, and you will have taken one more step toward a self-sufficient lifestyle. Pork is also very nutritious.

PIG
PROCESSING AND BUTCHERING

You may be wondering just what you will end up with after raising and butchering your own pig. A typical butcher hog, at 250 pounds live weight, will have a carcass weight (also known as "hang" or "hanging weight") of about 180 pounds, or 72 percent of live weight. This comprises 145 pounds (58 percent of live weight) of retail cuts and 35 pounds (14 percent of live weight) of fat, bones, and skin.

Retail cuts (also known as freezer weight) would be approximately:

- Ham: 45 pounds
- Loin chops: 37 pounds
- Cured or fresh picnic roasts: 13 pounds
- Boston butt roasts or slices: 14 pounds
- Bacon: 29 pounds
- Spareribs: 7 pounds

Additional meat is obtained by saving the feet, tail, neck bones, organs, and trim (used for sausage meat), approximately 25 more pounds.

You need to decide whether to take your pigs to a butcher or butcher them yourself. The decision depends on various factors, including your ability to process the hog, your personal feelings about killing an animal you have raised, your storage space, and the availability of the tools to get the job done.

Scheduling the Butchering

Fall is the traditional time of year for butchering on the farm. This coincides with the natural harvest cycle, and the cooler temperatures help meat chill quickly and thoroughly. If your meat will be processed by a butcher, you can have your hog slaughtered at any time of the year.

Several weeks before you intend to slaughter, evaluate your pig for readiness. Pigs fed to a weight of 225 pounds are ideal. Feeding for additional weeks to gain more fat is a waste of money and effort unless lard is your primary goal. A pig gains approximately 1.8 pounds per day from the size of 150–225 pounds. At this rate, a hog weighing 190 pounds will take about twenty days to go from its current weight to the ideal weight of 225 pounds. Knowing this, you can determine your slaughter date within a few days.

The exact instructions on home butchering are beyond the scope of this book. If you are interested in processing your own meat, refer to the Morton Salt Company's publication *Home Meat Curing Made Easy*. This booklet, originally published in 1941 but updated many times over the years, is the standard to which all home processors should adhere. Additionally, *Butchering, Processing, and Preservation of Meat* by Frank G. Ashbrook and *Sausage and Small Goods Production* by Frank Gerrard are excellent resources for the home processor and meat preservationist.

Selecting a Processing Facility

When evaluating a butcher, look for one that is licensed under the appropriate rating, such as Custom, State-Inspected, USDA-Inspected, or Organic, to accommodate your intended purpose for the meat. Speak with prospective butchers to obtain their ratings. Remember that if you intend to sell your pork, you will need to have it processed through an appropriately rated butcher.

All butchers have their own methods and recipes for curing ham and bacon and for making sausage and other cured meats. The best way to identify the butcher who processes the meats to your taste is to sample his product. By sampling the items you know you want to have, you can determine who makes the meats most pleasing to you. Ham, bacon, and sausage will be your primary cuts of cured meats. This step is imperative if you will sell your pork products. Your product reflects on you, not on the butcher. If your processor does not make a product that is pleasing to your customers, you will lose future sales.

When selecting a butcher, inquire as to packaging methods. If you are raising your pork strictly for home use, you may be perfectly happy with meat packaged in white butcher paper. If you sell the meat, you may prefer to seal the meat in clear plastic vacuum-sealed packages, which are sturdier when being moved around the freezer. Also, many consumers prefer to see what is in their packages. Vacuum-sealed packages cost more to produce, but look at it from your customer's point of view. Does the package look good, clean, and appetizing? Then it will probably be pleasing to your customers.

Some butchering facilities will allow the customer to be on site when processing an animal. Ask if you can be present when a pig is processed so that you can watch the method and understand how a pig goes from farm to freezer. If you are uncomfortable with what you see, or you feel that the facility is lacking in some way (such as cleanliness or staffing), look elsewhere.

Make arrangements with a butcher for processing your pig at least one month in advance of when the pig is ready. Although your processing will take place at the facility, you should become familiar with the various cuts of meat to give your butcher specific instructions on what you want in freezer cuts. Most butchers will assume that you want them to cure your hams, bacon, and sausage. If you intend to make your own cured pork products at home with your own recipes, notify the butcher that you want the fresh (green) hams and bacon for home curing, and be ready to pick them up as soon as the carcass is chilled and ready to be cut or cured. Keep in mind that your home-cured meats will be for your consumption only and not salable in most areas.

MEAT CUTS

Boston Butt	Blade Boston roast, boneless blade roast, blade steak, shoulder roll, cubed steak, shoulder chops, pork cubes, neck bones, fat back, lard
Foot	Pickled or fresh pigs' feet
Ham	Boneless leg, boiled ham, smoked ham, canned ham, ham slices, smoked ham: rump or shank
Hock	Smoked for soup bones, pickle, fresh hocks
Jowl	Smoked jowl or cheek, bacon
Loin	Blade, sirloin, rib, loin, butterfly or top loin chops, country-style ribs, back ribs, smoked loin chops, Canadian-style bacon, boneless top loin roast, tenderloin, blade loin, center loin, sirloin, crown roast
Picnic	Smoked arm picnic, arm roast, ground pork, fresh arm picnic, arm steak, sausage
Side	Spare ribs, slab bacon, salt pork, sliced bacon, lard
Trim	Sausage, pork cubes, ground pork

ADVICE FROM THE FARM
MARKETING YOUR MEAT

Sample a variety of cured products from your potential butcher. If you don't like the flavor of his products, your customers probably won't, either. Customers always relate the quality and flavor of the meat back to the farm. If your butcher isn't good, they will think it is the fault of the meat.

—Bret Kortie

Sanitation in the processing facility is the number-one thing to evaluate. Look for a federal- or state-inspected plant. Visit the site and look for cleanliness. When looking for a butcher, ask how many years he has been in business, what type of experience and training he has, and any specialties he may have. An experienced butcher will give you a higher quality and better-looking product. Your butcher should be someone who is easy to work with; you have to get along so that the person understands what you want in your processing.

—Doug Klarenbeek

About two days prior to slaughtering, move the intended butcher pig to the holding pen to ready it for transport. You can even load the pig onto the transport vehicle and allow it to settle in. Gently handled pigs that are not stressed, overheated, agitated, or frightened will produce much finer and mellower tasting meat. The pig can be washed down if necessary at this time. Remove feed from the pig at least twenty-four hours before slaughter but provide ample amounts of water. Feed withdrawal makes removing the viscera easier and minimizes the chance of contamination from the stomach contents because the gut will not be full of fermenting food.

When transporting your pig to the processing facility, take care to prevent the pig from overheating. Most pigs lie down when being transported. Unless they are excessively crowded or being transported with pigs they are not familiar with, the pigs will remain calm. Do not transport two or more boars that have not been housed together because they will engage in fighting, sweating, and other dominating behaviors. Give the pig time to walk off the trailer or truck at his own pace, without being chased or hurried. If you can deliver the pig the night before slaughter or very early that morning, the pig will have time to settle down.

Marketing Pigs as Meat

The first order of business in selling meat in the United States is to research and learn the applicable laws for your state. Most states do not allow the sale of meat that has been processed in another state unless it was handled through a USDA-inspected facility. If you intend to sell all of your meat within your state, you may be able to use a state-inspected facility for processing. Selling meat by the cut or package is usually governed by USDA standards and must be processed as such.

Selling a pig on the hoof is the easiest method of marketing your hogs. This means that the pig is sold live and delivered to the butcher or picked up by the customer. Pigs sold in this manner become the property of the buyer immediately, and all processing and transport are now the new owner's responsibility. Offering to deliver the animal to the local butcher may seal a deal for you. A lot of people living in town do not want to be introduced to their future food, nor do they have the vehicle for transport.

Assisting potential buyers in choosing the cuts of meat is another helpful service. First-time buyers are often uneducated about the meat that comes from a pig, thinking it all chops, bacon, and ham. Letting them know they have options may be of great benefit to them and will also establish a relationship of trust as you pass along your knowledge.

Talk to the butcher you will be recommending to your customers. Get a list of any specialty products, such as sausages, hams, crown roasts, and smoked meats. Be willing to sample many of his products so you will be able to talk about them with your customers. If your butcher uses a standard cutting form, get a copy of that also. Standard cutting forms are lists of the usual or most popular cuts of pork, the amount you want in your packages, and the size of the portions (that is, thickness of the chops). This will help you discuss the usual cuts of meat and make it easier for customers to decide which cuts they want from the pig they have purchased.

Marketing Organic Pork

If your desire is to market organic pork, you need to be certified by a recognized organic-certification organization. The certification process involves a lengthy application that will inquire about your production methods; sustainable practices on your farm; feed sources used; any fertilizers, chemicals, or supplements used; and your general husbandry methods. You will be required to comply with all standards as established by the USDA National Organic Program (NOP) as well as any additional requirements by the certification

DID YOU KNOW?

Crubeens is the English word for the hind feet of a pig. Served pickled, boiled, roasted, hot, or cold, crubeens are prime pub fare. According to food writer and historian James Beard, they have a delicious flavor, tender texture, and a pleasingly gelatinous quality.

Prosciutto is a cured Italian ham.

board. The best place to start your certification process is at the NOP website (www.ams.usda.gov/nop), where you can review all of the rules and regulations.

With the exception of producing certified organic feed, most small farms already comply with the bulk of the standards of organic certification. If you are able to purchase organically certified feed, you might consider applying for organic certification to boost the sales potential of your products.

Once you are certified, you will also need to find a certified butcher to process your products. A list of certified butchers and the standards by which they must process your meats are available through the NOP.

Selling Processed Pork Products

Selling standard products from your pigs is a perfectly legitimate way to offer pork of higher quality than feedlot-produced products. First, evaluate what cuts and quantities you will sell. Plan to offer a limited number of options in products. For example, offer a pork package that includes chops, hams, bacon, roasts, and one or two types of sausage.

This package simplifies relations with your butcher as well as with customers, as each pig will be cut and processed the same way. Again, check the regulations in your state for selling meat by the package.

If you want to get into the specialty market, no meat lends itself better to specialty products than does pork. Pigs have been used from nose to tail for centuries, and they have been creatively turned into delicacies in every nation that raises them.

Specialty products, such as sausages and hams, will likely need to be processed through a USDA butcher. All recipes for intended

ADVICE FROM THE FARM

MARKETING YOUR MEAT

When trying to sell your pigs, highlight the benefits of buying your pigs and why your pigs are special. People love a good story. Tell them one.

—Dr. Vincent Amanor-Boadu

Creating your pig production model takes time. Be willing to make short-term sacrifices for long-term goals. You may have to give up a few personal extras in order to get your facility producing, but it will be worth it.

—Angela Peters

Only the top 30 percent of animals should be sold for breeding stock; the rest should go for meat. Be prepared to stand behind the stock you sell. Sell only the best, the ones that will produce

the best offspring. Selling whole butcher hogs is the best way to sell meat; just deliver it to the locker plant and the butcher takes it from there.

—Josh Wendland

The US consumer is rapidly becoming aware of health issues associated with confinement hogs being fed growth hormones and given routine antibiotics. Being able to offer a healthy alternative, such as clean meat, is leading niche farmers to success. Advertise your meat as the wholesome product that it is, and you will attract discerning customers. Charge a premium price for this quality; clean meat has value, and you should be compensated for raising it.

—Bret Kortie

Some breeders choose to preserve Heritage or rare breeds like the Gloucester Old Spots.

Look for websites and magazines on topics related to what your farm does. For example, if you raise organic pork, advertise in magazines that promote organic and naturally raised products.

When posting an ad, make your pigs stand out from the rest. Tell about the way they are raised or anything special about them. If you can offer delivery or other services, be sure to mention them.

Purebred pigs can be marketed as breeding stock, as specialty meats, and as butcher pigs. By starting with purebreds, you always have the other two markets available. Be an upstanding breeder, and your reputation will soon follow. Remember the section on screening breeders? You want to meet all the requirements that you had for the person you bought pigs from.

Have the paperwork ready for any potential purebred pig buyer. Offer a receipt or bill of sale and the registration paperwork necessary for the buyers to register their pigs. If buyers come to your farm and you are sure that they will purchase a pig, have the pig already moved to a convenient location for loading.

products must be reviewed and approved by a USDA inspector prior to marketing the product. Labeling is also regulated by the government and needs approval before marketing.

To comply with the regulations of marketing under a brand or specialty label, seek the advice of a USDA-inspected butcher or the meat processing department of your state university. Colleges offer helpful information relating to marketing meat products in your state. Usually, they offer meat-quality testing and can help with recipe development as well.

If you raise purebred pigs or one of the endangered or Heritage breeds, you should advertise your pigs as such to draw in specialty buyers. Each breed's meat tastes different. Highlight the unique qualities of your breed to attract the customers seeking such qualities. For example, pigs with higher fat content are typically attractive to people who enjoy slow cooking and old-fashioned flavors. Diet-conscious individuals seeking lean meat could benefit from some of the breeds known for low fat content. All customers are different, and you should advertise your unique pork in a manner that will connect with those seeking the specific attributes your products have to offer.

Selling Breeding Stock

If you raise purebred pigs, get involved with the association for your particular breed. Typically, a breed association will offer a breeder directory, either in print or online or both, which can help you find customers. Most people looking for pigs will look closest to their home first.

Many websites and animal-oriented magazines offer ad space where you can tell people about your pigs. If cost is an issue, search out the free ones first and start there. You'd be surprised at how many people will find you through these avenues.

Other Options for Selling Pigs

Some farms offer piglets for sale at weaning age with an offer to raise the pig to butcher weight. If you have enough feed stored for the duration of the growing pig's life, you can easily calculate your feed and labor costs for such an endeavor.

Alternatively, you can sell weanling pigs to other farmers who wish to raise the pigs themselves. You will probably get only current market price for your pigs, so this may not be a profitable option. If you anticipate a drought or shortage of feed, or if the market is low for pork, selling your pigs at weaning time may be the best option. You can always sell a few and raise a few to gauge how this system will work for you.

Pigs raised to butcher weight can be marketed through livestock sale houses or directly to butchers. Butchers are often interested in working directly with farmers to obtain their pigs. By establishing a relationship with a butcher, you are almost guaranteed to sell all of your pigs as they become available, especially if you offer pigs that are raised differently from your competitor's pigs or have unique qualities such as fat or flavor. A butcher can tell you approximately how many pigs he can use in a certain time frame, allowing you to plan your production methods and numbers of pigs.

PART SIX

RABBITS

BY CHRIS McLAUGHLIN

WHY RABBITS?

Time and time again, rabbits have proved themselves an ideal livestock choice for small farms. In a world where farming is synonymous with land and acreage, hobby farms create an opportunity to live your rural dream in a smaller space. Even hobby farms located in suburban and urban neighborhoods will find small-scale rabbit keeping an easy endeavor.

MEET THE DOMESTIC

RABBIT

Raising rabbits demands little in the way of financial resources or specialized equipment, and the rewards are considerable. These petite creatures can offer companionship and the chance to participate in competitions. On a more practical level, they can provide meat and wool as well as top-of-the-line manure for your garden and compost piles (it's also an excellent medium for raising fishing worms!). You don't have to choose just one reason to raise rabbits. For example, a rabbit raised for fiber can also be a beloved family pet.

We brought home our first rabbit seventeen years ago, and we still consider rabbits a perfect fit for small-scale hobbyists. Rabbits are easy to house, handle, care for, and transport. They're also curious, affectionate, and extremely entertaining creatures. Keep reading to see why rabbits are so popular with small-scale farmers!

As you may have guessed, rabbits didn't start out as pets. In fact, the term *domestic* in the general sense of the word doesn't refer to being a pet at all. When animals are domesticated, they're kept and cared for by people for any number of purposes, be it for fur, meat, fancy (show), or—yes—companionship. The domestic rabbits we see in the United States today are a direct result of European settlers' shipping Belgian hares (they were actually rabbits) to America in the early 1900s for breeders and collectors. Later, in 1913, New Zealand Red rabbits were brought to America by sailors, who kept them on board as small and easy-to-care-for meat animals. Since that time, many more rabbit breeds have been imported from countries worldwide.

Rabbit History 101

According to geologists, rabbits have been hopping about the planet for 30–40 million years. It's said that the Phoenicians kept caged rabbits as far back as 3,000 years ago, and the Romans followed suit shortly thereafter. Eventually, people in southwestern Europe, on land that's now considered part of Spain and Portugal, saw the wisdom in raising (instead of hunting) rabbits for meat, and so began the heritage of the domestic rabbit.

RABBIT RANKING

In the animal kingdom, the class Mammalia holds a vast number of orders and suborders. Animals are classified according to their traits and habits to make it easier to observe and study them. Rabbits belong to the order Lagomorpha. This name comes from the Latin words *lagos*, meaning "hare," and *morphe*, meaning "form." The order has only two families: Leporidae (rabbits and hares) and Ochotonidae (pikas).

Wild rabbits, like this one, are descendants of European rabbits.

As people migrated from place to place by ship, rabbits either escaped at port or were released from their cages. These rabbits made themselves at home wherever they landed, which created wild populations of *Oryctolagus cuniculus* throughout Europe. Around the sixth century AD, French Catholic monks began raising rabbits for meat. Keeping rabbits within the monastery walls led to breeding them for various uses, sizes, and colors. These monks are still credited today with the first true domestication of the rabbit in Europe.

Although European settlers introduced the Belgian hare to America in the early 1900s, we know that Lop and Angora rabbits were already prevalent in the United States by the mid 1800s. In 1910, thirteen people formed the new National Pet Stock Association of America—an organization that would later become the American Rabbit Breeders Association (ARBA), which now boasts an estimated 30,000 members and recognizes more than fifty rabbit breeds.

The Rabbit Today

Depending on where you make your home, rabbits may be kept as companions or family pets and live indoors or outdoors in an enclosure. Other rabbit fanciers raise their rabbits in a "rabbitry" (a rabbit-specific area or enclosure where rabbits

SPLITTING HARES

Rabbit species are often confused with one another. These furry, hopping creatures all look very similar, but our domestic rabbit species isn't the same or even close to American wild rabbits (called cottontails or brush rabbits). In fact, they're cousins, at best.

The rabbits that we call companions here in the United States have all been domesticated from the European rabbit, *Oryctolagus cuniculus*, which live in family colonies and create underground systems of warrens or burrows that they use to hide from predators and to give birth. Cottontails are of the genus *Sylvilagus* and lead a different lifestyle. They live above ground and give birth above ground in depressions in the earth. Cottontails will, however, use burrows that were created by other animals or brush piles as temporary protection against predators.

Domestic rabbits need supervision outdoors because they are easy targets for predators.

fairs? Are you a knitter or spinner who uses rabbit wool and wants to make a little cash by selling your surplus fiber?

You'll save yourself a lot of time and trouble if you decide what you want at the outset instead of purchasing rabbits on a whim and slowly migrating toward what you'd really like to do with them.

The next step is to look at all of the recognized breeds on the ARBA website (www.arba.net) and then make contact with a club or breeder that specializes in the breed that catches your eye. These people can give you a better understanding of their breed and help you decide whether it's the right choice for you and your plans.

Whether you're planning on showing rabbits or not, I highly recommend attending a large rabbit show and perusing the aisles for a few hours. Even if not every breed is there, many of them will be. Talk to the owners, ask to pet their rabbits, and ask why they chose their breed. Stand around the judges' tables and listen to what they're looking for in a winning animal. Most judges are more than happy to answer questions either during a break time or after the judging.

First Things First

Before you jump into creating a rabbitry and purchasing animals, ask yourself some important questions and answer them honestly and thoroughly.

- How much space do you have? Enough for a few rabbits or an entire rabbitry? Ideally, you want your rabbitry to be on the coolest side of the property. Your answer will also be based on the size of the rabbit breed that you choose. For instance, the amount of cage space you'll need for five French Lops could house ten Netherland Dwarfs. Don't forget to add some space if you'd like a play yard where the rabbits can be turned out to run, dig, or nibble on grass.
- Can you provide protection for your rabbits against weather and predators?
- Will your rabbitry bother the neighbors? A large rabbitry could emit odors. If you think this may be an issue, do you have space in a more isolated area?
- Does your city allow rabbit keeping? Many cities and homeowners associations do not allow livestock keeping in

are raised and cared for) as a hobby for showing. Angora rabbit breeds are raised for both show and fiber because they produce wool that can be spun for knitting. Rabbits are small and easy to care for, and provide healthy meat, which also makes them a smart choice for breeding and raising for food. The savvy rabbit-raiser will research individual rabbit breeds and choose one that interests him or her as well as incorporates one or two of these uses.

The Right Rabbit

With fifty-plus rabbit breeds recognized by the ARBA and a vast array of coat colors and patterns, how do you decide which breed is best for your rabbitry? Well, start by determining what you'd like to do with your future livestock. Do you want to make money by selling them for meat, fiber, or fancy? Do you want to show your rabbits? Do you have children that are looking for a new hobby and can show them at local

A soft coat, gentle personality, and undeniable cuteness add up to popularity as a pet.

their communities. If yours does, it may require a license or permit for the hobby. Be sure to research the requirements of your community before buying any animals.

- How much help do you have for rabbitry maintenance? If you're relying on family members, which tasks are they capable of performing and willing to perform?
- Which facets of raising rabbits appeal to you? Companionship? Wool or meat? Manure to enrich your garden soil?

Rabbits as Pets

Rabbits have kept humans company as long as any domestic animal—and for good reason. They're small, gentle, soft, and extremely quiet. They're easy to keep and have very few specific needs. Pet rabbits don't require vaccinations and can be kept in even the smallest apartments. Rabbits make great pets, especially if you take the time to handle and socialize them.

Rabbit owners who interact with and watch their furry friends find that rabbits have individual personalities and can be very entertaining. (The juveniles are the most fun to watch, with their air-springing antics!) That said, rabbits are easily

---

ADVICE FROM THE FARM
THE RABBIT HABIT

With current trends in sustainable suburban agriculture, rabbits just make sense as the perfect new "livestock" for virtually anyone!

—Allen Mesick

Friends of ours in the 4-H group, who had been participating in the group for many years, suggested we try the "rabbit project." What started as a second-grader's first-year project for 4-H has evolved into a rabbitry with over twenty Netherland Dwarfs shown nationally through various rabbit breeding groups. As the mother

of the rabbit breeder, I would highly recommend the 4-H rabbit-breeding project to any parent looking for an educational program to teach children basic animal care skills and responsibility.

—Brenda Haas

I would like to stress how important it is to have the courage to ask questions when in doubt. Everyone started their rabbit raising hobby somewhere and got to where they are today because of the help others gave them. Most rabbit-raisers are willing to share their tips and tricks for successful raising if you ask. We raise rabbits for a hobby. It is important to enjoy the hobby, learn as much as you can, and have fun!

—Keelyn Hanlon

The Mini Lop is a popular and adorable lightweight breed.

startled and can scratch young children. Rabbits' legs and backs are also easily injured, making it important to supervise kids when they're handling them or caring for them.

Although rabbits prefer to have all four paws on the ground, once a rabbit is completely comfortable with you, it will relax its natural instinct and lie on its back in your arms. Now that's trust! Most rabbits are up for some lap time that involves petting and chatting. The best way to do this is to place a towel on your lap to prevent scratches from your bunny's toenails should it try to hop away.

Rabbits can become very fond of their caretakers, but this can only be achieved through consistent contact. Playtime also keeps rabbits physically and mentally stimulated.

Common Pet Rabbit Breeds

It's worth mentioning that any rabbit breed can make a great companion. It truly depends upon the individual rabbit's personality. That said, the majority of pet rabbits are in the lightweight and small categories. Popular rabbit breeds kept as pets include:

- American Fuzzy Lop (4 pounds)
- Dutch (5½ pounds)
- Dwarf Hotot (2–3 pounds)
- Himalayan (4½ pounds)
- Holland Lop (4 pounds)
- Jersey Wooly (3 pounds)
- Lionhead (2–4 pounds)
- Mini Lop (6½ pounds)
- Mini Rex (4½ pounds)
- Mini Satin (4 pounds)
- Netherland Dwarf (2 pounds)
- Polish (2–3 pounds)

RAISE A FRIENDLY RABBIT

Always remain aware that, first and foremost, your rabbit is a prey animal. As rabbit-raisers, the only way to avoid evoking rabbits' defensive behaviors is to gain their trust. But humans are predatory animals, and rabbits instinctively know this. Thankfully, hundreds of years of domestication are on your side. Rabbits bred to be pets are distant cousins to the wild rabbits in the world today, so it shouldn't take long for you two to become friends.

Your rabbit will be an affectionate pet if you put some time into handling and socialization.

Housing and feeding rabbits isn't complex in the least. It does, however, require
some common sense on the part of the person running the rabbitry. Of course,

PROPER HOUSING AND
NUTRITION FOR

RABBITS

The Rabbit Residence

Where your rabbits will call home is one of the most important decisions that you'll need to make. The following considerations may sound like a lot at first, but, as I said earlier, all you need is a little common sense to get it right.

Placement and Protection

If you're starting from scratch, the most logical first step is to decide where the rabbitry should be situated. One of the most important housing issues you'll face is safeguarding your rabbits from the weather. Because rabbits won't survive direct sunlight and aren't especially tolerant of heat, it stands to reason that they should be kept on the northernmost side of the house. You'll still need to provide protection from the sun, but this will be easier to do in a cooler area.

Even if you live in a cooler climate, you should err on the side of caution, as most cool places still gather some brilliant sun in the summer months. Rabbits do best in temperatures between 50 and 75° F. And as long as they're protected from rain, snow, and driving wind, they'll be fine in lower temperatures. But even in cooler temperatures, rabbits can't survive the summer sun bearing down on them. In fact, heat is the toughest thing for them to deal with. At more than 85°F, rabbits begin to get uncomfortable, even in the shade.

When the temperature rises, place a freestanding water mister or fan near the hutch or a frozen water bottle in the cage. If the rabbit continues to pant, and you don't have fans or a mister system, the rabbit should be brought indoors. If you live in an area that sees frequent high temperatures, and you don't want bunnies running around your home for the summer, be prepared to have a cooling system in place.

When deciding where to place your rabbitry, also think about what you'll need to maintain it. If you don't want to haul a hose or an extension cord a long distance, place your rabbitry near utilities. But if you're going to have electricity running to your rabbitry, make sure that you don't store your hay anywhere near the outlets or lightbulbs. If you're really concerned about potential fires, you can always add a smoke detector to your facility's decor.

Housing Basics

Some rabbit keepers make their own all-wire cages or hutches because, in the long run, it's a major money-saver. But those who don't mind the price and want cages right away can, of course, purchase them. Some suppliers are cheaper than others, so shop around for the best deal and keep an eye out for sales.

When I talk about a "rabbitry," I'm referring to the large structure that houses the smaller cages. Rabbit owners use a variety of buildings to house rabbits and their cages. The structure need not be elaborate; a simple lean-to can suffice. But if you use a structure that isn't fully contained, then you'll need something that can be pulled down over the front of the cages to protect the rabbits against sun, rain, wind, and snow. If a rabbitry has a very angled rooftop, it can help, but a tarp or something similar must be attached to the structure for added protection.

Your rabbit will appreciate some type
of bedding on the floor of its cage.

In addition to weatherproofing your rabbitry, it's vital that you provide protection from predators, such as raccoons, foxes, and weasels. But make no mistake, the "predator" that most often causes damage and fatalities in the rabbitry is the family dog—not just yours, but neighborhood dogs as well. The key is to secure your rabbitry against any and all possible predators to prevent heartbreak.

You can provide protection one of several ways. For one, you can house your cages inside a small barn or shed that has doors. Another option is to enclose the rabbitry with a chain-link dog run that has a gate and top. A garage or mudroom can easily be turned into a rabbitry. Or, if the cages are inside a simple plywood outbuilding, you can attach a cover to the front of the roof so that it can be rolled down and secured for the night. Removable sides for the cages themselves are also a good idea. I know a couple of rabbit keepers that even have removable wooden panels for their hutches.

When it comes to rabbit cages, you have more choices than you might think. First, you have the traditional rabbit hutch built with wood and welded wire. While we're on the topic, don't ever use a hutch that's been built with chicken wire; it's much too hard on rabbits' feet and is easily broken by predators and rabbits' teeth. People generally like the look of rabbit hutches, and you can paint them whatever colors you like. You can also purchase removable wooden fronts to keep out inclement weather.

The downside is that the rabbit inhabitants will probably chew on the hutch and soak the wood in urine if the hutch isn't constructed so that the rabbit can't easily reach the wooden frame. Hutches can also be costly to build or purchase depending on how many you need. On

the other hand, you can easily find used wooden hutches online or in classified ads.

You can opt for the all-wire "hutch" or cage. Many of these cages come attached to another cage. Some of them have slide-out litter trays. One of the most popular wire cage arrangements is the "stackable"—stacked three to four cages high on top of one another and secured by metal legs. Each stacked cage has an individual pull-out tray. Stackables take up the least amount of space in the rabbitry. There are also rabbit cages with solid bottoms, which are especially nice for rabbits living inside the home, but they work just as well outdoors, too.

A rabbit cage can be constructed on the ground as long as there's some sort of barrier to keep the rabbit from digging out and anything else from digging in. A wire cage sitting on bare earth with straw or similar bedding on the wire floor of the cage works well.

One of the drawbacks to having rabbits on soil is that they are more likely to contract worms. Most rabbit breeders only occasionally have to deworm their herds because they use elevated cages. Rabbits that are kept in contact with the ground, however, may need to be on a regular deworming program.

Build Your Own Cages

If you're planning on keeping three or four rabbits, cages are probably the least expensive way to house them. However, if you're planning on keeping a large rabbitry, or one that you know will grow, the most economical choice may be to build the cages yourself so that you can purchase the building materials in bulk.

Another great reason to build your own cages is that you can design them yourself, taking advantage of every square foot of space in your rabbitry. Once you're good at it, you can even create a side

HOW MUCH CAGE SPACE?

The rule of thumb for cage space is 1 square foot of space per pound of rabbit. So, a 6-pound rabbit needs a cage that's at least 2 feet wide by 3 feet long (6 square feet) to be comfortable. I personally prefer to offer a bit more room than that. Most cages are 18 inches high or a little taller, but for the smallest breeds, such as Netherland Dwarfs, you can purchase shorter ones.

Wire cages provide safe confinement as well as ventilation.

Baby-saver wire. This is a special wire that's used for the sides of cages housing a breeding doe; so expect to pay a little more for cages that are built with it. The wire is very efficient at keeping kits inside the cage should they accidentally come out of the nest box while they're still young. The bottom of baby-saver wire has ½ × 1-inch mesh while the top graduates to 1 × 2 inch. If you're breeding rabbits, it's worth the extra expense.

Outbuildings

You can use a potting shed, barn, garage, or any other existing outbuilding to house your rabbit cages. If you don't have something readily available, you can always build a shed structure specifically for the cages—somthing just large enough for the cages to slide into. The shelter must be weatherproof and critter-proof but with plenty of ventilation so that your rabbits don't suffocate or overheat.

The Importance of Ventilation

Ventilation is often overlooked when creating a rabbitry, yet it's one of the most important aspects affecting your rabbits' health. All wire hutches have ventilation in spades, but whatever structure houses the cages needs to be ventilated as well. Many prefabricated sheds or barns have to be modified to add additional ventilation. Placing strategic air vents in the rabbitry will allow breezes to blow through, creating better air circulation. If your building needs a little help despite good air ducts, add some fans. If you use fans, you'll need electricity in your rabbitry—an excellent idea in any case because lights are always helpful in the winter. But if your rabbits are right next to the house, you may be able to get away with an extension cord from an outside electrical outlet.

Provisions for Extreme Weather

While it is nice to have some heating in the rabbitry, you don't need to heat it too much, even in the coldest areas of the country. As long as the rabbits are protected from wind, rain, and snow, they'll do very well.

Heat is another story. Each rabbit will tolerate different temperatures, but generally as temperatures near 85°F, rabbits get stressed and become visibly uncomfortable. To ward off potential heat stroke, some breeders freeze 2-liter bottles filled with water and place them in the cages for the rabbits to rest against. However, some rabbits never get the hang of leaning against the bottles, so

business for yourself, selling cages to other rabbit enthusiasts to offset rabbit-keeping expenses. It's something to think about!

Wire Primer

All wire is not created equal when it comes to rabbit housing. Whether you're purchasing or building cages, there's a few things you should know about wire.

Wire specifics. The wire used for rabbit cages should be galvanized welded wire (not chicken wire or hardware cloth). If possible, choose wire that was welded *before* it was galvanized; your rabbit will be sitting on smoother and stronger wire.

Wire for the sides and top of the cage. With wire gauges, the higher the number, the lighter the wire. That means that 14-gauge wire is heavier (thicker) than 16-gauge wire. You can use 16-gauge wire for small rabbit breeds. Medium or large rabbits need 14-gauge wire cages. The mesh (the rectangular holes) should be 1 × 2 inches or smaller.

Wire for the cage floor. You want 14-gauge galvanized welded wire, but, for better support and to prevent your rabbit's feet from getting caught in larger mesh, the mesh should measure ½ × 1 inch. This mesh size will allow manure to drop down to the tray or ground below.

Rabbits tolerate winter weather
much better than they do the
heat of summer.

regular fans, misting fans, garden hose misters, and overhead tube misting systems are extremely useful.

Keeping Rabbits Indoors

If you plan on raising only a couple of rabbits, keeping them inside the home with the family has some nice advantages. Indoor rabbits have the opportunity to play more often with their families, making it much easier to raise a friendly rabbit and enjoy its company. Another plus is that, aside from when you bring rabbit outdoors to play, you don't have to worry about protecting it from predators.

Indoor rabbits are sometimes housed in cages at night and allowed to roam free in the home during the day. Some are only brought out of their cages to play. And some—because rabbits can be litterbox-trained—live cage-free inside the home. However you decide to keep your indoor rabbit, you need to take some precautions before you allow it to run around the house.

Rabbit-Proofing Checklist

- Relocate plants to a nonrabbit room because many are toxic to rabbits. Eating too much of any plant can give rabbits diarrhea and an upset stomach.
- Get down on all fours and look for any hidden nooks and crannies that your rabbit might be able to squeeze into and become stuck or lost. Domestic rabbits are burrowers by nature.
- Find all wires and cords that the rabbit could chew and either relocate them or cover them with aquarium tubing. Rabbits have to chew to keep their ever-growing teeth filed down, and

a rabbit can electrocute itself or severely burn its mouth by chewing on wiring.

- Lock up all toxic chemicals, including antifreeze, cleaning supplies, pesticides, and fertilizers, so they can't be reached.
- Place things such as trash bags and clothing out of your rabbit's reach. Should a rabbit ingest plastic or fabric, it can easily develop a gastrointestinal blockage.
- Choose an uncarpeted room for your rabbit to play in. After chewing, the next best thing to a rabbit is digging. It is part of their genetic makeup. Carpet fibers are notorious for causing intestinal impaction.
- Rabbits may also scratch at doors, so many indoor rabbit keepers attach Plexiglas to the lower part of doors to prevent damage.

Necessities and Accessories

In addition to the cage, you'll need some supplies for your rabbits' health and happiness.

RABBIT FEET AND WELDED WIRE

Welded-wire cages can cause a painful condition called "sore hocks" in rabbits; this is certainly an issue for some rabbits with very little fur on the bottoms of their feet. Rabbits are more comfortable with more support underfoot.

Removable hard plastic mats, made to secure to the cage floor and with thin slits to allow manure and urine to escape through, are a terrific solution. With this flooring, the rabbits are happier, and their owners can continue to use well-ventilated and easy-to-clean wire cages.

That said, wire flooring isn't the only way that rabbits acquire sore hocks. Rabbits sitting in their own urine and feces can also suffer this affliction. Solid flooring will only prevent sore hocks if it is cleaned thoroughly and regularly.

can be made of wood, cardboard, or other rabbit-safe materials.

Urine Guards

Many cages that are sold preassembled have what are called "urine guards." These are 4-inch metal strips that run along the inside of the cage to direct urine into the pan underneath the cage and away from the floor and surrounding area outside the cage.

Toys

Some rabbits enjoy having chews and toys in their cages, and nearly anything safe to chew and ingest is fair game: toilet-paper rolls stuffed with hay, apple tree branches, cardboard boxes, and grass mats. When you notice paper products becoming soiled, toss them. Toys that double as gnawing items will help keep rabbits' ever-growing teeth worn down.

Play Yard

Play yards for rabbits are the same as the ones used for dogs. They're portable fencing structures that allow the rabbit a little freedom and a change of scenery. I love having play yards around because it allows me to give my rabbits exercise while still keeping them confined in a safe area. Always place part of the yard in the shade or position an umbrella over half of it. Be aware of the sun shifting as the day goes on. And never leave your rabbits unattended in a play yard.

Running the Rabbitry

The first step to enjoyable and successful rabbit keeping is to have a sound management program. As the manager of your rabbitry, you decide what program will work best for your situation. Following are some considerations.

Food Dishes

Ceramic crocks can work well for holding pellet feed, but rabbits may flip them over, so the best crocks are the heavy ones. Rabbits may also sit in their crocks, ending up with manure and urine in their food. For pellet feed, I prefer a metal self feeder called a *J feeder* that attaches to the cage from the outside with only the bottom lip sitting inside for the rabbit to feed from. The bottom of the feeder has wire mesh so that the dust from the pellets falls through. The J feeders hold so much feed that you'll find yourself refilling them less often than you would using crocks.

Water Containers

You can offer water in a crock or in a water bottle designed to attach to the outside of the cage with only the nipple of the bottle sitting inside. If you have a large herd, you may want to use an automatic watering system with plastic tubing that runs through the rabbitry and connects to every cage. Small nipples sit inside the cages, similar to the water bottles. I prefer individual water bottles to a watering system because I can tell at a glance if my rabbits are drinking enough water. A rabbit that isn't drinking may be ill, or the nipple may not be working properly.

Hay Rack

A hay rack is nice to have inside the cages; it keeps hay fresh and off the floor.

Hiding Place

Although it's not necessary, rabbits feel more secure when they have a place reminiscent of their wild ancestors' warrens. These havens

CLOTH CAUTION

Don't put towels, blankets, pillows, or anything cloth inside your rabbit's cage. Rabbits will chew and ingest cloth but won't be able to vomit it up. This can cause a gastrointestinal obstruction that can only be rectified by surgery; unfortunately, most rabbits don't make it to surgery in time.

Indoor rabbits benefit from supervised time for outdoor exercise.

Daily Management

- Replenish food and water every day. If you're using water bottles or crocks, you may have to refill the water twice a day in hot weather.
- Look at every rabbit, keeping an eye out for problems or health issues, such as unusual behavior, loose stools, watery eyes, or runny noses.
- Handle the rabbits daily if possible. Your rabbits will be friendlier, and you can keep a closer eye on their body condition.
- Check on any litters if you have does raising kits. Lookout for illness and change wet bedding in the nest box.
- Spot-clean cages with a handheld scoop. This simple task makes weekly cage cleaning simple and fast.

Weekly Management

- Handle your rabbits. If you aren't handling them daily, at least do it weekly. Check their teeth for malocclusions or breaks.
- Make repairs to cages and other equipment. You'll thank yourself later for staying on top of the small things.
- Clean all cages weekly (at least). This is one of the best ways to raise healthy rabbits, and it keeps the fly population down.
- Check your calendar or hutch cards for does that are preparing to kindle (give birth) in the next week so that you're prepared.
- Thoroughly clean nest boxes that are removed from cages, or add nest boxes to cages with does that'll be kindling in the next week.
- Check your supplies and make note of when you will need more feed, hay, bedding, and so on.

WATER

Water is a top priority when it comes to proper rabbit care. Cool, fresh water is a must *at all times*. In fact, rabbits won't touch their food if they're out of water. If a rabbit goes off its feed, the first thing to check is the water supply. Is the crock or water bottle empty? If you're using water bottles or a watering system that delivers water through a nipple, check that the nipple is letting the water flow freely.

Monthly Management

- Give the entire rabbitry a once-over. Do you need to add fans or misters for expected hot weather? Extra straw bedding for a cold front?
- Check the litters again. Which litters have outgrown the nest box? Is it time to separate them by sex? Any runny eyes or noses?
- Check and restock your first-aid kit.
- Trim every rabbit's toenails and check every rabbit's teeth.
- Fill out pedigrees for new kits, update hutch cards, and update your expense book.
- Mark upcoming shows on your calendar and make sure that you have everything you need to prepare for those shows.

The Importance of Record Keeping

If you have only one or two rabbits, record keeping may seem like overkill. But as your rabbitry grows, you'll be amazed at just how easy it is to confuse sexes, ages, and breeding dates. Specific details are also extremely important if you plan on making money with your rabbits.

In rabbit keeping, a *hutch card* is attached to the front of a rabbit's cage, and it contains applicable information pertaining to that rabbit, such as name, age, breed, which buck bred the doe, litter due dates, how many kits were born, how many kits survived, and so on.

Although hutch cards are very handy for on-the-spot info, I keep detailed records in a hard file or on a computer program. Pertinent information includes breeding dates; which doe was bred to which buck; birthdates of litters (transfer those dates to individual rabbit cards *on the day* they are separated); pedigrees for each rabbit, filled out completely; rabbitry expenses; and any income from your rabbits.

Keeping accurate records show you which rabbits are producing the highest quality litters, which rabbits do the best at shows, which pairings work and which don't, and how much you're actually making

Be a rabbit ambassador so that everyone loves your rabbits as much as you do.

You always have the option of using old-fashioned sticky flypaper ribbons. They work pretty darn well, but dust and debris in the air shorten their shelf life, and woe unto the person that gets his or her hair caught in it! Another fly-control system that many rabbit-raisers rely on is a battery-operated dispenser that automatically releases metered doses of pyrethrin, a pesticide derived from the pyrethrum flower, every so often. This capitalizes on an effective organic gardening method that won't harm your rabbits.

Be a Good Rabbit Ambassador

The best way to achieve peace, acceptance, and even support for your rabbitry is by being considerate of your neighbors. Don't overwhelm your property with livestock, keep healthy and happy animals, and follow excellent husbandry practices.

If the sight and smells of your rabbitry annoy the neighbors, they may complain to the authorities. Of course, you want to be on top of your city's zoning laws pertaining to rabbit keeping. Even if you abide by all of the city's regulations, your neighbors' complaints will still cause a stir. Aside from the fact that you'd like to enjoy a harmonious relationship with your neighbors, you also want to be a model ambassador for rabbit keeping.

Feeding Your Rabbits

There are as many opinions on feeding rabbits as there are rabbit keepers. While hobby farmers are looking to meet their rabbits' nutritional needs, they're also looking to do it simply and

with your endeavor. Computer programs made especially for hobby farmers allow you to keep track of your rabbits' lineage, tattoos, show records, and pedigrees and your own business expenses. Many rabbit keepers feel that this is the best way to go as far as accuracy and simplicity.

Controlling Flies in the Rabbitry

Flies are attracted to rabbit urine and manure, and they show up practically overnight. Fly control is best addressed before you bring home your rabbits so you can launch a preemptive strike. Flies aren't just a nuisance—they also can spread disease. A rabbit with loose stool is a target for fly strike, which is almost a sure thing if you have a fly problem. A fly strike is when a fly lays eggs on a rabbit, and then the larvae eat the rabbit's flesh when they hatch. Not a pretty picture!

The first way to keep flies to a minimum is to clean the cages as often as possible. Spot-cleaning in between full cleanings keeps manure from piling up. The best place for rabbit manure is the compost pile, but mix it in rather than just dropping it on top of the pile. This will help keep flies from hovering around the compost pile. Red worms can also help decompose the manure, creating wonderful compost in the process.

Flytraps alone won't make a big dent in the fly population—not to mention that the traps smell terrible! Fly parasites and natural predators are another option; you can order them online. These parasites hatch and kill the fly larvae before they can grow into adult flies. I've found this approach to be helpful, but, once again, not as a sole preventative.

HORTICULTURAL LIME

For cages that have no drop pan underneath, if you're not raising red worms directly under the cages, spread horticultural lime there; this is said to make the ground inhospitable as a breeding area for flies. Use only horticultural or "barn" lime, which is gray in color—hydrated lime is pure white and not safe to use in this manner.

Pellet feed contains balanced, complete nutrition for your rabbits.

economically. Keep in mind that you want to provide optimum-quality feed. You don't have to stop at pellets, however. Your rabbit can enjoy grass hays and other goodies, as well. Your best bet is to come up with a plan that works for you and your rabbitry and modify it later if necessary.

Commercial Pellets

High-quality pellet rabbit feed should make up the majority of your rabbits' rations. These cylindrical pellets include everything your rabbits need—protein, phosphorus, calcium, and trace minerals from varying ratios of alfalfa and other hays, such as timothy or oat. I've found that my rabbits prefer some brands over others due to the quality and freshness of the hay. In a pinch, I've purchased an off-brand feed only to have my rabbits turn their twitchy noses up at it.

A nice pellet feed will look rich and smell fresh. Those of lesser quality will look incredibly light and dry and not have much scent when you stick your nose into the bag. That said, if you choose a pellet feed that's made of timothy, it may be lighter than alfalfa pellets, and this is perfectly normal and fine. In fact, timothy pellets will be higher in fiber, which is a good thing. However, you might be limited to what is available in your area unless you order your food online.

I follow a general rule of thumb for feeding adult rabbits: give them as much as they'll eat in a day. If you find that there are always pellets left in their bowls, they're getting too much. But if they jump on top of your hand every time you fill their bowls, then you may be offering too little. I've had great luck with this guideline, and I never have to measure my rabbits' feed.

It's important to keep pellet feed as fresh as possible for your rabbits, and this includes keeping out unwanted critters such as mice

and rats. Feed should tightly sealed and kept in a cool, dry place. For me, nothing is more effective (or lasts longer) than old-fashioned metal garbage cans. Absolutely nothing can chew through them, and they seem to keep feed the driest, too.

Certain circumstances will call for a change in portion, a special mix, or an extra nutritional supplement. For instance, you might have to adjust your rabbits' feed when the weather turns cold because they'll need to consume more calories to keep warm. A nursing doe will need more calories to produce milk for her kits. A mature rabbit needs less food than a juvenile rabbit that's still growing. Angoras need a pellet feed that's extra-high in protein to produce their long wool. In addition, Angora owners sometimes feed their rabbits special mixes that help pass ingested wool through their systems to avoid blockages (Angora rabbits ingest a *lot* of wool). You might give show rabbits a special mix to help them gain a little weight as well as to encourage a nice coat. These are just a few examples of special circumstances you may encounter.

Hay

Most rabbit raisers supplement their rabbits' diets with free-fed grass hay. Rabbits are foragers by nature and are healthier and happier when grass hay is added to their daily diet. Most rabbit keepers recommend a filled hay rack for each cage. I always keep oat hay around for my rabbits—it's like magic when it comes to firming up loose stools, and it makes great bedding, too.

Vegetables, Fruit, and Treats, Oh My!

This is a controversial topic. Some people feel that their rabbits' "natural" diet would consist entirely of vegetables. This simply isn't

---

DID YOU KNOW?

When switching brands or quality of pellet feed, be sure to make the switch gradually. Rabbits' systems are delicate, and even a basic switch can throw off their intestinal balance.

A HAPPY, HEALTHY HOME

Sanitation is the key to your rabbits' good health, so cleaning and maintaining is a daily commitment. You cannot let droppings and urine build up, or you risk leaving your animals open to a host of diseases.

—Sami Segale

Proper feeding and nutrition are fundamental in providing your rabbits with the right start. I prefer a 16-percent alfalfa-based pellet fed every night. I also like to give a handful of oat hay once a week to keep their digestive systems happy and running smoothly.

—Keelyn Hanlon

I owe my rabbits' good health to keeping a clean environment and maintaining comfortable temperature control in the rabbitry as well as keeping them on a restricted diet. We use pine wood chips in the catch pans under each cage. Temperature is controlled in the rabbitry by the structure's insulation, fans (not trained directly on the rabbits) in the warmer months, and a dehumidifier in the winter months. Our rabbits enjoy a snack of timothy hay once a week and are on a strict diet of 2–3 ounces of Select Series PRO Formula Rabbit Food as a complete feed.

—Brenda Haas

true. Most wild rabbits aren't sitting in a voluptuous garden, happily munching fresh veggies all season. The majority of wild rabbits forage on grasses and other greens, shrubs, bark, and twigs. They may or may not occasionally run into something a bit juicier in the way of vegetables and fruit. Therefore, a vegetable diet for rabbits is not replicating the diet of their wild cousins.

That said, rabbits are herbivores and can benefit from nutritional vegetables, even on a daily basis. In fact, rabbits can eventually enjoy 1–2 cups of veggies per day, as long as the vegetables are introduced very slowly and don't include watery lettuces that can bring on diarrhea.

Now we're getting to the biggest problem associated with feeding rabbits too many vegetables too quickly: diarrhea. Diarrhea isn't a disease, it's a symptom, but it can kill a rabbit very quickly. Some rabbits can be cured of it overnight if offered oat hay or old-fashioned oats (*not* oats with added sugar), but some cases are hard to cure, and a rabbit can die before its diarrhea clears up. Baby rabbits have an extremely hard time surviving a bad case of diarrhea.

Being an avid gardener, I do enjoy sharing some of my garden bounty with my rabbits, but I only offer vegetables in very small amounts. I also vary the types of vegetables I offer to maintain a good nutritional balance. Because I'm cautious, I monitor those rabbits with sensitive digestive systems and personalize their snacks. I have one little blue doe named Blizzard that gets diarrhea if she has anything more than a tiny carrot. In fact, she tends to get diarrhea so easily that I always add some old-fashioned oats to her food, which works beautifully.

If you want to offer a handful of veggies to your rabbits as part of their daily diet, do so *very* slowly and watch for signs of watery stool. If stools are getting loose, back off a bit with the veggies and greens and offer the affected rabbits some oat hay or old-fashioned oats.

You can give rabbits fruit sparingly, with the same rules of caution and moderation. I give my rabbits apple or pear pieces once in a while, but steer clear of sugary, watery fruits like peaches; these loosen stools almost immediately.

Other treats that rabbits really dig are dandelions (if they're from your lawn, be sure that they are free of herbicides or pesticides), dry bread, parsley, carrot tops, grape leaves, and plantains.

Why Your Rabbit Is Eating Manure

On occasion, you may catch your rabbit eating manure. People tend to be concerned by this; after all, it does seem like a problem. This behavior is called *caecotrophy,* and it is an important part of the rabbit's digestive process.

Rabbits' bodies produce two kinds of feces: the hard, round, black (or dark brown) droppings that you typically find in or under the hutch and *caecotrope,* which is softer, greenish, and mucousy and looks like a cluster of grapes. The rabbit's cecum (a sac that contains healthy bacteria and other organisms to help the rabbit digest food) produces these softer feces, and the rabbit ingests them directly from its rectum. Rabbits use the microorganisms in the caecotrope to help them digest their food properly. This is a nutritional requirement for rabbits; without it, they become malnourished.

WHAT ABOUT WEIGHT?

Make it a weekly habit to run your hand over each one of your rabbits' bodies. A rabbit should feel filled out without having a large stomach or fat pads on its shoulders. You should be able to feel the hip bones, but they shouldn't be protruding or sharp; likewise, the spine should not be obvious, and you shouldn't be able to easily feel a rabbit's ribs.

Rabbit keepers need to learn and practice proper handling techniques for many reasons. First, handling rabbits correctly is the only way to ensure the safety of both the rabbits and the keeper. Second, handling is the primary way that we teach rabbits that they can trust us. Mishandling and therefore unintentionally frightening them only fosters mistrust of human beings. Good handling practices engender calm and secure rabbits, in turn making them easier to handle. Last but certainly not least, all living creatures deserve our respect and kindness. Period.

Understanding where the rabbit is coming from in terms of instinctual behavior should be foremost when trying to catch it, take it out of its cage, turn it over, or carry it around. If you put yourself in your rabbit's place, you'll gain insight on how to handle it in a nonthreatening way.

A RELATIONSHIP WITH YOUR

RABBITS

How Rabbits Behave and Why

One thing you need to remember is that rabbits are at the bottom of the food chain. There's a reason they produce so many offspring—rabbits are dinner for many other animals, and reproducing in vast quantities ensures the survival of the species. Rabbits themselves are herbivores, solidifying their place on the chain—the only thing they hunt for is a good patch of grass.

In addition to rabbits' rapid reproduction, their eye placement also tells us that they're prey and not predators. A rabbit's eyes are set on the sides of its head, which allows for a nearly 360-degree view around it. This comes in handy when everything is trying to eat you. A predator's eyes are placed together at the front of the head, creating binocular vision so they are able to focus on their prey as it runs off.

Humans, as you might guess, are predators. You're a carnivore, or omnivore (generally speaking), and your eyes are set on the front of your face. Although domestic rabbits have been around people for thousands of years, instinct tells them that it's still a strong possibility that you might eat them. Now all you have to do is convince your rabbit that you're one predator that isn't going to hunt it down.

Instinctive Ability

Rabbits are incredibly aware of their surroundings and are ready to take off at a moment's notice should they sense that a predator is near. They have a highly developed sense of smell and excellent vision, even in low light, which makes sense because they're most active at dawn and dusk. They're also tuned in to very low sounds—lower than we humans can hear. Nature has programmed rabbits to be on high alert.

Rabbits prefer to keep all four paws on the ground at all times, which can be frustrating for people trying to handle them. From the rabbit's point of view, a rabbit whose feet aren't on the ground isn't able to hop away and is probably about to become some predator's meal. Therefore, a rabbit that hasn't been socialized with humans will struggle, twist, and jump away from those trying to hold it. Other rabbit defenses include biting with long teeth and scratching with powerful hind legs and sharp toenails. In fact, when a predator has a wild rabbit on its back, this fluffy critter is able to tear open its attacker's abdomen.

Excavation Expertise

Domestic rabbits are programmed to dig—period. They'll dig out of enclosed areas that don't have a bottom. They'll dig in flower beds, on carpeting, and on furniture. There's no point in trying to stop this behavior because they can't help it. Their ancestors needed this skill to dig warrens for protection and raising families. The best way to keep this natural behavior in check is to supervise rabbits in your home and be sure that if you aren't supervising them in an outdoor run, there's a floor to their confined area. You can also encourage your rabbits to dig

DID YOU KNOW?

Rabbits are often said to be nocturnal animals, but they're actually *crepuscular* animals, meaning that they are most active and prefer to eat at dusk and dawn.

only in appropriate areas by providing them with a designated space to enjoy their excavating urges. A large terra-cotta flowerpot filled with soil works well, as does a child's sandbox.

Another instinctual behavior that rabbits aren't going to give up is chewing. As mentioned, a rabbit's teeth grow continuously throughout its life. Your rabbit *needs* to chew on something. If you don't provide something for it to chew on, it'll find something on its own. This might be its wire cage (which can cause malocclusion) or perhaps your phone's cord. Baseboards, carpeting, electrical cords—they're all fair game and all extremely harmful for the rabbit and for your home. Keep your bunny's teeth busy with grass hay, wood blocks, and apple branches to save you both some trouble.

Domestic Rabbits in the Wild

If your domesticated rabbit escaped its hutch, it would instinctually dig a burrow, but that's as far as its survival skills would go. The sharp instincts necessary for a rabbit to survive in the wild have been watered down by generations of domestication. This is why seasoned owners cringe when someone reports that he or she has "set a rabbit free." Although the person feels he or she has done the right thing, the freed rabbit won't survive for very long.

One thing working against domesticated rabbits in the wild is coat color. Fanciers have bred many colors in rabbits for people

---

HARES AND RABBITS

To set the record straight, both the jackrabbit and snowshoe rabbit are actually hares, whereas the Belgian hare is really a breed of rabbit. Like rabbits, hares (of the genus *Lepus*) also belong to the family Leporidae, but there's quite a bit of difference between the two. One of the major differences relates to their young. Baby hares are called *leverets* and are born with more survival advantages than rabbits. Newborn leverets are *precocial,* which means they come into the world completely furred and with their eyes open. They can hop around about an hour after birth and begin nibbling the grass around them within a week. Rabbit *kits* are *altricial,* which means that they're completely hairless and blind at birth, obviously requiring more care.

A mother hare spreads her litters out in several different depressions on top of the ground. This tactic leaves the little leverets more exposed than the babies of tunnel-digging rabbits. But because leverets are capable of seeing and moving around when they're born, they're able to hop about, hide, and return to their birthing area when the mother hare is around to nurse them.

Hares have longer back legs than rabbits, making them stronger and faster. Their ear length is also exaggerated when compared to rabbits' ears. Unlike rabbits that live in colonies in the wild, hares are solitary; they pair up only for mating and then go their separate ways. In addition, hares rely on their speed to escape predators, whereas rabbits tend to hide in burrows or brush piles. Finally, hares have never been domesticated as pets.

The instinct to dig and burrow for safety is innate in all rabbits.

to enjoy, but these unnatural colors do not necessarily blend in with natural surroundings. Coat color can make a domestic rabbit a beacon to every predator in the area, including hawks, foxes, coyotes, raccoons, and even pet dogs. Some domestic rabbits have the *agouti* coat color, a grizzled brown closest to the color of their wild ancestors, but there are other essential survival skills that a domesticated rabbit simply does not have. For example, domestic rabbits are heavier than wild rabbits, which means that they can't escape a predator as quickly.

Even a wild cottontail rabbit has a lifespan of only about a year—possibly three, if it's very, very clever. If a freed domestic rabbit survives as long, it's due to sheer luck and nothing else. The moral of the story? Never let a domestic rabbit loose in the wild!

What Rabbits Like and Dislike

Now that you've read a bit about rabbits, you should have a better idea about what they like and dislike. They certainly prefer to keep their feet on the ground, but we need to handle rabbits as part of their daily care, and we also enjoy their company and often want to hold them in our arms. If you exhibit your rabbits, they'll be picked up and placed on their backs throughout their lives. It stands to reason, then, that a gentle, consistent approach to these tasks is necessary in order for rabbits to become familiar with them and learn to trust us as handlers. Rabbits do seem to enjoy stroking and face rubbing, and they are most relaxed with a regular routine in the rabbitry.

As far as temperatures, rabbits seem to enjoy the same, or a bit cooler, climates as humans so. It bears repeating that rabbits can take the cold extremely well as long as they have protection from the elements, but they can't tolerate high temperatures for very long.

Many rabbits enjoy sharing a cage with a friend. Remember, our domestic rabbits are related to European rabbits, which live in colonies. Unfortunately, providing companionship isn't always

as easy as it sounds. For example, two intact bucks (male rabbits that are not neutered) won't make good cage mates. They'll fight for territory and does and—prepare yourself—end up literally tearing off each other's testicles in the process. If you'd like bucks to live together, they'll need to be neutered to create harmony.

Does can often live together without being spayed, but some are aggressive and refuse to share cage space. You can try to combat this by placing them in a home together that did not belong to either doe previously. Obviously, you shouldn't keep an intact buck in the same cage as an intact doe unless you plan on breeding them. You also should not pair up two does that just don't want to play nice or pregnant does. You do *not* want to house two does together when one (or both) has a litter. This situation can end in disaster if the mother doe becomes protective of her kits or if her roommate sees the kits as intruders.

While I do have a pair of does living together, I also have another extremely aggressive doe that has never gotten along with other rabbits. My solution is to arrange my rabbitry so that the rabbits are close in proximity to each other for company but are not actually sharing a cage.

Aggressive Behavior

Although they may look cute and cuddly, rabbits are completely capable of showing aggression. They may growl, lunge, bite, and scratch with both their front and back feet.

With a suddenly aggressive rabbit that hasn't displayed aggression in the past, your first step is analyzing why its personality has changed. The rabbit may be injured or uncomfortable. Give it a good physical once-over, looking for fur

DID YOU KNOW?

Rabbits, hares, and pikas (all from the order Lagomorpha) have something in common with rodents: their teeth continue to grow all of their lives. Both rodents and lagomorphs must chew constantly to keep their teeth worn down.

Rabbits enjoy time with carefully chosen companions.

or ear mites, lacerations, or soreness, especially in the leg(s) or paw(s).

If your rabbit randomly nips or lunges at you, and you're sure that it's physically healthy, begin responding to bad behavior with a sharp yelp or "Ouch!" If you were attempting to give it a treat and it lunged to get the treat, remove the treat. Don't reward the rabbit with the object of its desire, or you'll reinforce the bad behavior. Conversely, try to use positive reinforcement whenever possible, giving your rabbit a treat or some special attention as a reward for good behavior.

Whatever you do, don't ever hit your rabbit for any reason. The honest truth is that the rabbit won't learn a thing from being hit other than to become more fearful and, therefore, more aggressive. If you're not comfortable handling an aggressive rabbit, ask a more seasoned rabbit-raiser to help you. Turning bad behavior around isn't easy if the rabbit makes you nervous.

The Alpha Rabbit and Aggression toward People

Like all animals (humans included) rabbits living together in the wild instinctually develop a hierarchy within the colony. The alpha rabbit is the pushiest within the group. Dominant behavior allows it to be, for example, the first one in line for food. This, of course, won't sit well with you because you're not willing to be bitten or scratched. Try using positive reinforcement to coax the rabbit out of its aggressive behavior toward you over time.

Another reason a rabbit will show aggression toward humans is fear. Fear can result from an experience with mishandling, abuse, or being chased by people or pets. In addition, if the animal isn't socialized properly and has no experience dealing with people who want to pick it up and cuddle it, it may react fearfully to these types of advances. This is why it's so important to socialize your rabbits from a very young age. The best way to retrain a scared animal is to handle it frequently for short periods of time.

There is one form of aggression that you won't want to correct or minimize. Very often, you'll find that a normally friendly doe becomes dominant and protective when she has new kits. You can't really blame her for that. In fact, you can be proud that your doe is being a good mom.

Food Aggression

To be clear, "food aggression" could technically be called "human aggression," considering it's the human that is attacked when food is brought to the cage. However, the behavior is about the food, not the human.

Rabbits that literally bite (and growl and lunge at) the hand that feeds them are acting instinctually. For a wild rabbit to survive, it has to be the first one to the food, especially in the winter when there may not be many green things to munch on. A rabbit that defends the food it finds is a rabbit that stays alive through the coldest months. *You* are aware that you're bringing the food to the rabbit (not stealing it). But to your rabbit, the action of putting your hand into the cage (your rabbit's domain) and quickly pulling it out in reaction to the lunge looks a lot like a rival rabbit going for the food and then retreating from the aggression. As a result, your rabbit feels that it successfully defended its food.

Unfortunately, this may just be a part of your rabbit's personality that you'll have to live with—there's no bunny therapist to convince it that it doesn't need to defend anything. But you can try some of the following tricks.

Instead of using a food bowl exclusively, place multiple bowls of your rabbit's food in different spots around the cage. The idea is to get the rabbit to feel comfortable and realize that it has food all over the place and doesn't need to defend what you're bringing in. (Or, the extra food can serve as a distraction while you put another food bowl inside.)

If you prefer to feed from a single food bowl, try placing it into the cage in a different spot every time. Moving the bowl from a spot that the rabbit associates with defense may help it snap out of its aggressive routine—or it may just give you an extra second or two to dodge an attack.

Putting your fingers near a rabbit's mouth may invite a bite, even from a typically nonaggressive rabbit.

If all else fails, you might consider using a J feeder instead of a bowl or crock. If your rabbit lunges, your hand is outside the cage, so there's no fear of getting bit.

Handling Rabbits

If you like rabbits enough to raise them, you'll more than likely find them irresistible when it comes to just holding and cuddling them for fun. And if you're going to show or sell your rabbits, training them to be handled means that they'll be better behaved on the show table and in front of potential buyers.

If you've just acquired your rabbits (or you've just added a new rabbit to the rabbitry), let them become used to their surroundings for several days or more before handling them. To get them familiar with you, place them on a towel on your lap or on the floor and do

Hand-feeding your rabbit treats, such as carrots, may help it associate your hand with *bringing* food, which could lead it to be kinder to you during feeding time. You can use a long carrot to keep your hand out of striking distance, if you want.

SOUNDS LIKE A RABBIT

Rabbits have long been considered quiet pets. For the most part, rabbits communicate with people through their body language. Running, leaping, and flopping over onto its side all point to a bunny doing the happy dance, for example. But, the longer you're around rabbits, the more likely you are to hear from them, too. Rabbits make many of these sounds at a very low level.

Cluck. It's a lot quieter than the clucking sounds that chickens make and means that the rabbit is satisfied with what it's munching.

Grind teeth. It's hard to confuse teeth-grinding with purring, even though the sound is made the same way. If your rabbit is grinding its teeth, it's in a lot of pain and needs medical attention.

Growl. A growl usually precedes a lunge and possibly a bite. If the rabbit perceives that you are a threat, it'll have no qualms growling and lunging at you. Forewarned is forearmed.

Hiss. The hiss is used to ward off other rabbits.

Hum. I think all rabbits hum it on occasion, but most rabbit keepers associate it with an unaltered buck wooing his ladylove.

Purr. Purring for a rabbit is lot like purring for a cat in that they're both sounds that signal happiness and contentment. However, a cat purrs with its throat while a rabbit makes the sound by lightly rubbing its teeth together. It's a very soft sound, but one you'll want to listen for.

Scream. The sound of a rabbit screaming will send chills down your spine for two reasons. First, it sounds eerily close to a terrified child. Second, rabbits only scream when a predator is chasing them down or they're dying. It's never a false alarm when a rabbit screams.

Snort. Snorting can come before or along with growling.

Whine or whimper. Rabbits will make this noise if they don't want to be handled, although I most often hear this sound from a pregnant doe that has been put into a cage with another rabbit, especially a buck. Either she doesn't want a cagemate, or she doesn't want a buck making useless advances.

Some rabbits may bite the hand that feeds them as a defense mechanism.

some calm petting. Do this a couple of times before you begin handling them regularly.

Picking Up and Carrying Rabbits

Always remember to put yourself in your rabbit's position. Again, rabbits prefer all four feet on the ground at all times. Your job as the rabbit handler is to pick the rabbit up and hold it in a way that makes it feel as secure as possible. It may kick

HARE'S LOOKING AT YOU, KID

Rabbits, with their curious, kind, and fast-footed habits, have captured the imagination to the point where we've placed them in the center of mainstream media by way of folklore, literature, advertisements, and entertainment. Check out some of the most famous rabbits:

Br'er (Brer) Rabbit—This tricky rabbit is linked to both African and Cherokee cultures but was adapted by Walt Disney in the animated feature *Song of the South*.

Bugs Bunny—With his world-renowned and charming catch phrase, "What's up, Doc?," Bugs Bunny is perhaps the most famous rabbit anywhere. He was created in 1940 by Looney Tunes and Merrie Melodies, which became Warner Bros. in 1944.

Bunnicula—He is the vegetable-juice-sucking title character from the 1979 children's book written by Deborah and James Howe.

The Hare—This is the guy who took too much for granted as he snoozed and lost the race in Aesop's fable *The Tortoise and the Hare*.

Harvey—This is James Stewart's 6-foot, 3½-inch tall invisible rabbit friend from the 1950 movie *Harvey*.

Hazel, Fiver, and Blackberry—All of the characters in the 1972 book by Richard Adams, *Watership Down*, are rabbits.

Leo—He is the main character is Stephen Cosgrove's 1977 children's book *Leo the Lop*, part of the Serendipity series.

Peter Cottontail—He lives in the 1914 children's book *The Adventures of Peter Cottontail* by Thornton Burgess.

Peter Rabbit—Peter is the adventurous bunny in Beatrix Potter's *The Tale of Peter Rabbit*, which was first privately printed in 1901.

Rabbit—He is the gardening enthusiast from the *Winnie the Pooh* stories written by A.A. Milne.

Roger Rabbit—He is Gary Wolf's main character from the 1981 novel *Who Censored Roger Rabbit*? and was later transformed into a cartoon character for the 1988 movie *Who Framed Roger Rabbit*?

Thumper—He was a deer's best friend in the 1942 Walt Disney classic animated movie *Bambi*.

The Trix Rabbit—Remembered by baby boomers for the ads featuring the tagline "Silly rabbit, Trix are for kids!," the General Mills cereal-box mascot was created in 1959 by Joe Harris.

The Velveteen Rabbit—In Margery Williams's 1922 children's book of the same name, this toy rabbit becomes real thanks to the love of a child.

The White Rabbit—Alice follows this rabbit down the hole in Lewis Carroll's 1865 novel Alice's Adventures in Wonderland; in this same book, Alice visits with the March Hare and the Mad Hatter.

Uncle Wiggily—He starred in the 1910 children's book *Uncle Wiggily* by Howard Roger Garis.

Be sure to fully support your rabbit's body when picking it up.

and guides the lifting action. If you want to learn this handling style, it's best to learn it hands-on from someone who is experienced with the technique because performing it poorly can hurt the rabbit.

Whenever you're carrying your rabbit, make sure that you're able to safely put the rabbit down if it becomes frightened or begins to squirm too much. If you think your rabbit may fall out of your arms, simply drop to one knee. Your knee will offer a little more stability as you attempt to get the rabbit back under control; plus, it's easier to set the rabbit down in this position. In the worst case, it's a much shorter distance for the rabbit to fall. If your rabbit jumps or falls from a distance, it can break a tooth, a leg, or even its back. For this same reason, never leave a rabbit unattended on a table.

those strong hind legs and thick, long toenails at you, and when those nails connect with your skin, it's incredibly painful. The smartest idea is to wear a long-sleeved shirt when handling a rabbit, at least until you're certain that it's no longer afraid of you.

The best way to practice lifting and carrying your rabbit is by using a large, flat table that's about waist-high. Place a towel on the table so the rabbit has something to grip when it's put back down. To pick up a rabbit, put one hand under it, just behind its front legs. The lower part of its chest will rest in your hand. Put your other hand underneath the rabbit's rump. The idea is to actually lift it with the hand that's under its front legs and chest while the hand on its rump supports its weight. Now, bring the rabbit against your body for extra support, with its head facing toward your elbow. This is called the *football hold*.

During the first session or two, the rabbit may get away from you and, in the process, leave a scratch. Although it's painful, try to remember that the animal isn't scratching you out of spite or aggression; it's truly afraid and wants to protect itself. The more often you handle your rabbit, the faster it'll relax and realize that it is a common practice without danger. Practice picking up the rabbit and using the football hold every day. Make these practice sessions short; picking the rabbit up a few times over the course of ten or fifteen minutes is plenty. After a couple of days of practice, walk with the rabbit around the table or a short distance away from the table and back again. Then set your rabbit down and start over.

Another way to pick up a rabbit is by the extra skin on the back of its neck. Rather than putting your hand under the rabbit's front legs and chest, grasp a fair amount of skin just behind the ears. Your other hand is once again placed under the rabbit's rump for additional support. However, with this technique, it's the rump hand that does the lifting. The hand at the nape of the neck just secures

Turning Rabbits Over

Thanks to the rabbit's distaste for being belly-up, you need to take your time teaching it to be turned over. Again, you're going to practice with a waist-high table. When turning your rabbit, you'll use one hand to control its head and one hand to support its hindquarters. Flatten the rabbit's ears against its neck and back while reaching around the base of its head with your fingers. You should have a firm grip without actually squeezing. A variation of this hold is to put your index finger between the rabbit's ears (close to the base of the head) and gently but firmly wrap your fingers around its head so that they point at its jaw. In a smooth motion, use the hand that's holding the rabbit's head to lift it up and slightly back, while at the same time tucking its rump toward you.

If the rabbit cooperates with you, the hand that was holding the hind end can now inspect the rabbit's body and teeth. But a rabbit that is new to being turned over will more than likely wriggle free a couple of times. If this happens, try to gently guide it back to its normal position and try again. Don't attempt to force the rabbit into submission, or you could break its back.

If the rabbit begins to struggle, use your free hand to grip the loin area; it may settle down. The real key to performing this technique correctly is learning how to have the proper grip on the rabbit's head, which prevents it from wrestling away from you. After you've practiced this hold successfully a few times, you'll know when you have it right.

Children should always be supervised when handling a rabbit.

Another way to turn your rabbit over is to begin with the animal resting with its four paws on the front of your body. From this position, you can grasp its head and rump in the same way as you would from a table position, but now you can bend gently at the waist to let it rest on its back. This type of turnover doesn't feel quite so drastic to the rabbit, and it may feel more secure starting with this technique. However, if you plan on competing in showmanship in 4-H, you will have to teach your rabbit to let you turn it over from a table position.

Transporting Rabbits

You may need to bring your rabbit to a show, move it to another rabbitry, or just take it to the vet. Rabbit carriers are handy little cages made especially for rabbit transportation. They're usually made of the same wire mesh that wire hutches are made of and come complete with a detachable bottom tray. Carriers are designed to hold one to four rabbits, allowing just enough space for the rabbit(s) to turn around. Miniature travel water bottles are usually attached to the fronts of the carrying cages. Rabbit carriers can be purchased wherever rabbit cages and hutches are sold.

Rabbits may also be transported in small dog or cat carriers. If you choose to use one that isn't especially designed for rabbits, choose one made of hard plastic as opposed to one made of cloth so that the rabbit can't chew on it. Also, choose one with plenty of ventilation, as rabbits can easily become overheated. A cardboard box meant as temporary traveling carrier can be equally dangerous for a rabbit on a long ride.

ADVICE FROM THE FARM
RABBIT WRANGLING

When designing your rabbits' homes, stimulation should be kept in mind. Rabbits are actually quite smart and can benefit from toys, ramps, cage proximity to other rabbits, the occasional treat, and regular interaction and handling. It is nice to design the rabbitry as a place you can hang out and enjoy your rabbits' company. Daily "rabbit pen time" can be a welcome break from a busy household.

—Sami Segale

The Netherland Dwarf (raised primarily for show) is known for its even temperament because most breeders raise this breed in the "cage" environment. This breed, like any other animal, needs to be raised with plenty of TLC. If you don't hold the bunny, then the animal won't want to be held. The same principles apply when training this breed for show. I have noticed that once a doe has been bred, she does get a little "nippy" if a buck is around. You need to pay attention to her behavior or just keep her away from the bucks if you don't want this aggression.

—Brenda Haas

Handling is an important part of raising rabbits because it is the best way to keep a close eye on your herd. Your animals should be comfortable being handled by you so that tasks such as nail trimming, grooming, evaluation, and judging are safe for you and for the animal. It is just as important that you be confident and safe in your handling of the animals so that the animals will feel safe.

—Keelyn Hanlon

Transporting by Car

If you're traveling in your car with rabbits, be sure that they aren't sitting in direct sunlight, and never leave a rabbit unattended in the car. While en route, keep the temperature of the car comfortable for you, which should be fine for the rabbits. If the air conditioning is on, let it blow directly on the rabbits. If outside temperatures are extremely low, keep your rabbits cozy by adding bedding such as hay or straw to the carrier.

When traveling with rabbits, most people think to bring food and hay, and I always suggest bringing extra water from home. Often a rabbit will refuse to drink because the water offered at its destination has a different flavor or odor than the water it's used to. Generally speaking, rabbits won't drink while traveling in the car. However, once you reach your destination, they'll appreciate fresh water.

Transporting Rabbits by Plane

Transporting rabbits by plane is not uncommon, but you'll need to consider a few things before you head for the airport. Contact whatever airlines you're thinking of using and ask questions. Find out their rules about rabbits in the cargo area. Although most airlines do accept rabbits, some cargo areas are more comfortable than others and are temperature-controlled. If not, the airline may not let rabbits board the plane in extremely hot or cold weather.

If you're flying with only one or rabbits, the airline may allow your rabbit(s) to travel in the cabin with you. Many airlines permit small dogs to travel in a dog carrier underneath the seat, so it doesn't hurt to find out if the same offer applies to traveling rabbits.

Whether you're shipping rabbits to someone or traveling with them, they'll need a United States Department of Agriculture (USDA) health certificate stating that the rabbits are healthy and disease-free. You can obtain a health certificate from a licensed veterinarian; it is required by all airlines. Health certificates are time-sensitive documents that come with an expiration date—usually around thirty days. Make the appointment with your vet a couple of weeks ahead of your flight.

To fly, you'll also need a proper carrier, which is usually one of the all-wire rabbit carriers or a basic dog crate. The carrier is required to have your name and address on it (as well as the recipient's name and address if you're shipping the rabbits). The

airlines expect you to provide food and water inside the carrier for the rabbits, and they will check that you did.

Permanent Long-Distance Moves

You can get your rabbits from one part of the country to another in a few different ways. The first is to ship them in advance to another rabbit breeder in the new location so that this person can care for them until you arrive. The advantage of this is that you won't have to worry about moving them and you (and possibly a family) all at the same time. The disadvantage is that the food (and water) will more than likely be different unless you ship some of each as well. The rabbits will also have to adjust to a whole new environment of sights, sounds, and smells without you. Then, after you arrive at your new home, they will have to transition all over again. You can also ship them directly to your new home if someone is there to set up rabbit cages, pick them up and care for them until you arrive. This solution will at least keep the moves to a minimum.

Another option is to travel with your rabbits to the new home. If you're traveling by car, it's best to use carriers that are twice the size of each rabbit to give them some wiggle room. Purchase an extra bag of rabbit feed to bring along with you. This will make it easier to transition them to a different feed should your brand not be available in that area, combining a little of the old feed with the new feed until your rabbits get used to the new stuff.

What's up, doc? As it turns out, not much. When it comes to rabbits and veterinarians, the two don't often meet. Rabbits are not difficult to care for, they usually stay healthy, they have no vaccination requirements, and they have life spans of seven to ten years. Of course, if an animal shows signs of illness, such as lack of appetite, unusual lethargy, weepy eyes, a runny nose, or obvious distress, then you need to visit a vet. Most veterinarians consider our domestic rabbits to be exotic animals, so find a vet in your area who specializes in exotics.

HAPPY AND HEALTHY

Often, rabbit breeders prefer to learn everything they can about their animals to avoid some or all of the costs of veterinary care. This may not be entirely possible, but most of the rabbit-raisers I know are well versed in rabbit ailments and treatments. Whether you have just a couple of rabbits or a full rabbitry, it's prudent to learn simple, safe home care. If you have a large rabbitry, you can save a bundle by knowing the basics.

Because rabbits are prey animals, their natural instinct is to mask signs of illness as long as they can. To do otherwise in the wild would mean ending up as dinner. Domestic rabbits have the same instinct, which is why many rabbits don't end up in the vet's office. By the time anyone realizes that something serious is going on, the rabbit is dead (or nearly so). Know your rabbits. If you handle your rabbits often, you'll be more likely to notice when something's off.

An Ounce of Prevention

All living creatures, no matter how well they're cared for, can contract an illness or have an accident. To minimize problems, it's important to provide the healthiest and safest environment possible for your rabbitry. Most rabbit keepers have stories about averting common health issues by simply taking the time to prevent them.

Start with the most basic necessity—your rabbits' home. The best cages or hutches are spacious, solid, and secure against all predators. They protect rabbits from inclement weather but still have plenty of ventilation. When it's warm outside, rabbits may need fans or misters around the rabbitry or frozen water bottles to lean against. When the temperature drops, you can offer them extra bedding, such as straw. If the hutches aren't housed in a barn-type unit, drop tarps around the entire rabbitry for cold-weather protection. Like all other warm-blooded animals, rabbits burn more energy in the winter in order to stay warm, so give them some extra food. You should also provide extra nesting material for nest boxes in cold weather.

Hutches, as well as any structure built around them, should be thoroughly and routinely cleaned; cleanliness is paramount in maintaining the health of your rabbits. Disinfect the cages from time to time to help prevent the

DID YOU KNOW?

While some over-the-counter medications have been used on rabbits for decades and may be perfectly safe in the appropriate doses, there are times when you'll need a prescription medication. This means a trip to the vet's office. Of course, a good exotic-animal veterinarian would never prescribe something that's harmful to rabbits, but it's good to be aware of what's safe and what isn't. Something as simple as the wrong antibiotic could have adverse and even fatal consequences; for example, penicillin and amoxicillin are lethal to rabbits. Antibiotics that are safe for rabbits include sulfademethoxine and enrofloxacin. Never use another animal's leftover prescription medication without first talking to your vet. Most of the time, if an illness requires an antibiotic, you'll need to see a vet to obtain it.

YOUR RABBIT FIRST AID KIT

The best way to avoid illness and disease is to start with healthy rabbits and use good husbandry practices in the rabbitry. That said, things can go wrong even under the best of circumstances. Having the right medications and supplies handy will allow you to treat an ailing rabbit immediately, which will increase the likelihood of a full recovery. The majority of minor rabbit illnesses can be treated at home. Here are some things you should have in your rabbit first aid kit:

- Antibiotic cream (the kind without added pain relief, which can sting before it helps)
- Beneficial bacteria (probiotics)
- Corticosteroid or salve
- Cotton balls and swabs
- Hydrogen peroxide
- Ivermectin (or other wormer, such as fenbendazole)
- Miticide (or cooking oil)
- Nail clippers
- Papaya tablets

- Plastic medicine dropper
- Rubber gloves, disposable
- Saline eyewash
- Sharp scissors, small
- Sterile cotton pads
- Styptic powder
- Sulfademethoxine
- Tetracycline ophthalmic ointment
- Towel
- Tweezers

I also like to keep an index card in the kit with the phone numbers of my vet and of the most knowledgeable rabbit raiser I know (just in case the situation is more than I can handle). I can't tell you how many times a rabbit breeder has come to my rescue in an emergency.

spread of disease. You can use a bleach and water solution (one part bleach to five parts water) with a wire brush on the cages and then rinse them well and let them dry completely in the sun. Many breeders disinfect a doe's cage before her litter makes its way out of the nest box.

Anything that makes your rabbits nervous can inflict damage in unexpected ways. Dogs shouldn't be allowed to roam loose around the rabbits, even if the rabbits are locked safely inside their hutches. A wandering dog can make a rabbit so nervous that it may race around its cage, flinging itself against the sides, sometimes hard enough to break its back. It's worth your time and effort to make your rabbits' home as stress-free as possible.

Obviously, nutrition plays a big role in health. The number-one thing that rabbits need is clean, fresh water, and plenty of it. The right amount of fresh, quality food and a reliable feeding schedule will also go a long way in keeping rabbits issue-free. I offer not only fresh pellet feed but also fresh hay daily. If you're planning on switching feeds, you need to do it gradually or your rabbits may get diarrhea—sometimes literally overnight. As we know, rabbits become dehydrated from diarrhea very easily and it's often hard to turn the situation around. This is the perfect time to remind you that, thanks to rabbits' sensitive digestive systems, fresh greens have no place in young rabbits' diets. Adults may deal with *some* green treats just fine, but chances are that a baby won't.

Remember that the best thing you can do for your rabbits' health is to observe and handle every one of them every day. This may sound time-consuming, but it'll go surprisingly fast after you get to know your rabbits and your rabbits are comfortable with you. Handling your rabbits will let you feel each one's body and coat condition.

Observing each rabbit for a couple of minutes every day will give you a lot of information about its normal habits. Know how your rabbits normally move, eat, drink, and behave when they see you. When you know each rabbit well, just a glance at the full dish of a food-loving rabbit will trigger warning bells. You'll be amazed at what early signals you'll catch and what disasters you'll avoid simply by spending a little time in your rabbitry.

Health Problems and Solutions

You can easily treat some conditions, such as sore hocks and the first signs of diarrhea, at home. But others, such as hindquarter paralysis or wryneck, are going to need the attention of a veterinarian. If you're not sure what you're dealing with or you're uncertain of how to handle a situation, please seek the attention of a veterinarian. This section shouldn't take the place of advice from your veterinarian.

Noninfectious Health Issues

Many noninfectious health problems are simple to cure. At the very least, animals with these conditions don't need to be quarantined because the disease can't spread to their furry neighbors. Here are some common noninfectious health issues and their respective treatments.

Corneal ulcers. Corneal ulcers are lesions on the eyes, often resulting from a scratch or from very dry eyes. The eye's cornea may look bluish, and you may notice that the rabbit is squinting or tearing. It's also possible that the ulcer may develop a secondary bacterial infection. Temporarily place the rabbit in a dark area to lessen the irritation to the eye. You may want to consult a veterinarian about medication.

temperatures in areas with high humidity. Watch out for quickly rising temperatures.

Affected rabbits should be relocated to a cool environment, such as an air-conditioned house or a cellar. They should remain there for as long as possible; certainly until it cools down outside. Try wiping the inside and outside of the rabbits' ears with a cool, wet cloth or ice cubes wrapped in cloth. The ears are a rabbit's cooling system; blood vessels in the ears are closest to the skin, and cooling them will expedite cooling the rabbit.

If the rabbit is overheated to the point that it looks lifeless, every second makes a difference. Fill a container or sink with room-temperature water (cold water could shock its system, and it could die from the shock). Place the rabbit's body gently into the water, supporting it the entire time. Don't let the rabbit's head go underwater—just soak it up to its neck. Afterward, take it to a shady, quiet place to recover.

Hindquarter paralysis. A rabbit that is dragging its hind legs around due to lack of motor control has dislocated or fractured his spinal vertebra. This sort of trauma can occur if the rabbit gets suddenly excited or frightened and gives a big thrust with its back legs. Often, owners never find out what happened, but it's usually a grave prognosis. Most rabbit-raisers will have the rabbit *euthanized to* end its suffering. However, you should certainly seek immediate professional help for your rabbit.

Hutch burn (urine burn). If the skin on the inside hind legs or genitals of a rabbit appears to have been scalded, it is likely a result of hutch burn caused by the rabbit's own urine. Sometimes the urine guards in the cage are at an angle that causes the urine to splash back onto the rabbit. Hutches that are dirty and wet can also cause the condition. A secondary infection can follow if the problem is not addressed. Correct the angle of the urine guards, clean and disinfect

Diarrhea (nonspecific). First, rule out the possibility of a bacterial infection. If the rabbit seems fine otherwise, it could be that the flora in the intestines is simply off balance. Often, a rabbit being treated with antibiotics will have diarrhea. However, nearly every intestinal illness and parasite can cause diarrhea as well.

Diarrhea requires immediate treatment. If the rabbit is on antibiotics, discontinue them until you can figure out the reason for the diarrhea. Add healthy bacteria to the rabbit's gastrointestinal tract by giving it a probiotic. Many rabbit breeders have found fresh oat hay (or old-fashioned oats) to be helpful for drying up diarrhea.

Fur chewing. If a rabbit is chewing the fur off of its body (or its neighbors' coats), it may be bored or nervous or not have enough fiber in its diet. It could also be coming down with enteritis (inflammation of the intestines) or could be experiencing a behavioral issue. Try giving the rabbit in question some magnesium oxide and adding more fiber to its diet with hay or other high-fiber foods. Provide some toys for its cage to alleviate possible boredom, and keep it away from other rabbits until the problem is resolved.

Heat stroke. A rabbit that is stretched out and breathing heavily may be suffering from heat stroke. Sometimes it will hold its head high and mouth open to aid in breathing. The fur around its mouth will be wet. If a rabbit in this condition isn't tended to immediately, death is almost certain. Rabbits are at risk for heat stroke when temperatures climb to 85°F and higher and possibly at lower

SWITCHING FOODS

Purchase the new food well before you run out of your current food. Start by replacing just one quarter of your rabbits' daily rations with the new pellets for a few days. Increase the new feed and decrease the old feed a little more every few days until the new feed fully replaces the old.

the cage, and use an antibiotic ointment on the affected areas of the legs and genitals.

Malocclusion. Also known as *buck* or *wolf* teeth, malocclusion occurs when the rabbit's top or bottom teeth grow extremely long and possibly crooked. When this condition is severe, the teeth interfere with the animal's ability to eat and cause wounds in its mouth. Some malocclusions can be addressed; others can be fatal.

If the malocclusion appears after a rabbit has been pulling on cage wire for a long period of time, try putting the rabbit into a wireless hutch to see if its teeth will grow back properly. Most of the time, however, malocclusion is a hereditary condition. If the elongated or crooked teeth are inherited, there isn't much you can do except clip the teeth regularly for the rest of the rabbit's life so it can eat properly. If the condition is so severe that the rabbit is constantly in pain due to sores or malnutrition, it may be best to have it

SIGNS OF PAIN OR ILLNESS

Below are common signs to watch for in the rabbitry. Not every symptom by itself means that a rabbit is ill, but each one is a red flag that tells you to look deeper into the situation.

- Lack of interest in food or water
- Consuming food slower than usual or dropping food when eating
- Teeth grinding
- Extreme weight loss
- Unusually aggressive behavior
- Manure caked on underside
- Sitting unusually (such as leaning back on its hind feet)
- Lethargy or looking depressed
- Extremely soft droppings or diarrhea
- Difficulty moving (moving very slowly or not at all)
- Limping
- Head tilting (loss of balance)
- Discharge coming from the eyes or nose
- Dried and caked fur on the inside of front paws
- Facing the corner of the cage, hiding, or sitting in a "hunched" position (if this is unusual behavior for the rabbit)
- Rapid breathing or signs of labored breathing
- Mouth wet with saliva
- Grunting or whining when moving or being handled
- Dullness in the eyes

It's easy to see how an Angora's fur could cause a blockage if ingested.

euthanized. It's considered quite unethical to breed a rabbit with this health issue.

Slobbers. Excessive salivation will cause a rabbit to have a wet face, chin, or dewlap (the flap of skin under the chin of some rabbit breeds). The saliva may end up causing severe irritation of the chin, neck, and front legs. Slobbers is a symptom of another problem, such as an abscessed tooth or contaminated hay. A tooth problem (which is usually the cause of slobbers) requires treatment. If contaminated hay is the cause (it may be moldy or soiled with urine or feces), remove and replace it with fresh hay. Be sure that the rabbit isn't suffering from heat stroke, for which one of the symptoms is a wet mouth.

Sore hocks. This condition occurs when the fur on the bottom of the rabbit's feet wears away and sores or ulcers develop. It's most often seen on the hind feet, but it can happen on the front feet as well. This is very painful, and the rabbit will look uncomfortable while sitting and try to rock back on its heels to avoid the sores touching the ground. The rabbit may also lose weight and look lethargic.

Sore hocks can be exacerbated by wire flooring, but, contrary to popular opinion, it isn't the primary cause of the condition. Rabbits with a very thin layer of fur covering the bottoms of their feet, nervous animals (that thump a lot), and wet conditions on solid flooring can also cause sore hocks.

To remedy the condition, place a solid, dry surface inside the cage for the rabbit to sit on. Treat the affected areas of the feet with astringent and follow up with a corticosteroid. Rabbits that are chronically affected by sore hocks might need to be placed permanently in a hutch with solid flooring. But be warned: if its cage isn't kept dry, the condition will worsen.

Wool block (fur block). Wool block is a physical blockage (usually of fur and other debris) in the stomach or small intestine. Rabbits ingest wool and fur when they groom themselves or nibble on their neighbors' coats. The wool mixes with the undigested food in their stomachs or intestines and becomes a firm ball called a *trichobezoar*. Signs of a rabbit with wool block include lack or loss of appetite, weight loss, intermittent diarrhea, very small stools, manure strung together by fur, and a complete lack of stool. If any or all of these symptoms are present in a molting rabbit, it may have this ailment. When wool block is caught early and treated, the blockage can pass through the rabbit's system without a problem, but the key word here is *early*. The rabbit may not survive if the blockage is left untreated.

Although wool block isn't limited to longhaired rabbits, the condition is predominantly associated with the wool breeds. Rabbit-raisers with wooly animals should take preventative measures against this ailment. To help prevent a trichobezoar from forming:

- Make sure that your rabbits' feed is high in fiber.
- Offer grass hay or oat hay for roughage as often as possible, preferably every day.
- Offer papaya enzymes (which break down wool fiber) one to two times a week in tablet form or whole form. Also try putting pineapple juice—which also breaks down fiber—in the water bottle.
- Keep longhaired rabbits groomed to eliminate as much loose wool as possible.
- Try not to overfeed a rabbit, as this could cause a digestive system backup and encourage wool block.
- Always have fresh water available for the rabbits to drink.

Rabbits can't vomit, so any blockage has to be pushed out the opposite end. If the blockage is severe, take your rabbit to a veterinarian. Surgery may be necessary. However, if symptoms have just appeared and your rabbit is still leaving some droppings, there are a few home remedies that you can try before taking a trip to the vet. Try to break up the blockage by treating it with the following home remedies. Try two home remedies for a day or two and then

THOSE PEARLY WHITES

Rabbits in the wild eat some particularly tough and abrasive plants, which quickly wear down their teeth. To remedy this situation, nature designed their teeth to grow about ½–¾ of an inch a month.

add another treatment. If it isn't working, bring your rabbit to the vet for more intense treatment. Don't wait too long because wool block can destroy the intestines quickly. "Too long" can vary between animals, but I'd say three days with home treatment and without stool is a big red flag.

- One third of a banana with the skin still on.
- ½–1 teaspoon of stool softener on a banana if the rabbit will eat it. If it won't, use an eyedropper to administer the stool softener. If using an eyedropper, put the rabbit on someone's lap with its feet resting on the person's legs. Try to insert the dropper at the side of the rabbit's mouth. Don't turn the rabbit over and force the medicine down its throat; it may aspirate the fluid into its lungs.
- A dropper or two of mineral oil per day over several days.
- 1 teaspoon of meat tenderizer mixed with mashed banana.

Wet dewlap. As mentioned, the dewlap is the big fold of skin just under the chin of a rabbit. Dewlaps are more pronounced in certain breeds than in others and are usually associated with does. Wet dewlap is indicated, obviously, by the fur on the dewlap becoming wet and matted. If this goes unnoticed, an infection can occur, and the skin will become irritated, turn green, and emit a foul odor. This condition is usually caused by the rabbit dragging its dewlap through a water crock while drinking. The nice thing is that you can easily cure wet dewlap: either replace the crock with a water bottle or raise the crock a couple of inches. Apply antibiotic ointment to the irritated area (you may want to clip the fur off the wet area before applying the ointment).

Wryneck (torticollis or otitis media). You'll recognize wryneck right away—the rabbit carries its head tilted and often at an abrupt angle. This can accompany loss of balance or even falling over when the rabbit tries to hop. Wryneck is actually a symptom of inflammation of the middle ear. Usually, the inflammation is due to an upper respiratory infection or a bacterium called *Pasteurella sp.* Your vet will run tests to discover what is causing the inflammation and will prescribe a medication to correct the problem. Wryneck isn't usually simple to cure. Sometimes the medication works, sometimes it doesn't.

Bacterial Diseases

The following diseases are infectious; that is, they're transmittable from rabbit to rabbit. Remember that the best defense is a good offense. In this case, adhering to excellent husbandry practices and rabbitry cleaning is a must. In addition, use a quarantine ritual for all newly acquired rabbits.

A healthy rabbit's eyes are bright and clear.

Abscesses. Although abscesses themselves aren't infectious, the reason the abscess is present in the first place might be. Rabbits can develop an abscess due to a scratch or bite from another rabbit, but abscesses can also pop up if the rabbits have the bacterium *Pasteurella sp.* in their bloodstream.

An abscess shows up initially as a lump somewhere on the rabbit's body, usually around the shoulders and head. Upon further inspection, you may notice that the skin over the bump is red and warm. Eventually, the abscess will burst and drain without any help from you. However, this leaves the wound open for even more bacteria to collect, and germ-filled pus from the abscess may spread to other parts of the rabbit. To prevent this, manually drain and treat the abscess before it bursts on its own. If you're uncomfortable doing this, then a rabbit breeder or vet is the next best bet.

Cleaning abscesses is easiest when you have another person to help you.

1. Have someone hold the rabbit so you can wipe medical disinfectant on the abscess bump. If you want to, clip some hair away from the area so that you can get a better view.
2. Take a brand-new razor blade (or other sterilized sharp instrument) and make a slight cut in the center of the raised lump. It might seem like it hurts, but if you've ever had an abscess (or a pimple), then you know that the pressure from the trapped pus is actually relieved when it is expressed.
3. Use clean cotton pads or tissues to gently press on the sides of the abscess to push all of the pus out.
4. Use the disinfectant to wash the wound and the area around it.
5. Put an antibiotic ointment on the wound (the human kind is fine, but not the type with added pain reliever). There's no need to bandage the wound—the rabbit will either pull it off or it will fall off on its own.

The big feet of a
Holland Lop.

for several days will clear up a bacterial eye infection, but if it's a blocked tear duct (which won't respond to the ointment), it'll have to be opened by a vet.

Eye infection (sore eyes). Sore eyes is a bacterial infection most often seen in kits still in the nest box. If the kits' eyes aren't opening around day ten, and they seem stuck shut, take a closer look. Take a warm washcloth and press it gently to the eye for a couple of minutes; the eye will loosen and pus will be underneath. With immediate treatment, the infection can clear up completely. But if the infection is severe, the rabbit could be blinded. Carefully flush the eye out with water and apply some tetracycline ophthalmic ointment a couple of times a day. Thoroughly clean and sanitize the nest box, and be sure to practice excellent nest-box sanitation at all times.

Enteritis complex. Enteritis is inflammation of the intestinal tract, which can result in a pot-bellied rabbit with diarrhea. Infected animals are often found sitting with their feet in a water crock.

6. Wash your hands well with soap and water and throw away the cotton pads or tissues used to clean the wound.
7. Apply some ointment every day to the abscess until it's completely healed.
8. Proper sanitation practices in the rabbitry will help prevent the infections that cause abscesses. Also watch that cage mates or breeding pairs don't fight, and be sure to separate littermates after weaning so that intact males don't start fighting over territory.

Foot abscesses look like little nodules under the skin on the toes, feet, or legs. Typically, they are caused by something directly irritating the foot. The nodules are abscesses and can become severely infected if harmful bacteria enters the site of the open wound.

Soak the affected foot in an iodine solution several times a day. If the abscesses aren't draining by themselves, they may need to be lanced. Keep cages disinfected and surfaces smooth without anything protruding into the hutch that could pierce or snag the abscesses.

Conjunctivitis (weepy eye). Symptoms of conjunctivitis include redness or irritation of one or both eyes, possibly with some discharge on the lower eyelids and matted fur at the corners of the eyes. Weepy eye can be caused by bacteria and is often associated with an upper respiratory infection. On occasion, a blocked tear duct is responsible. Ophthalmic ointment applied two to four times a day

QUARANTINE PRACTICES

Quarantine practices in rabbitries are valuable for preventing the spread of disease. Forgoing the simple practice of isolating newcomers from the rest of the herd until you're certain they're disease-free can be a costly mistake. Quarantining is a free and uncomplicated practice that could save your entire herd. Simply house the new rabbit in an area that's well away from the main rabbitry for about four weeks. The amount of time is up to you; however, it can often take a couple of weeks for the rabbit to start showing symptoms—better safe than sorry! When you're on rabbit-feeding (or cage-cleaning) rounds, work on the newcomer's cage last. This way, if it does end up sick, you didn't carry its germs to the rest of the herd.

THE HEALTHY RABBIT

At-home rabbit healthcare should focus primarily around prevention rather than treatment. By practicing appropriate husbandry, most rabbit breeders will drastically reduce or completely eliminate incidence of disease. If a problem occurs in the rabbitry that involves a group of rabbits, the breeder should immediately look to his or her husbandry techniques first and troubleshoot all potential areas of concern.

—Jay Hreiz, VMD

Supplements that can be added to the rabbit's diet to help improve fur condition are cod liver oil, linseed oil, dry bread, sunflower seeds, and pumpkin seeds. To improve the body condition, try adding some calf manna, raisins, sweet corn, apples, apple twigs, comfrey, and sweet feed.

—Linda Hoover

If a rabbit isn't eating, try simethicone (infant gas drops). If you have weepy eye problems with a doe, try putting her in a hutch with another friendly doe. They clean each other's eyes, and this sometimes solves the problem.

—Betty Chu

Because death can often be the outcome of this disease, the best chance a rabbit with enteritis has is to see a vet. It needs broad-spectrum antibiotics, a high-fiber diet, and reduced stress. This condition requires aggressive treatment.

Mucoid enteritis. Refusing to eat, grinding teeth, and producing jellylike stools are all signs of mucoid enteritis. Another telltale sign is that the rabbit has a potbelly and, if you pick it up and *gently* shake it, you'll hear what sounds like water sloshing around inside its abdomen. This type of enteritis is often referred to as "weanling enteritis" because it's typically seen in young rabbits that are being weaned (5–8 weeks old).

Unfortunately, the prognosis is not good. Usually, in rabbits this young, they go from looking just fine to dying within twenty-four hours. Keeping rabbits on a high-fiber diet will generally help avoid mucoid enteritis. For rabbits transitioning from their mother's milk to solid foods, have grass and oat hay in the hutch for them to eat as they come out of the nest box.

Pasteurellosis (snuffles). An upper respiratory infection, snuffles is often a precursor to other complications, such as pneumonia. A rabbit with snuffles will sneeze and produce a white nasal discharge. Because no medication can effectively cure pasteurellosis, a rabbit-raiser should work diligently to prevent it.

The affected rabbit may simply live with the snuffles symptoms showing up every so often until it dies, or it may die right away. That said, you can give affected rabbits antibiotics to prevent secondary infections. This isn't a bad idea, because it's often the secondary infection that does the most harm. Make sure that the rabbitry has good ventilation and disinfect water bowls to help keep bacteria to a minimum. Be sure to isolate any rabbit with snuffles from the rest of the herd.

Pneumonia. A rabbit that is having trouble breathing and has bluish lips and a discolored tongue may have pneumonia. This is an infection of the lungs caused by a bacterium or virus, and it often follows on the heels of snuffles. Remember to quarantine any animal showing signs of

pneumonia. The affected animal needs to see a vet immediately.

Tyzzer's disease. Tyzzer's disease, an infection of the bacteria *Clostridium piliforme,* causes acute diarrhea and a rapid decline of health. This disease can kill a rabbit within three days. It also shows up most often in weanlings and is hard to distinguish from mucoid enteritis without a necropsy. Some rabbit owners have had success administering metronidazole to affected rabbits when symptoms arise. You should isolate the sick rabbit immediately and take it to the vet. Both rodent control (mice carry it) and good sanitation practices are key in preventing this illness.

Vent disease. Vent disease is actually the rabbit form of syphilis. It affects both bucks and does, and its symptoms begin with inflamed, possibly scabby, lesions on the rabbit's genitals. You may also find scabs on the affected rabbit's nose and mouth that ooze a yellowish discharge. Affected rabbits may refuse to breed, or a pregnant doe may miscarry her litter. Penicillin can be injected intramuscularly or applied topically on the lesions by a veterinarian. Be extremely careful about letting other breeders use your buck, and always check both bucks and does for lesions and scabs before breeding. Don't breed infected animals, and treat any animals that have been in contact with a sick animal.

Parasites

Parasites, by definition, are creatures that feed off of a living creature for their own survival. Rabbits have their fair share of critters that use them as hosts.

Coccidiosis. Coccidiosis shows up in two ways: the intestinal form and the hepatic or liver form. Both types show the same symptoms: diarrhea, trouble gaining weight, a potbelly, and poor condition of the rabbit's fur and flesh. Affected rabbits may also be prone to secondary infections. Rabbit breeders use amprolium or sulfadimethoxine to combat *Coccidia* in their rabbit herds. Again, practicing good rabbitry sanitation is paramount.

A rabbit's big ears can be an inviting spot for mites.

Cuterebra fly (warbles). The cuterebra fly, or bot fly, often creates lumps around a rabbit's neck and shoulders and possibly other areas. At first glance, these may appear to be abscesses, but upon closer inspection, you should see breathing holes in the center of the inflamed lumps. The fly has penetrated the rabbit's skin and laid its egg there to feed off of your rabbit. While it's growing, it continues to feast on your rabbit until, one day, a mature flying insect crawls out of the hole. It's obviously in your rabbit's best interests to remove the larva before too much feeding takes place.

Removing the fly larva is exactly the same as draining an abscess—with one main difference: you must not smash or rupture the insect while it's in the rabbit because doing so can send the rabbit into shock and may lead to its death. If you've never removed a warble before and don't have the help of an experienced person, a vet visit may be in order. This is just one reason why you need to keep flies down in the rabbitry.

Ear mites or ear canker. Symptoms of a rabbit with a mite infestation include head shaking and scratching, which can lead to cuts and bleeding where the rabbit has scratched. The rabbit may also have yellowish scabby material on the inside of the ears. Isolate rabbits infested with ear mites and treat them for three or more days with either a mitacide or olive oil in the ear canal.

Encephalitozoon cuniculi. This parasite can be the cause of neurological problems in rabbits and can cause wryneck similar to that caused by *Pasteurella*. It can also affect the spinal cord, heart, kidneys, and brain. The parasite is sloughed off when the rabbit urinates and is transmitted from rabbit to rabbit via contact with an infected rabbit's urine. This makes it easy for a doe to pass it on to her kits. *E. cuniculi* is difficult to diagnose without a necropsy. Unfortunately, this means that by the time the symptoms appear, the internal damage has been done. There isn't a treatment for *E. cuniculi* at this time.

Flystrike. Flystrike occurs when a rabbit has wet, loose stool or doesn't clean itself properly. Flies see wet feces and urine as an invitation to lay their eggs. The eggs are usually laid in the rabbit's genital area (check the side pouches next to the rabbit's sex organs). Once the eggs are laid, they quickly hatch into maggots and begin burrowing into the rabbit. They eat the rabbit's flesh as they grow and can make their way deep inside a rabbit's body. The rabbit then goes into shock from the maggot infestation and dies soon afterward. This whole process can take only twenty-four hours.

Flystrike has nothing to do with the rabbits getting less than optimal care. It just takes one day of pasty stool and a single opportunistic fly for flystrike to happen. You'll need to remove all of the maggots from the affected area for the rabbit to recover. Hold the affected area under running water to make the maggots surface. Once they surface, use tweezers to remove every maggot. Be gentle because usually the area is swollen and sore. Afterward, dry the area so that flies aren't attracted to the moisture. Keep your fly-stricken rabbit indoors until it's fully recovered.

This technique works well for me most of the time. If flystrike isn't caught right away, though, there isn't much you can do. If you can't bring yourself to get near the maggots, a trip to the vet is in order. If it's a bad strike, you should see your veterinarian anyway. A vet may also be able to give you a product that kills fly eggs before they have a chance to hatch. Note: Never use a household fly repellent on the rabbit cages and certainly not on the rabbit itself!

Fur mites. The symptoms of a fur-mite infestation are simple: noticeable fur loss on the head, neck, face, and base of the ears. Assuming this isn't due to a natural molt, isolate infested animals first. Depending on the severity of the infestation, a simple application of kitten-strength flea powder may do the trick. If the mites are persistent, you can give ivermectin orally or topically or inject it subcutaneously.

Pinworms. Pinworms are internal parasites that won't show any clinical signs unless there's a major infestation. Sometimes you can see the pinworms by the rabbit's anus or in the droppings. You can also bring a stool sample in to a vet for positive identification. Piperazine is the usual dewormer for pinworms in rabbits. You can use ivermectin, but ivermectin is most effective on roundworms.

Many people who keep a rabbitry intend to raise rabbits for show, fiber, or meat. These activities usually involve enlarging and improving the herd through breeding. The most valuable thing you can offer your breeding program is a rabbitry full of healthy and happy rabbits. Neither overweight nor underweight animals are desirable for a breeding program meant to improve a rabbit line. The pair chosen to breed should be free from any abnormalities, health issues, or what are considered "faults" for the breed and the show table.

BREEDING LIKE

RABBITS

Rabbit Reproduction

Does don't come into heat (estrus) domestic dogs and cats do; rather, they are *induced ovulators*, meaning that they release an egg only *after* breeding with a buck. So, it's possible for a doe to accept a buck for breeding at any time of the year. However, there are certain times when does are more likely to accept a buck's advances. Look for signs of the doe's acting restless or rubbing her chin on the feeder and water-bottle nipples. If her genitals are inspected, the vulva of a doe that's ready to breed will be a deep reddish color (as opposed to a pale pink).

A doe has two uterine horns, each with its own cervix. The uteri both usually carry kits from breeding with a single buck, but It is possible for a doe to become pregnant by two different bucks. This makes the litters of each horn due on two different dates. Unfortunately, in these cases, it is likely she will lose both litters due to malnourishment or other developmental issues.

Choosing Rabbits to Breed

The first thing you should think about when choosing a pair of rabbits to breed is how their physical characteristics complement each other. If you've never chosen a breeding pair before, it's in your best interest to seek guidance from a seasoned rabbit breeder in your breed. You need to pick a pair of adult rabbits so that you're sure they've reached physical maturity—this is especially important for the doe that has to carry and nurture a litter.

Once you've chosen the buck and the doe, look them over for physical health issues. You're looking for signs of illness, disease, and malocclusions. You want clear-eyed rabbits at a healthy weight. Make sure that both of the buck's testicles are descended.

The Best Time to Breed

The health and well-being of your rabbit and her kits depends not only on good timing but also on your own practical scheduling considerations. Does give birth approximately thirty-one days after being bred. Will you be available for the kindling (birth)? If you're a beginner, will your mentor rabbit keeper be available to help you?

Weather is also a consideration. For those living in cold regions of the country, kindling in freezing temperatures may not be optimal and actually makes it harder to save kits that are accidentally born "on the wire" (on the floor of the cage rather than in the nest box). It is also

BREEDING AGES

Rabbits may be capable of breeding at three months old, but their bodies are still immature. Following are general guidelines for the optimal age to breed variously sized rabbits.

Small breeds
(e.g., Netherland Dwarf, Dwarf Hotot): five months old
Medium breeds
(e.g., Dutch, Mini Lop): six months old
Large breeds
(e.g., French Lop, Flemish Giant): eight months old

How can you tell a buck from a doe? Determining males (bucks) from females (does) is called sexing and is easy when you know what to look for, especially with adult rabbits. Young rabbits can be trickier, but you get better at it with practice.

With one hand, hold the rabbit's head and the scruff of its neck while the other hand cradles the rear end. Place the first and second finger of the hand holding the rear end on either side of the rabbit's tail. With an adult male, you'll be able to see the testicles. When sexing babies or youngsters, the testicles will not have dropped yet, so you'll need to inspect further. Use your thumb to gently press down in front of the rabbit's genitals. If the opening is round and you see a protrusion coming out of it, this is a buck. If the opening is more of a slit with no protrusion (and no testicles), the rabbit is a doe.

Start practicing with young kits at about three weeks old and double-check your findings a week or two later for accuracy. If you still aren't confident that you've properly sexed a litter, have a more experienced breeder sex them for you. Mark the insides of their ears with a B or a D using a fine-tipped permanent black marker.

harder to keep the litter warm enough inside the nest box. Typically, a doe will pull enough fur from her abdomen to keep her babies warm, but on occasion, she'll fall short. This is something to think about.

Finally, you'll want to think about how timing affects your rabbit's purpose. Some breeders choose not to breed a doe in the hottest part of summer because a juvenile rabbit's ears tend to grow longer in the heat (to help them keep cool). This may not seem important, but if your rabbit is destined for the show table, this could make its final ear length too long and disqualify it from competition. Rabbit raisers interested in specific shows will also concentrate on how old the rabbits will be for competition. Many want the oldest junior animals competing in the junior classes; the older junior rabbits are more physically mature, which betters their chances of winning the class.

How to Breed Rabbits

For the most part, bucks and does are very capable of producing offspring on their own. That said, rabbit breeders have found that a few pointers can help you avoid problems and get the does successfully bred.

Always take the doe to the buck's cage for breeding. Does are highly territorial, but a buck is always willing to make an exception for a lady visitor. Some breeders are comfortable leaving the doe with the buck for a couple of days or a week to ensure successful breeding, but I don't recommend this for a couple of reasons. First, if the doe isn't receptive to the buck's advances, you may not be around to break up a fight. Second, while you can still get a general idea of when to expect the kits, this approach is inexact because you don't know when the actual breeding took place.

REBREEDING

The most humane time to rebreed a doe that has just had a litter is when her kits are six weeks old. At six weeks, the kits will be weaned, and the doe will have some down time to recuperate before raising another litter.

The buck will chase the doe around the cage a couple of times before she'll let him mount her. If you've never watched rabbits mate, you're in for a quick education. The doe raises her hindquarters up, and the buck mounts her, breeds her, and then squeals and falls over. This all happens quickly, mind you, so don't take your eyes off them. Bring the doe back to her cage after she's bred. To be sure that enough semen is deposited to create a litter, place her back into the buck's cage in a couple of hours.

Occasionally, the doe isn't ready to breed and won't raise her hindquarters for the buck. Instead, she'll turn and attack him. In my experience, this is a quick attack-and-retreat situation. Once the doe and buck pull apart from an attack, take the doe out before she decides to attack again. There are teeth involved, so wear thick gloves. You can bring the doe back to the buck's cage in a day or two to try again.

Even more occasionally, the buck won't be interested in breeding. If your stud just sits there, dazed and confused, try placing him on the doe's back. It's amazing how quickly they catch on with this simple maneuver. If he still is uninterested, return the doe to her cage and let her scent hang around his place for a bit. Then, bring her back over for another rendezvous, and he'll more than likely have figured out his mistake. If he still won't breed, have her visit another buck.

Keep Good Records

Whether you're breeding rabbits for profit or for show, I can't emphasize enough the importance of good record keeping; no other practice will keep you more informed about what's happening in your rabbitry. Immediately write the breeding date on the hutch card as well as in your permanent records. Going back later and guessing is risky because you could easily get the kindling date wrong, resulting in the potential loss of kits.

Is She Pregnant?

There's a rather nonscientific method of testing a doe to see if she has "taken" (gotten pregnant). About ten days after the breeding date, place the doe back into the buck's cage. Usually, if she has taken, she will run around, avoiding the male, possibly grunting

Large breeds, like the Flemish Giant, take longer to reach breeding age.

comes time for the doe to kindle. Still, we rabbit raisers can do our part in preparing for the pitter-patter of fluffy feet.

Caring for the Pregnant Doe

During pregnancy, does don't require much in the way of special care. In fact, most breeders will suggest that you don't even increase the doe's feed until the second half of the pregnancy (day 16). Even then, breeders usually offer just a slight increase because obese does can have a harder time with pregnancy and giving birth. Pregnant does tend to drink more water, so be diligent about keeping plenty available. If you have an automatic watering system in your rabbitry, be certain to check the nipple in her cage daily.

The All-Important Nest Box

Nest boxes are extremely important for the kits' survival. Our domestic rabbits' descendants burrowed into the ground and created special warrens in which to deliver and nurse their kits. For the domestic rabbit, we offer her a nest box with shavings at the bottom and straw packed inside. Place handfuls of nesting materials such as straw or oat hay inside the cage so that the doe can build her nest inside the box the way she likes.

or squealing. While this test can be helpful, some does will allow breeding again anyway, so this isn't a guarantee.

You can also weigh her on the breeding date, and then weigh her again in ten days. Assuming you're feeding her the same rations as before the breeding, if she gains somewhere close to a pound, it's another good indication that she might be pregnant. A more reliable test is to palpate the doe two weeks after exposing her to the buck and see if you can feel the marblelike embryos in her belly. Place your hand just in front of her groin and gently feel around the sides of the belly. Once she's far along, you can actually feel movement.

Preparing for Kindling

Of course, Mother Nature will be the main birthing coach when it

---

BREEDING DOES IN TANDEM

One of the best pieces of advice about breeding I've ever received is to breed does in tandem. That is, never breed only one doe in the rabbitry if you can help it. I've tandem-bred every time, and it has saved many litters for me. Tandem breeding is especially important if a doe has never kindled and raised kits before. The idea is to breed at least one doe who has kindled and successfully raised a litter to weaning. If you don't have another doe that can be tandem-bred, ask another rabbit raiser when he or she is breeding and if you can breed according to his or her schedule just in case. Here are some possible reasons to tandem breed:

- A first-time mom might not have any idea what to do with the kits and might scatter them all over the cage. The doe usually shows no interest in the kits after that, and the kits are left to perish unless they are given a foster mom.
- A new mother may become frightened while kindling and start eating her babies. As gruesome as it sounds, it's a common occurrence in nature. Try to rescue the remaining babies and have a more seasoned doe foster them.
- A doe may give birth to more kits than she can feed. A couple of the babies may begin to weaken from not getting as much milk as the others. In this case, having another mother foster a couple of the more robust babies will help save the smallest ones. A clueless doe with her first litter may have no problem with subsequent litters. Give her a second chance, and decide from there if she's fit to be in your breeding program.

Tiny newborn kits on a bed of straw.

It is the doe's instinct to construct a tunnel-shaped nest. After utilizing the nesting materials, she will pull fur from her belly, sides, and neck and under her chin to make the nest not only cozy but also extremely warm. After the doe kindles, she'll pull more fur and place it over her kits.

The question of when to put the nest box into the doe's cage is tricky. Some breeders say to wait until a day or two before the kindling date so that the doe doesn't soil the box before kindling. Others say to put it in a little sooner just to be on the safe side. I fall into the second school of thought—better safe than sorry. In addition, only some does like to make their nest days early; most build it just hours before the kits arrive. Be sure to place the nest box inside the cage in the back corner opposite where the rabbit leaves most of her droppings. You definitely don't want the doe's box over her potty area.

Kindling

A quiet environment is best for a pregnant doe that's days from giving birth. Does preparing to kindle are focused on nest building, and their bodies are preparing for labor, so the less commotion, the better. This is the case especially with does that are kindling for the first time. As previously mentioned, an anxious or fearful doe may resort to eating her new young or refusing to feed them.

Pregnant does don't generally make a big production about giving birth, but they do show some signs that clue you in to the upcoming event. If you know what the doe's eating habits are, you may notice is that she will eat less or go off her feed entirely a day or so before the kits arrive. The second clue is that she'll hop around, gathering the nesting material that you provided to create a cozy nest. Whether she does this on the same day as kindling or a day or so before depends on the doe. As mentioned earlier, she'll also begin to burrow in the straw bedding and pull fur from her body and line the nest with it. Not only does the fur help keep her kits warm, the missing fur on her abdomen allows for easier access to her nipples.

Although does give birth around day thirty-one, it could be a little earlier or a little later. About twenty-eight days after breeding, approach the hutch quietly when the doe isn't sitting in her nest box and peer into it—don't use your hands at this time. Watch for a minute or so and see if the fur she's placed inside the nest is moving up and down. If so, you can be sure she's already kindled.

Does can kindle at any time, but most seem to favor the quiet early-morning hours. Often, the babies will already be there when you visit the doe's hutch in the morning. Sometimes a doe will kindle a little later, and you'll catch her in the act. The safest course of action is to stay back so that you don't disturb her or interfere with the birth, although a couple of instances may require you to intervene; for example, if the kits have been born on the wire (outside the nest box). If you catch them while they're still warm, simply place them deep into the fur inside the nest box. If they're cold, you'll need to warm them up before putting them into the nest box.

Does and Kits

You've successfully bred a well-matched rabbit pair, you've taken great care of your pregnant doe, and now there's a brand-new litter of kits in your rabbitry. Here's what you can do to help your doe-mom raise a happy and healthy litter.

Caring for Nursing Does

Veteran rabbit breeders have their own home concoctions for nursing does. A favorite remedy for does that have just given birth is bread soaked in warm milk and offered to the doe right after kindling. Most rabbit-raisers allow a doe to have as much pellet feed as she will eat after kindling as well as offer her daily nutritional snacks, such as small pieces of carrot and apple. Don't overdo the added snacks.

Caring for Kits

Rabbits have been kindling (giving birth) since long before breeders came into the picture. However, that was also before they

The doe pulls fur from her body to keep the kits warm.

nurse. If the doe has eleven babies and you have another doe with only a few, you may want to give a couple to the doe that has fewer so that all of the babies have the best chance at reaching nipples to nurse. Second, if there are any stillborn babies, they should be removed from the nest box. Third, if the doe is inexperienced or has had a history of not caring for her young, look at the kits' tummies each day to be sure that they're full and round, indicating that they're getting plenty of milk. Another reason, purely curiosity-related, is to check out the coat colors.

If you've never handled newborn kits before, there are some things to take into consideration. Personally, I like to distract the doe before I go near the nest box—not necessarily because she may be protective and therefore aggressive (although she might), but because I want her to see me as nonthreatening. I feel that bringing my doe a goodie and letting her begin to munch on it before I remove the nest box to inspect it conveys my message of peace. You can offer her an apple slice, a carrot, or another favorite treat and then remove the nest box to check on the brand-new arrivals.

There are two schools of thought on handling kits from birth. One is that the doe may become stressed due to the human scent on her kits and either kill them or refuse to care for them. The other is that

became domesticated and dependent on our care. When we keep rabbits in cages, we place them in a setting that doesn't necessarily allow nature to take its course. More bunnies survive in captivity, and it's hard for does that are trying to wean their kits to get away from them! So, we've placed ourselves in the position of becoming caregiver, nursemaid, weaning attendant, and sometimes foster parent to our rabbit herds.

The Early Days

Does that have raised litters before usually have it down pat. That said, many breeders like to peek into the nest box for several reasons. First, it's nice to know how many kits the doe will have to

ABOUT NEST BOXES

All rabbit breeders have a favorite type of nest box. Some feel that all-wood boxes keep the bunnies warmer; these should be carefully sanitized between litters using a solution of one part bleach to five parts water. Other breeders prefer the galvanized-steel type with a sliding plywood bottom. The upside to this type is that it's easy to clean and easier to disinfect than solid wood.

Many nest boxes include a small shelf toward the top of the box. This comes in handy when a litter is old enough to chase their mother around. Does will often jump onto the shelf and take a well-deserved break from their rowdy youngsters.

Another popular type of nest box looks like a wire mesh basket. The bottom and sides can be lined with cardboard, which is discarded as it becomes soiled and changed between litters. This type is used more often in mild weather because they're obviously not as warm as the other types.

Look at as many varieties of nest box as you can find. Ask other breeders which types they prefer and why. Nest-box size is determined by the size of the rabbit. For large rabbit breeds, the box should be 20 × 12 inches, and 10 inches high. For medium breeds, the box should be 18 × 10 inches, and 8 inches high. For small breeds, the box should be 14 × 8 inches, and 7 inches high.

if the doe knows you well, she'll have no problem with your touching them. If the doe is unfamiliar with you, or she hasn't been in your rabbitry for very long, put some vanilla extract (from the kitchen) on your finger and wipe it onto the doe's nose. This way, the kits will all smell the same as far as she's concerned. If you want to err on the side of caution, gently push the fur and nesting material aside with the back of a wooden spoon to get a good view. Count the babies and be sure that your doe has enough nipples to feed them. Remove any placentas, dead kits, or bloody fur. Afterbirth and dead babies will attract flies, ants, and who knows what else.

Stillborn kits are quite common, for numerous reasons, especially in a doe's first litter. Rather than focus on kits that didn't survive, focus on the healthy babies—they need you! Dwarf breeds sometimes give birth to what are referred to as *peanuts*. These are kits that are half the size of their brothers and sisters and very often aren't fully developed. Some breeders feel that they won't make it and pull them out of the nest box immediately, but I leave them with their siblings. Many don't make it, but I've had my share survive, too—especially if I give the wee one to a doe that has only a couple of kits in her litter.

Foster Care

Sometimes a doe will give birth to an extremely large litter or will refuse to care for or feed her kits. Here's where you will congratulate yourself if you had the foresight to breed more than one doe at a time. While it's possible to hand-raise kits, it usually doesn't work out very well. Baby rabbits are extremely difficult to hand-feed and often end up aspirating the substitute food into their lungs, which

results in pneumonia. The best thing to do is give the kits to another doe to raise. Simply remove them (or some of them, in the case of an enormous litter) from their original box and place them into the foster mom's nest box. If you feel more comfortable, use the vanilla-extract-on-the-nose trick so the foster mom doesn't realize that you've added more mouths to feed.

Warming Up Cold Bunnies

Once in a while, a doe will kindle on the wire, or a baby will be attached to a nipple and won't let go until its mom has hopped out of the nest box. If the kit is basically warm but has wandered, place it back into the nest with its siblings, and it'll warm up quickly. Check on new litters periodically throughout the day just in case a baby ends up on the wire. Kits are born blind, deaf, and very naked. They become chilled quickly, and the exposure can kill them in minutes.

That said, don't ever immediately assume that a baby on the wire is dead. Kits will become very still when their temperature drops; this preserves their energy, which keeps them alive for the longest time possible. Survival time has everything to do with how old it is, the temperature outside, and how long it's been on the wire.

There are a couple of different techniques for warming up a baby rabbit. The first thing I do is put it under my shirt, against my skin. This begins to warm it up on the spot, and I often soon begin to feel the motions from its feet moving. You can also turn a heating pad on low and wrap a towel around it. Put the heating pad and towel with the bunny inside a shoe box so that the heat surrounds it. It's better to warm the bunny up slowly. In fact, don't give in to the urge to turn the pad on high; the kit's skin is thinner than paper and can burn easily.

The baby rabbit has a "5" drawn inside its ear for temporary identification.

litter are completely different from those of the foster babies. If they're the same, mark the fostered babies' ears with a permanent marker *before* you place them into the new nest box. More often than not, you'll have to mark the babies' ears every couple of days.

Continuing Care

Raising baby rabbits is best left in the capable hands of the doe, barring any unusual complications. It's important to note that rabbits only feed their babies once or twice a day. The doe also sits over the babies for a couple of minutes each day and cleans their bottoms (which encourages them to eliminate) while she's in the nest box. Other than these times, the kits stay in their fur-lined nest box and mostly just sleep and grow.

The most important thing you can do for your kits is to keep the cage and nest box clean. After the doe has been visiting the nest box for a week or so, she may soil it, which is unhealthy for the babies (especially if they become wet from urine). Find a small container (just large enough to hold the kits comfortably) and gather up all of the clean fur you find in the nest. Carefully take the kits out of the nest box and put them onto the fur in the temporary container. You can now clean out the entire nest box.

Get rid of all soiled or damp materials and add fresh straw. Form the bedding into a nest shape and make a hole in the fresh straw. Now line the hole with the clean fur you saved, reserving a little to place over the babies. You can even gently pull some fur from the doe's belly to add to the nest. Chances are, once the nest box is placed back into her hutch, she'll add some herself. Add the babies to the fur-lined nest and cover them with the extra fur.

When the kits are about ten days old, their eyes will begin to open. They may open one at a time; that's fine. If it seems to you that a

Another technique is to put warm water in a bowl on the counter. Now place the baby into a storage baggie (leave the top open and hold it securely), and place the storage baggie so that the baby is lying in the baggie in the warm water. The idea is to not let the water touch the kit but to let the warm water heat its skin through the plastic bag. Yet another method is to put a lot of hand towels into the dryer, and when they're warmed up (not burning!), hold the kit in your hands with a towel wrapped around him. When the first towel cools, wrap a new warm towel from the dryer around him. Do this until the bunny pinks up and is quite warm without the towels and then put it back in the nest with its siblings. Before returning a chilled bunny to the nest box with the rest of the litter, it really needs to have completely warmed up. If it's still cool at all, the other bunnies will wriggle away from it; without their body warmth, it could become chilled again and perish.

Fair warning: sometimes a kit can be brought back from the brink only to gape and gasp and not pull through. You've done your best, but some kits are just too far gone to be brought back completely. This is the not-so-fun part of rabbit raising.

In a foster situation, don't forget that you want to remember which kit belongs to which mom. You're lucky if the markings of the original

WORD TO THE WISE: ANGORA WOOL

If an Angora doe is in full coat when she pulls fur for her kits, it often comes out in long strands. These fibers often wrap around the delicate necks and feet of the newborns, which can be dangerous for their circulation and possibly life-threatening. Before your Angora kindles, you can trim her coat short (about the length of a puppy's coat) so that when she pulls fur, the strands are short. You can also gather the fur she's already lined her nest with, cut it into short pieces, and place it back into the nest box.

In breeding, try to always breed faults to good points. If one [rabbit] has weak hips, do not breed it to another with weak hips, as that is all you will produce. Try to match up any faults in one rabbit to the good points in the one that you are breeding it to. This is not as easy as it sounds. You will often find yourself doing a balancing act when trying to match up faults with strong points. For example, a buck with a good crown and fair body could be bred to a doe with poor ear carriage and a good body. Ideally, you should always attempt to breed rabbits with the fewest faults.

—Chris Zemny

Reputable breeders keep records on their animals, make responsible breeding choices, and have an effective plan to deal with the results. I also think it is important to be ethical when selling animals. If an animal isn't something you would be proud to have represent you on the show table, it isn't ethical to sell it to someone looking for a show rabbit or to start their own breeding program.

—Keelyn Hanlon

Some first-time does may not birth in the box, so it's a good idea to use a sheet of wood to cover almost half of the cage floor. If the doe does give birth outside of the box, gently move the kits to the box as soon as you notice them.

—Brenda Haas

bunny's eyes should be open and they aren't (within a couple of days of all the others' eyes), gently run a cotton ball moistened with warm water over the eyelids to see if that helps coax them open. If not, the kit may have an eye infection; look for crusty material gluing the eyes shut. This is common with kits and one reason why it's important to keep the nest box as clean as possible. If you notice that one kit is having an eye problem, be sure to check the eyes of the entire litter.

Eye infections can be simple to cure if caught early; otherwise, they can result in permanent blindness. Use the moistened cotton ball on the crusty eye until the eyelid is clean. After that, use clean hands to gently pry the eyelids apart so that you can see the eye. Sometimes this simple procedure clears up the problem. If the rim of the eyelid is red, use a couple of drops of eyewash. Check the eye again in several hours and again the next day. If you see pus coming from the eye at any point, you'll need to put tetracycline ophthalmic ointment into the eye to battle the infection.

Treat an infected eye with the wash-and-ointment treatment for several days, and you should notice positive changes. Once the eye is healthy again, it may be as good as new, or there may be some cloudiness over the cornea. Cloudiness means that the rabbit is either partially or totally blind in that eye.

As the litter grows, you'll certainly notice that the hutch needs more frequent cleaning than when it housed just the doe. Remember, if you keep up high sanitation standards, you'll avoid a large number of health issues.

Weaning Kits

The first step in weaning kits from the doe is to remove the nest box. When the kits are about sixteen days old (give or take), they'll begin to hop in and out of the nest box where they've spent their entire lives until now. It's only fair to warn you that this is when beginning rabbit keepers often get sucked into the rabbit-raising world for life. There's

nothing cuter than kits between the ages of two and four weeks old.

If you've been giving the doe greens, stop once the kits pop out of the nest box. You don't want the youngsters' digestive systems to deal with greens and treats yet. You should, however, have plenty of oat hay in the hutch to help the kits' digestive systems transition smoothly from mother's milk to pellet feed.

You typically remove the nest box from the hutch when the kits are three weeks old. If you live in an extremely cold region, you can wait an extra week or two. But remember that the kits will now begin eliminating in the box, and this will not only make them wet but also encourage bacteria growth. Remember to clean the box out regularly. A nice layer of straw as bedding on the cage floor will help keep the young ones warm, too.

Another reason to leave a nest box in for longer than the kits need it is actually to help the doe. The youngsters will attempt to nurse all day and will, in general, pester her to no end. If the nest box is the type with a small shelf over part of it, she can escape them for a time.

Kits at three weeks old are ready to be handled by their caregivers. Handling these youngsters every day will make them easy to handle throughout their lives and will get them on the path to becoming good companions and show rabbits. Young rabbits are fragile creatures that are injured and frightened easily, and their moves are fast and sudden. You can show older children how to handle kits properly, but kids younger than ten should only be allowed to pet them while someone older holds them.

Somewhere between week five and week eight, wean the kits from the doe and separate them from each other. The general rule of thumb is that the smallest breeds are weaned closer to the five-week mark and the largest breeds around the eight-week mark. From a human perspective, this may seem sad, but from the doe's perspective, I assure you that it isn't. Her body works very hard to produce milk for kits, and she'll be grateful for the relief that weaning brings.

As kits grow older and larger, you'll separate them from their mom and each other.

Some breeders wean the kits all at once, but I always remove half of the litter at a time. This allows the doe's milk production to reduce gradually and lessens the chance of her teats becoming engorged.

You can take half of the litter out of the doe's cage in groups organized by sex or by number. If you separate them by sex, be sure to double-check your conclusions a week or two later. Young rabbits have been known to "change sexes" due to an initial misreading of their sex. The first group can be housed together in a temporary cage, or they can be placed into their own cages at weaning time. A few days later, you should separate the second half from the doe as well.

Many breeders will tell you that the youngsters can be caged together up to four months old. My advice is to resist the urge to house littermates together for more than a couple of days after weaning; otherwise, days tend to lead to weeks, and weeks lead to does being bred by their brothers and bucks fighting to the point of attempting to remove each other's testicles. Rabbits grow up fast.

If you'd like to bide your time as far as having to provide cage space, you can always keep the doe kits with their mom for a bit longer. At about three to four months old, the young does may start to get territorial, but this arrangement usually works for a while.

Pedigrees, Tattooing, and Culling

Weaning time is the perfect time to do other important tasks for the new kits, such as filling out pedigrees, tattooing, and culling. As far as pedigrees, you should fill them out for all of your stock. Yes, some may go to pet homes where papers are unnecessary, but most people still enjoy seeing where their rabbits came from and keeping track of their rabbits' ages. Pedigrees also have your rabbitry information on them, which ensures that potential buyers can reach you if they

have any questions or if they want to recommend you to someone else who wants to purchase a rabbit. Pedigrees are mandatory if you plan on selling purebred show or breeding animals. In any case, weaning time is a great time to get this paperwork done.

Part of filling out the pedigree is assigning a number, which will be tattooed inside the left ear, to each rabbit. Weaning day is also the perfect time to tattoo the new litter. For pet and meat rabbits, tattooing is unnecessary.

Now for culling. What exactly is culling, and how is it done? Somewhere along the line, people got the impression that culling means killing those rabbits that you don't want to keep in your rabbitry. That it is not what is meant by culling your herd. Culling is selecting which animals will be kept in your breeding program, which will be kept for show, which may be sold, and which will be used for meat. Culling is simply deciding where a rabbit is best suited because every rabbit is born with different qualities.

Proper culling is how your herd will become a strong representation of the breed, fiber, or meat standard. In essence, you'll judge each rabbit for its intended purpose. If you breed French Lops for the show table, you may look at a young Frenchie to make sure that it's healthy and has no inherent disqualifications, such as a malocclusion. You also want to compare it to the French Lop standard (as well as to your other show French Lops) to see how it holds up. A beautiful French Lop may not have the best ear placement for a winning show rabbit, but it could do well in youth shows or as a pet, and you may want to cull it from the rabbitry for either of those purposes.

It's wise to have a cage or several cages in an area of the rabbitry to temporarily house the rabbits you won't be keeping. This makes it easy for you to see at a glance what you have available when a phone call comes in from someone looking to purchase. Before you let any rabbit leave the premises, double-check that it hasn't changed over time and grown into a rabbit that you'd like to keep. You wouldn't be the first breeder to let go of a great show rabbit because it was slow to come into its own. In fact, most experienced rabbit-raisers will tell you that it's very difficult to decide which rabbits are worth keeping and which aren't when they're eight weeks old. Most people wait a couple more months to decide,

Small-livestock keepers have long known that rabbits can easily be kept for both pleasure and profit. While I don't think anyone claims that they're rolling in riches thanks to rabbit keeping, a variety of rabbit products and by-products can help offset your hobby costs as well as some household bills. Rabbit fanciers can sell their livestock as show animals, as breeding stock, as pets, and for meat. Consider also the fiber, manure, and fishing-worm markets. You'll be able to strike a balance between your monthly budget and the creatures you love to keep.

MAKING MONEY WITH

RABBITS

Show Rabbits

More than 800,000 rabbits are shown at rabbit shows in the United States each year. So, yes, there's a market for show rabbits, and there are fanciers out there who show the breed you're raising. To acquire a good reputation for breeding high-quality rabbits, you'll first need to purchase quality rabbits of your own, and then you'll need to begin breeding rabbits that are excellent representatives of the breed. Therefore, marketing and selling show stock may take a little more time and money to show a profit than some other rabbit-raising ventures. Still, in my experience, showing rabbits and breeding for the fancy is the most rewarding and certainly the most fun opportunity.

Rabbit shows are inexpensive and easy to enter. Just attending a show seems to get the fever going in just about anyone who's ever considered owning a rabbit. There are opportunities to make great friends and to take home a wealth of information. In addition, many people make rabbit shows a family affair.

If you'd like to raise show rabbits for profit, contact a local or national club devoted to your chosen breed. This is the best way to find the most reputable breeders in your breed. I do need to warn you: as in every other area of life, there are unscrupulous people who want your money, with little concern about the quality of their rabbits. To protect yourself, know as much as you can about the breed and the standard for that breed beforehand. Consider any breeder who insists that you buy at that show or on that day as a red flag. That isn't to say that a reputable breeder can't have what you're looking for at the show, but most will have no problem inviting you to come by at a different time to see what they have available. Trust your gut.

As a breeder of show rabbits, your goals should be, first and foremost, to breed rabbits of superior quality to what you currently have in your rabbitry and to be a reputable ambassador for your breed. The show-rabbit world is smaller than you can imagine. Rabbitries that rise to the top and prosper are those that are completely fair, honest, and helpful.

WHAT TO PACK FOR RABBITS THAT YOU'RE SELLING

· Any comment cards (cards from shows that have the judges' notes on them) that pertain to the rabbits you are selling
· Pedigrees
· Poster board (or dry erase board) and markers to advertise that you have rabbits for sale
· Small baggies filled with rabbit food as a transitional feed for whoever buys your rabbits

You can raise fiber rabbits to sell raw or spun wool.

I'd like to impress upon you the responsibility that you have to your fellow breeders, to rabbit owners who will come after you, and, mostly, to the rabbits themselves. As a breeder, your goal should be to end the genetic line of negative inherited traits if they arise. I'm not talking about mismatched toenails or ears that are too short (although you want to better those traits as well). I'm speaking of issues such as malocclusion and other physically disabling traits that negatively impact the life of the affected rabbit. Malocclusion can also be caused by a rabbit continually pulling on its cage wire, but it is often the result of bad breeding.

On the flip side, don't ever sell a rabbit that you know has faults or is ill to an unknowing buyer. Don't ever falsify papers or lie about an animal's age or its show record. And don't even consider taking advantage of a 4-H kid. Your name will spread like wildfire through the show circuit in your area and beyond. A reputation of questionable ethics is hard to shake.

Advertising is one of the most important aspects of any business. Most people end up selling their show stock through word of mouth. Breeders display whiteboards and hand out cards at shows to advertise what they have available and might even offer show specials. They watch for who wins at the shows and approach the winners to buy rabbits that will improve their own stock. Make yourself known to local 4-H groups, and advertise on bulletin boards at farm-supply stores. Create your own website, complete with pictures of your stock, and try to get your link on other breeders'

websites as well as on club sites. Anything you can do to get your name out there and in a positive light will come back to you in sales.

Rabbit Fiber (Wool)

The first thing you should know about raising rabbits for wool is that you won't get rich doing it. Still, if you're a spinner or a knitter and are involved in the fiber world in any way, you may find a nice little market for the by-product that your pet or show rabbits have to offer. Just as with other fiber, such as that of sheep or alpacas, rabbit wool can be dyed and is considered top drawer in the fiber market. Plush and silky Angora rabbit wool is one of the softest fibers in the world. The calm, docile Angora makes a wonderful pet, and if you're going to have one sitting on your lap, you might as well collect or spin its wool!

Angora rabbit wool comes one of two ways: raw or spun. Raw fiber is shorn or plucked from the rabbit while the rabbit is molting. Prices will vary depending on the quality of the wool, the length, and the percentage of guard hair. The softest part of the rabbit's coat is the undercoat, but there are longer, rougher hairs on the outside called *guard hairs*. The fiber that fetches the highest prices has high density (thickness), good crimp (waviness), and a minimum of guard hairs. If the rabbit has won awards for its coat, you can charge more, too. Plucked wool may bring a higher profit than shorn fiber, but many Angoras today are being bred to hang on to that long fiber, making it necessary to shear it.

FUZZIES AND WOOLIES

Both the American Fuzzy Lop and the Jersey Wooly produce true wool. Because these two breeds are of smaller stature and their adult coats contain many more guard hairs than those of the other Angora breeds, they are less profitable for wool production. Still, if you raise Fuzzies or Woolies, their wool can also be spun.

In addition to harvesting and selling raw fiber, you can spin your wool into skeins that are ready for knitters to use. Yarn and craft shops and knitting clubs alike will be interested in local, humanely harvested Angora wool. You can also dye rabbit wool to add variety to the product. If you are a knitter, you have yet another opportunity: you can make and sell garments from the Angora wool you've harvested and spun from your own rabbits.

Many Angora breeders bring their raw or spun wool and knitted products to rabbit shows to sell. Fiber festivals, farmer's markets, and specialty shows are other possible markets for rabbit wool. The Internet offers instant marketing opportunities. Angora rabbit raisers often market their products on their own websites or in online stores such as www.Etsy.com or www.MadeItMyself.com. Become involved with the National Angora Rabbit Breeders Club to get started.

RAISE A RARE BREED

Today, the American Rabbit Breeders Association recognizes more than fifty rabbit breeds. Some breeds' numbers are dwindling, and these rabbits need help staying in existence. Some rabbit enthusiasts enjoy raising rare breeds for conservation purposes as well as for showing or meat. It's personally satisfying to know that your hobby is making a difference in a breed.

The Livestock Conservancy (www.livestockconservancy.org) is a nonprofit organization working to protect more than 150 breeds of livestock and poultry from extinction. Endangered breeds are categorized based on their populations. Here are the Livestock Conservancy's definitions for three of its categories and the rabbit breeds that currently in them:

Critical

Fewer than 200 annual registrations in the United States and an estimated global population of fewer than 2,000. The American Chinchilla is on this list.

Threatened

Fewer than 1,000 annual registrations in the United States and an estimated global population of fewer than 5,000. The American, Belgian Hare, Blanc de Hotot, Silver, and Silver Fox are on this list.

Watch

Fewer than 2,500 annual registrations in the United States and an estimated global population of fewer than 10,000 as well as breeds that present genetic or numerical concerns or that have limited geographical distribution. The Beveren, Giant Chinchilla, Lilac, and Rhinelander are on this list.

Rabbits can make lovable pets for people of all ages.

Rabbit Meat

While European countries such as France, Spain, Italy, and Germany have long been utilizing this lean and nutritious meat, rabbit has barely begun to touch the meat industry in America. Meat rabbits are easy to raise, and the market for them has huge growth potential here in the United States. Certain breeds, such as the New Zealand and Californian, are considered the most profitable when it comes to raising rabbits for meat, but you need to decide which breed is right for you.

The most profitable way to sell rabbit meat is if you do the processing (butchering) and dressing (cleaning and packaging) yourself. However, this requires more than just basic knowledge. Before processing rabbits at home, you first need to explore and understand any restrictions and regulations regarding this activity.

Not everyone can stomach processing rabbit meat. Many meat-rabbit breeders opt to send their young meat animals to outside processors, who then return the processed meat to be marketed and sold. If you can sell your live rabbits directly to someone who will process and sell the meat themselves, you can cut out the cost of the middleman. Selling live rabbits to a processor brings in the least amount of money, but it's also the easiest way to run your business. Check out the list of processors on the American Rabbit Breeders Association website (www.arba.net).

Researching regulations and planning your business approach *before* you begin breeding rabbits for meat are very important. Rabbits are usually sold as fryers at 3½ pounds or more, which is around ten to twelve weeks old, so that's not a lot of time to do all of the footwork involved in complying with laws and getting the word out.

Figure out if you have a market for rabbit meat in your area—again, preferably *before* deciding to start a rabbitry for meat rabbits. Otherwise, you may end up looking for pet homes for a lot of rabbits.

About the Pet Rabbit Market

Whether or not it's ethical to get into the pet rabbit market is debatable among rabbit breeders. If you breed and sell pets responsibly (in small numbers), this may be a good outlet for you.

The most frequent argument against selling pet rabbits is the same as the one against breeding cats or dogs—many unwanted rabbits wait for homes in animal shelters everywhere.

Let's talk about culling, which forces us to look at selling pet rabbits from a practical and humane standpoint. When a rabbitry has culled those rabbits that are deemed unfit as show or meat animals, the fate for the rabbits that didn't make the cut is usually euthanasia. Obviously, selling the rabbits as pets instead is a humane alternative. In these cases, a rabbit breeder will typically find homes through word of mouth rather than sell them to a pet store. The breeder might ask friends and neighbors to spread the word, advertise in classifieds, or even spread the word at local shows.

Some people will breed extra litters specifically to sell as pets to the general public. However, in an attempt to retain some control over who purchases their animals, as well as to educate potential pet owners, these breeders sell pets directly from their rabbitries. Usually, breeders offer lifetime rabbit-care support to those who take home their rabbits.

Breeding rabbits to sell to pet stores is frowned upon. It isn't that the breeder in this case is a "bad" breeder, but rather that the rabbits are treated solely as products—not as living creatures—by the pet shop industry. As a responsible rabbit breeder, your first concern should be where your offspring will end up and how well they will be cared for.

Rabbit Manure

Rabbit manure is a natural by-product of your rabbitry and is precious as far as gardeners are concerned. Seasoned

Red worms have an appetite for rabbit manure and are the perfect choice for composting.

worms to fishermen. The investment for this type of business is minimal—small containers for the worms and maybe some marketing efforts. Choose containers that you can poke holes in, and label the containers with your contact information should someone want to purchase more worms.

If you'd like to stand out in a crowd of fishing-worm raisers, you can place worms into an interim container and fatten them up using a special supplemental diet (as opposed to just manure) a couple of weeks before putting them on the market. One such diet consists of five parts chicken layer mash, two parts wheat or rice bran, one part agricultural lime, one part wheat flour, and one part powdered milk.

Market your worms by contacting local fishing-supply, sporting-goods, and convenience stores near popular fishing spots. Be sure to let fishing clubs, summer camps, and youth organizations (Boy Scouts, Girl Scouts, 4-H) know that you're in business. Get permission to hang flyers at nearby marinas and piers.

gardeners will tell you that rabbit manure is, hands down, the very best manure for creating magnificent compost. It's the most nutritionally balanced of any herbivore's manure, and it produces extensive positive effects on the garden bed.

As a rabbit keeper with manure to spare, you can tap into this market by advertising online, placing flyers on bulletin boards at nurseries, and contacting garden clubs and community gardens. Spreading the word locally is key because you won't want to ship this particular product. There's no "going rate" for rabbit manure, and how much you can sell depends on how many rabbits you have in your rabbitry! If you don't have a compost pile (although I encourage you to start one) and aren't raising worms, then you may be thankful to just have the manure hauled away for free by zealous gardeners.

The Rabbit–Red Worm Connection

If there were ever a perfect companion to rabbits, red worms are it. While all worms can be used for composting, the ones with the voracious appetite for rabbit manure are red wigglers (*Eisenia fetida*) or red tigers (*Eisenia andrei*). These are surface-dwelling worms that munch on the debris left on the ground, quickly turning it into nutritional castings for gardens. When worms are raised in rabbit manure, the castings are mixed in with the composted material from the manure. Just to be clear, this actually creates vermicompost (compost that has worm-composted organic materials in the mix) as opposed to pure worm castings (without other composted materials). Vermicompost is exactly what worm farms produce from kitchen scraps.

You can use the vermicompost that's created under your rabbit cages in your own garden, but you can also sell the extra red

PART SEVEN

SHEEP

BY SUE WEAVER

WHY SHEEP?

Sheep are to hobby farms as diamonds are to gold: they make a good thing better.

Be they pets or profit makers, sheep should be part of every small-farm scene.

They are inexpensive to buy and keep, easy to care for, and relatively long lived,

making them great investments. Given predator-proof fencing, minimal housing,

good feed, and a modicum of daily attention, sheep will thrive.

SHEEP

FROM THE BEGINNING

New shepherds can learn to care for sheep in a relatively short time, which makes them attractive even to first-time farmers. Sheep make delightful pets; they'll mow your yard, come when called, and, with training, maybe even pull a cart.

Farmers looking to turn a profit in sheep can do so. Hand-tamed miniature lambs fetch respectable prices as pets. Clean fleeces from many breeds are wildly popular among handspinners. Niche-marketed lamb sells readily to ethnic and organic markets, and sheep's milk cheese is popular with gourmets around the world. In addition, raising livestock qualifies landowners for lower agricultural land tax assessments in many locales.

You may have reservations; even I haven't had sheep all my life. For more than fifty years I was blissfully content with a husband, horses, and dogs. Then came hobby farms—then guineas, chickens, and an ox. I had always liked sheep, so why not give them a try? I was smitten, totally and thoroughly captivated by our first woolly pets. Now sheep are a part of me. I've been entrusted with the care of these beautiful creatures. I'm sure I've been blessed!

In the early days, there were majestic wild sheep called *mouflons*. About 11,000 years ago—probably near Zawi Chemi Shanidar, in what is now northern Iraq—a hunter stopped fiddling with his spear point, kicked a log onto the fire, and said to a friend, "Wouldn't it be smarter to snatch some lambs and raise them here in camp?"

So humans and sheep formed an alliance. People protected sheep from wolves, bears, and mountain lions; sheep reciprocated by developing wool. About 3,500 BC, women puzzled out how to weave sheep's woolly covering into fine, sturdy cloth that kept wearers toasty in the wintertime and cool under the blazing summer sun. Decked out in woolen garments, men said to one another, "We don't have to stay here on the Mesopotamian plains where it's always pretty warm; we could go out and conquer the world!"

Sheep were already out in the world. Domestic sheep reached parts of Europe by 5,000 BC, having been carried west by intrepid Neolithic farmers. Swedish farmers began raising northern short-tailed sheep between 4,000 and 3,000 BC. Between 1,000 BC and 1 AD, Persians, Greeks, and Romans labored to develop new and better sheep. The Romans brought their revamped woollies along (a walking food supply) when they conquered Europe and North Africa; by 50 AD, the Romans had erected a wool-processing plant near Winchester, England.

Historically, the greatest sheep was the mighty Merino. Some researchers think it sprang from a genetic mutation some 3,000 years ago; others believe it was developed during the reign of Queen Claudia of Spain (41–54 AD). Whatever the case, income from the Spanish Merino wool trade transformed Spain into a world power and financed its New

DID YOU KNOW?

Ancient Chinese sheep were used to pull carts, carry burdens, and give children rides, especially in Xiongnu, where youngsters rode giant sheep as part of a hunting game. In Tibet and Nepal, thousands of sheep are still employed as pack animals.

The Corriedale is a dual-purpose sheep.

include the Black Welsh Mountain, Cheviots (American Brecknock Hill Cheviot, Border Cheviot, Keyrrey-Shee, North Country Cheviot), and Scottish Blackface.

Dairy and Meat

Farmers raise sheep not only for wool but also for dairy products and meat. The American dairy sheep of choice is the high-producing East Friesian Milk sheep. An increasing crowd of American sheep owners are clambering aboard the sheep-dairying bandwagon. Rich, mild-tasting sheep's milk is higher in protein, calcium, and fat—as well as vitamins A, D, and E—than is cow's or goat's milk, making it an ideal cheese-making medium. Many of the world's great cheeses are made from ewe's milk. These include Greek Feta, Kasseri, and Manouri; Spanish Iberico, Manchego, and Roncal; Sicilian Pepato and Ricotta Salata; and that perpetual French favorite, genuine Roquefort cheese.

Medium- and long-wool breeds are sometimes called *dual-purpose sheep* because they are raised for meat as well as wool. During the twentieth century, breeders created additional dual-purpose varieties by crossing Merino or Rambouillet fine-wool sheep with medium- and long-wool sheep. These breeds include the Corriedale, Cormo, Romeldale, and Targhee. The British long-wool and luster sheep are also bred for both meat and wool.

Scarce and Endangered

Commercial breeders generally raise big, meaty breeds, such as the Suffolk and Hampshire; wool factories prefer the Merino and Rambouillet. Many producers crossbreed for added productivity. As a result, a great many of yesterday's sheep breeds are becoming scarce and endangered.

The Livestock Conservancy publishes a priority list of threatened breeds. Listed as "Critical" are the Florida Cracker, Gulf Coast Native, Hog Island, Leicester Longwool, Romeldale/California Variegated Mutant, and Santa Cruz. Breeds considered "Threatened" are the Black Welsh Mountain, Clun Forest, Cotswold, Dorset Horn, Jacob (American), Karakul (American), Navajo-Churro, and St. Croix. Breeds on the "Watch" list are the Lincoln, Oxford, Shropshire, and Tunis. Breeds being monitored as

HERITAGE SHEEP

Want to keep sheep and also make a difference? Raise one of the endangered or rare breeds listed on the Livestock Conservancy's Priority List.

The Livestock Conservancy was founded in 1977 to preserve rare farm animals and promote genetic diversity in livestock. It acts as a clearinghouse for information on these subjects and registers breeds that don't have American registries of their own. It also assists in the conservation of Heritage-breed cattle, asses (donkeys), goats, horses, pigs, sheep, and poultry.

Sister organizations include Rare Breeds Canada and the Rare Breeds Survival Trust (UK). Australia, New Zealand, and many other countries support preservation efforts of their own. With their help—and yours—we can bring these endangered breeds back from the brink of extinction.

This crossbred ewe has a Scottish Blackface sire and Black Welsh Mountain dam.

"Recovering" are the Barbados Blackbelly, Shetland, Southdown, and Wiltshire Horn. There is also a "Study" category, under which no sheep breeds are currently listed.

Buying the Right Sheep for You

So you've decided you want to keep sheep. Although shepherding is far from complicated, it makes sense to learn all you can about sheep before buying any. Purchasing the sheep is just the first step. You'll need to find the best way to transport your new animals and to determine where they should be kept.

What to Buy

What sheep are best for you and your farm? Is it hot where you live? Cold, wet, arid? Are your pastures lush or rocky? Do you have time to spend with your sheep, or would self-reliant sheep work best? Know what you need before sheep shopping.

Of course, there's nothing wrong with bringing home a few "easy sheep" (whatever you can find locally) to gain sheep-keeping experience, as long as they're healthy and you enjoy them. But to save yourself a heap of heartache, do all the homework before launching a business or investing in pricey registered sheep.

Choosing the Breeds

At least five dozen sheep breeds are reasonably and readily available in North America, each developed to meet specific needs. Make a list of the qualities you're looking for, and then go shopping.

Purebred, Registered, or Crossbred?

Purebred sheep are those whose parents, grandparents, and beyond were all the same breed. Registered purebreds have papers to prove it. Purebreds "breed true," meaning that their offspring inherit their parents' looks as well as many other traits, including elements of their behavior and personalities.

Crossbred sheep have parents of two different breeds; sometimes the parents are themselves crossbred. Because the parents are genetically dissimilar, crossbred sheep benefit from *heterosis*, or *hybrid vigor*, which translates into hardier, faster-growing animals. A crossbred often proves more productive than either parent.

If you plan to market breeding stock, show at breed shows, or participate in Heritage or rare-breed preservation efforts, you'll need registered sheep. These sheep cost more—often much more—than do crossbreds or unregistered purebred sheep. In addition, although their breeding-quality offspring fetch higher prices than those of crossbred lambs, the market (meat) lambs and commercial wool of registered stock do not. Registered sheep take some of the guesswork out of breeding, especially for shepherds who know their breed's bloodlines and how best to use them. Registered flocks are generally more uniform-looking than crossbred or grade (unregistered) herds. If this is important to you, buy registered sheep. If you simply want productive sheep, think crossbreds.

Availability

When choosing a breed, consider availability. You could opt for a breed common to your area (in which case, you wouldn't have to travel far to purchase seed stock or replacements) or something unique (making yourself the only regional source of stock for others who might like to own this breed).

If you decide on something unusual, you'll have to buy your foundation animals from afar. Although transportation may be an issue, keep in mind that horse and livestock transporters will haul sheep for a fee and that lambs of miniature sheep breeds can be

Observe the sheep's behavior and assess its overall conformation.

shipped between major airports in large dog crates. However, a single sheep is a stressed sheep; when shipping sheep, always buy multiples if you can.

For locally common breeds and crossbreds, a number of sources exist. Watch newspaper classifieds and ask at feed stores, veterinary practices, and county extension services. In addition, peruse registry directories and breed-journal ads.

Don't buy your first sheep at auction! Livestock sales are the farmer's dumping ground. Except for slaughter lambs, most sheep sold at auction are culls: aged ewes with broken teeth, ornery rams, and sheep afflicted with hoof rot (or worse). Even healthy sheep passing through sale barns are exposed to sheep that are not. Experienced shepherds occasionally bag a jewel in the rough, but until you can recognize the many signs of potential pitfalls, you're better off purchasing sheep directly from a breeder.

DID YOU KNOW?

Many of today's sheep breeds have long, illustrious histories:

- The Black Welsh Mountain dates to the pre-Middle Ages.
- The Border Cheviot was first recognized in 1372.
- Cotswold sheep are from pre-Roman-conquest times.
- Icelandic sheep were carried to Iceland by Viking settlers in the ninth and tenth centuries AD.
- The Jacob was possibly mentioned in the Bible and has been raised in Britain for more than 350 years.
- Karakul sheep were pictured in Babylonian temple carvings.
- Navajo-Churro sheep came to the New World from Spain circa 1493.
- The Shetland arrived at the Shetland Islands via Viking settlers.
- The Tunis has existed since the pre-Christian era.

Again, when you do buy, try to buy two. Sheep are gregarious, and a solitary sheep will be frightened and lonely. One possible exception is a bottle lamb (or adult former bottle lamb) that accepts humans as its flock.

Talking to Breeders

When you've narrowed your choices to a few favorite breeds or crossbred types, contact registries or peruse online directories to find breeders in your area. Visit as many as you can. Shepherds love to talk sheep, and you'll learn so much by picking their brains.

Producers' websites are also great sources of valuable information. Find them via registry websites, search engines, and online breeder directories. In addition, breed-specific books and journals are essential for newbie shepherds, both for care information as well as bloodline data and health-issue information you might not find elsewhere.

The best way to obtain information, general or specific, about sheep is to work with a mentor. If a local shepherd befriends you, consider yourself blessed. Barring that, you can meet loads of helpful, experienced shepherds online. Ask one or more friendly souls to coach you; most will be happy to oblige.

Selecting the Sheep

You've selected your breed and contacted a breeder. Now you're off to see some sheep. When assessing the sheep, you must consider several factors, including conformation, health, and teeth, as well as sex-specific qualities, such as udder condition.

328   HOBBY FARM ANIMALS

Notice the absence of front top teeth, which is normal in sheep.

Conformation

When evaluating purebred sheep, assess their type. Type is what makes sheep of a given breed resemble one another. It's outlined in each breed's standard (a registry-generated guide to that breed's ideal). If you understand your breed's standard before you go shopping, you'll select better sheep.

Quality sheep of all breeds have traits in common. You'll want straight, sturdy legs, set one in each corner; such sheep have wide, strong hindquarters and broad chests. They have deep bodies, well-sprung ribs, level backs, and fairly straight underlines. They're wide and firm at the haunches, and their backs are nicely fleshed. If fleece conceals their basic structure, explore prospective purchases with your hands—that's the only way to know for sure what's under there.

ADVICE FROM THE FARM
CHOOSING AND BUYING SHEEP

Do your homework. Locate a reputable sheep breeder and then tell the breeder what you are looking for. Let the breeder select the animals that he or she feels will work for you. I was raised on a ranch and have been around animals all my life, but I bought my present sheep sight unseen from breeders who I researched via websites and chat lists. I knew they were top-level breeders and were producing the quality of sheep I wanted. They were also the breeders who welcomed my questions via e-mail and took the time to respond. And I joined all of the Yahoo! groups that discussed the breeds I was considering.

—Lynn Wilkins

Talk to old-time sheep buyers if you can. They have worked all classes of sheep for years and know just about all there is to know about these animals. They can tell you which sheep are most likely to be a profit for you, and this is important.

—Connie Wheeler

If you really like a breed, get it. I think if you care for your animals, as in feeding and healthcare, then they'll do well no matter where you are located. It might cost you a bit more to have them shipped to you, but you'll have gotten the breed you really wanted, and, in the end, you'll take care of them better for it.

—Laurie Andreacci

Never be afraid to ask questions. And remember, we're all in this together, so help out your fellow sheep breeders.

—Laurie Andreacci

Rams can be more trouble than the average small-scale sheep owner wants to deal with.

Health

Any sheep you buy should be healthy. Healthy sheep are alert and interested in their surroundings. Unhandled sheep will be on guard but not frantic, and tame sheep may beg for attention. Avoid droopy, disinterested sheep that cough, wheeze, or just plain strike you as sickly. Healthy sheep have reasonably tidy backsides, nondrippy eyes and noses, and bright pink mucous membranes.

Raggedy, moth-eaten fleeces bespeak sheep riddled with lice, mites, or wingless biting flies called *keds*. Limping sheep may be suffering from a dreaded contagious disease called foot rot. Unless you know for certain that they're not afflicted with such a malady, don't buy them!

Teeth

A sheep has a pad of hard tissue but no front teeth in its upper front palate; eight incisors grace the front lower jaw. Lambs grow eight baby incisors, which are replaced by two larger permanent teeth each year, beginning at the age of one. They start at the center and continue outward, so you can easily tell a sheep's age through year four. Once they've emerged, a sheep's teeth continue to wear and spread farther apart; by the time they're eight or nine, most sheep have lost or broken some of their incisors. Because they can no longer efficiently graze, older animals require specialized feeding, so don't pay top dollar for "broken-mouthed" or "gummer" (toothless) sheep.

By the same token, unless a sheep's lower teeth align properly with its hard dental palate, it can't efficiently grasp and rip off grass. If its upper palate juts out farther than its teeth, it is "parrot mouthed"; if its incisors protrude beyond its dental pad, it is "monkey mouthed." Either way, this sheep is a problem feeder and one you should probably avoid.

Sex-Specific Factors

When buying a ewe, inspect her udder. Either reach under her or ask her owner to set her up on her hindquarters, which is better still. She should have two symmetrical, widely spaced teats set on a warm, soft bag. Sheep shearers in a rush sometimes zip off teats.

Lumpy udders are a mark of mastitis. You might have to bottle-feed a problem ewe's lambs; if that's not an option, refuse her.

If you aren't lambing-savvy or have chosen a breed known for lambing problems, pick experienced ewes. Ewes reach peak productivity between four and six years of age. However, large-scale breeders often cull six- to eight-year-old ewes that can work perfectly in a hobby farm setting. They'll teach you loads in exchange for a smidgen of extra care.

When buying pets, forgo the rams. Rams are unpredictable, and a surly one is trouble on the hoof. If you can't resist that winsome ram lamb, think castration. Altered males (called *wethers*) make wonderful pets. However, if you need a breeding ram, buy the best one you can afford. He'll influence your entire lamb crop, so he should exemplify the best of his breed. He must have two large, smooth, roundish testicles suspended in a free-hanging scrotum; never buy a ram with a single testicle. In sheep, size counts. Measure a ram's scrotum at its widest point. Miniature and small-breed rams should tape close to 12 inches, and large ones at least 14–15 inches.

If the ram is horned, make certain that you can place two fingers horizontally between his face and horns. Sheep chew with a side-to-side motion; if his horns crowd his jaw, he could starve.

The Sale

When buying expensive registered stock, request references and check them out. In every case, ask for detailed health and production records.

Breeding prospects must be identified with a number that proves its participation in the USDA's scrapie eradication program.

If you'll be crossing state lines with your purchases, arrange for health papers (you'll probably have to foot this bill). Read the registration papers carefully. Are they up to date and transferred to the present seller's name? Do the identification numbers match those tattooed on the sheep or printed on ear tags? Make sure you get the right sheep!

Sheep breeders are required to participate in a federal program called the National Scrapie Eradication Program, directed by the U.S. Department of Agriculture (USDA)'s Animal and Plant Health Inspection Service (APHIS). All sexually intact sheep and all wethers older than eighteen months of age are assigned unique numbers and must be permanently identified with these numbers via ear tags, tattoos, or microchips. Don't buy sheep from a noncompliant herd! Find out more at www.eradicatescrapie.org.

Transporting Your Sheep

Hauling your sheep can be as simple as boosting them into the topper-sheathed back of your pickup or as complex as hiring an animal transporter to truck them across the country to your farm. To make the move as smooth as possible, do everything you can to make your sheep comfortable. Bed their conveyance so they can lie down en route. Don't use sawdust, wood chips, or finely chopped straw; all work their way deep into fleeces, reducing their value. Longstem straw or grass hay works well; it's soft and cushiony, and sheep can nibble it if they want. On long trips, stop and provide drinking water every few hours. Also ensure that your mode of transportation is a safe one. Sheep can jump higher and push harder than many people imagine.

Sheep don't tolerate stress. Keep things low-key, and—unless it's absolutely necessary—don't haul a sheep by itself. Conversely, don't crowd sheep into a space that's too small. Sheep need room to move around. Ensure adequate ventilation, without drafts.

Holistic shepherds sometimes dose each sheep with probiotic paste or gel before and after hauling. This boosts immunity and reduces tummy upsets. In addition, spritzing their faces with Bach Rescue Remedy will help steady your sheep's nerves.

Have your facilities ready to receive the sheep on arrival. This includes starting them off with the same feed that their former owner fed them. Switch feeds gradually to prevent nasty digestive upsets.

New sheep should be quarantined for at least three weeks and dewormed. If their vaccination history is uncertain, revaccinate. When quarantine time is up, introduce new ewes gradually, perhaps by penning them in an enclosure adjoining the existing flock's pasture. Rams are another story.

Housing for hobby-farm sheep is the essence of simplicity. Sheep are happier and healthier when kept outdoors, so provide your flock with well-maintained pastureland. Your friendly county extension agent can help you choose the right pasture species for your locale and teach you how to plant and maintain it. America is a huge country. Minnesota pasture grass species don't thrive in the Ozarks, nor do western forage grasses do well in the Deep South.

SHEEP

FEEDING AND HOUSING ADVICE

Provide plenty of water in your pasture, and keep it sparkling clean. Sheep won't drink enough to satisfy their needs if their water supply is alive with green algae or filled with their flockmates' feces. Don't just dump and refill the trough; scrub it out. A weak chlorine bleach solution helps chase away stubborn algae deposits and disinfect the trough. Feed stores sell sheep-appropriate fiberglass and metal water troughs, but we use injection-molded plastic horse-stall cleaning baskets. Heavy-duty plastic cattle and horse mineral-lick tubs and pans make dandy, easily cleaned sheep waterers, too. Pastured sheep need shelter from sun, hail, and possibly snow and freezing rain. Trees work fine as sunshades, but your sheep need manmade shelter to withstand harsh storms. If you don't take your sheep inside at night, make a point of checking them at least twice a day. Count noses. Walk around each sheep, checking for illness or injuries. Patrol all fences and associated structures on a weekly basis, scouting for downed wire, exposed nails, loose tin, wasp nests, and any other accidents waiting to happen.

Lush, green pasture provides essential nutrition for a grazing flock. Given their choice of victuals, sheep choose 40 percent grass and 60 percent browse, meaning they'll cheerfully rid your meadows and open woodlots of tender tree sprouts, brambles, and briar roses.

Provide pastured animals with a high-quality, free-choice, granulated sheep mineral mix. Place it inside their shelter, or buy or make a covered container to install outside. Feed a little "sheep candy" every day. A scant handful of corn, pellets, or sweet feed per head keeps them happy to see you, and happy sheep are a whole lot easier to handle.

If your sheep don't keep weeds and saplings in check, occasionally mow the pasture—or buy a few goats to add to the flock. Constantly grub poisonous plants out of your pastures. Many are bitter, and well-fed sheep won't eat them—although there are notable exceptions. It takes only a nibble of certain species to dispatch a lamb. It's best to err on the side of caution.

Fencing, Shelter, and Stalls

Sheep also need fencing to keep them safe from predators, simple shelters to protect them from the elements, and pens and stalls for lambing and times of quarantine. Wherever you keep your sheep, you'll need to make sure that they have all of the amenities. Fencing and shelters alone often are not enough to adequately protect your flock. For additional protection, consider engaging the help of a guardian animal (discussed later in this chapter), whether a specially trained dog, a llama, or even a donkey.

DID YOU KNOW?

In days of yore, shepherds in Europe were often buried with a tuft of wool in their hands. This practice sometimes signified their devotion to their charges. At other times, it simply showed that they were shepherds—thus excusing occasional lapses in church attendance because they couldn't leave their flocks during lambing.

Such an enclosure will keep the sheep in but may not keep predators out.

Fence Me In

In most cases, a sheep fence is more important for keeping things out than for keeping them in. Sheep are incredibly easy to fence because one nose-first encounter with electric wire convinces them that they want to stay put. Keeping predators at bay is another story.

One solution is to coyote-proof your sheep's nighttime fold with tightly stretched, all-the-way-to-the-ground woven-wire fencing surrounding the area. Reinforce it with strands of electric fence along the top and outside bottom. Should you leave the farm during the day, the sheep can go into this enclosure, too. Dogs can be far more lethal than coyotes, and dogs tend to strike during daylight hours.

Some variables to consider when selecting fencing are soil base (for example, rock, sand, or clay), lay of the land, proximity to roadways or neighbors, breed(s) of sheep, local predators, and costs of fencing materials and labor. It's beyond the scope of this book to discuss the hundreds of types of fencing available, so talk to your county agricultural agent and peruse the fine agricultural bulletins available online from university resources.

Shelter

All that a small flock (twenty-five or fewer sheep) really needs is access to simple shelters, erected with their open sides facing away from prevailing winter winds. They're inexpensive, portable, and easy to clean, and they can be constructed by any reasonably adept home carpenter in a day. Be sure to allot enough space in your shelters for the number and type of sheep that you own. Most breeds are fine with 8 square feet; 12 square feet is adequate for ewes with lambs.

Whenever possible, place shelters on a southern slope with well-drained soil. Keep them close to your house to discourage predators and to make midnight-lambing time checks a little bit easier on a sleepy shepherd. Available space in an existing barn or shed can be used as well if the sheep are outside for at least part of the day. Indoor accommodations must be well ventilated but draft free.

Pens and Stalls

You'll need an indoor spot to erect your lambing *jugs* (small pens where ewes are confined with their newborn lambs for a few days postpartum) and a quarantine, or "sick sheep," stall. The areas needn't be elaborate, just big enough for a single sheep or a ewe and her lambs to be comfy, dry, and out of drafts.

When choosing bedding, select absorbent material that is unlikely to work its way into the sheep's fleece. For example, wood shavings quickly imbed themselves in fleeces and ruin them, and the fine dust particles in sawdust are hard on sheep's respiratory systems. Dust-free long-stemmed wheat straw is the bedding of choice, but other types of long-stemmed straws work well, too. Other workable options include peanut hulls, ground corncobs, and the hay that your sheep inevitably pull down and waste.

IS IT SAFE TO EAT?

There are few things more heartbreaking than losing animals because they ate poisonous plants that you should have grubbed out of their pasture. As shepherds, it's our obligation to make pastures as safe as possible for our woolly friends. Unfortunately, it can be hard to know which plants and brush to remove. Flora varies widely from place to place, so contact your county agricultural agent to identify mystery plants; don't assume that they're all safe to eat.

As ruminants, sheep get most of their nutrients from forage.

Principles of Feeding Sheep

Only a few classes of sheep routinely require concentrates: ewes during flushing, late-gestation and lactating ewes, hard-working rams, geriatric or underweight sheep, and an occasional lamb.

Concentrates must be sheep-specific, meaning whole or cracked grains or commercial products labeled for sheep. Although the subject is complicated, suffice it to say that sheep are easily poisoned by too much copper in their diets. Swine and poultry feeds are high in copper, as are some beef and dairy cattle rations. Even some goat mixes contain more copper than some breeds can handle. It pays to look for the words "sheep feed" on the label. For the same reason, you shouldn't expose sheep to mineral licks or loose minerals designed for other species. Sheep require minerals, but in sheep-specific form.

No matter what some folks insist, sheep cannot safely digest spoiled or moldy feed, including musty hay. Consider that mycotoxins—poisonous compounds produced by molds—could be in moldy feed. Moldy feed can trigger bloat, too.

Measure feed, especially concentrates. Don't feed a lot today and half of that amount tomorrow. When group-feeding concentrates, don't let bossy gluttons hog it all. They can bloat or get acidosis and founder while meeker flockmates literally starve.

Sheep are creatures of habit. Try to feed them about the same time every day. They prefer to eat during daylight hours and in the company of other sheep. If you must switch feeds, do so gradually, over a 10–14-day period, to allow sheep's rumens to adjust.

Hay There!

Forage is the basis of most sheep diets. Consider hay, for example. Most hobby farmers have to buy it. In some parts of the country, acquiring decent hay is an expensive proposition; if you live in a place

---

ADVICE FROM THE FARM

HOUSING AND FENCES

The ideal fence would be made of the kind of nonclimb fencing that equestrians use, although this is quite expensive compared to regular "field fence." An electric wire placed 12 inches away from the fence and about 12 inches off the ground will keep sheep completely away from the fence. It's also a good idea to spray vegetation killer under and on both sides.

—Connie Wheeler

It's good to have a hilly pasture where your sheep get lots of exercise going up and down hills to feed and water.

—Connie Wheeler

Although we do have barns and other sheltered areas, most of our sheep don't use them except on extremely rainy, windy days. They'll stay out in the rain most of the time. Hair sheep have a thicker skin to compensate for their lack of wool.

—Connie Wheeler

My mantra is, "Always close all of the gates, all the way, all the time." Even if you're just popping into the pasture to grab a bucket or something. It's tempting to not always bother. Sometimes you get away with it, but sometimes not.

—Lisa MacIver

These ladies think that a repurposed plastic tub makes a fine feeder.

where quality hay is grown, you are at an advantage.

Small square bales usually work best for sheep; plus, they're easier to handle and store in a hobby-farm situation. Because you can dole out exactly what they need, the sheep waste less when fed from smaller bales. Large round bales are acceptable if they've been stored under cover and are free of mold. However, your sheep will pull hay down and then defecate and lie in it. You also can expect a lot of debris-contaminated fleece at shearing time.

Nutrient-rich hay is green. Quality alfalfa is dark green, but if it's a strange lime green, it has been treated with propionic acid preservative. Authorities claim it's safe, but if you're not convinced, avoid it. Well-stored grass hay is light to medium green. The outsides of both legume and grass bales become bleached if exposed to light in storage, but inside the bales, the hay should still be green.

If you see white dust, blemishes, or black spots, or if there is a musty smell, the hay was cut and stored while still damp. Even if you don't spot actual mold, do not feed this hay. Hay that's pale to golden yellow wasn't put up promptly and got sun-bleached in the field. It's nutrition-poor and apt to be dusty.

Quality hay is made from leafy plants mown just before blooming. It has thin, flexible stems that easily bend without breaking. Coarse, stemmy hay is low in nutrients and unpalatable. Leaves contain most of the plant's protein; stems are mainly hard-to-digest, low-energy cellulose. As plants mature, they grow more stem than leaf. Stemmy hay made from older plants is no bargain, and hay containing mature blossoms and seed heads was mown well past its prime.

Refuse hay containing sticks, rocks, dried leaves, weeds, and insect or animal parts: your sheep can't eat the sticks and rocks; the weeds may be toxic, and their seeds will infest your property; and baled insects and desiccated animals can cause serious disease. Don't feed your animals uncured new crop hay, either. Age new hay fresh from the field for 5–6 weeks before feeding it. Feeding uncured hay can easily trigger bloat. Nutritious hay put up dry and carefully stored holds its nutrient content for several years. When it is still green inside, old hay can be a bargain.

Whenever possible, buy tested hay from a reputable hay dealer. It takes a lot of guesswork out of the equation. Be extra-cautious if you buy hay at auction. Some sellers place their best bales on the outside, where buyers can see them, and tuck junky hay deep inside the stack. Inspect the hay you plan to buy. Agree to pay for a couple of bales so you can open them and see what they look like inside. If the hay doesn't pass muster, you aren't obligated to buy the rest.

If you can have your hay delivered and neatly stacked in the barn; it's extra money well spent. Buy only the amount of hay you that can store under cover. Stack it on pallets to get it off the floor; if you don't, the bottom layer will rot. Don't let barn cats defecate or have kittens on it; cats are primary carriers of toxoplasmosis. Finally, to minimize waste and prevent disease, use those feeders. Never feed hay off the ground.

The Amenities

Your sheep need something to eat out of. There are countless styles of grain and hay feeders to choose from. Large, sturdy plastic grain feed pans are tough, portable, and easily cleaned. Wall-mounted steel rod racks work well for hay. A great portable hay feeder for just a few sheep is a horse-style nylon hay bag. You'll find durable multiple-sheep feeders at farm and feed stores, or order them from sheep-supply catalogs.

Another option is to build your own feeders. If you request item #900000 from Premier's website (www.premier1supplies.com) or catalog, you'll get free downloadable or hard-copy plans for its double-sided and single-sided flock feeders.

Sheep shouldn't eat out of standard big-bale feeders. When they bury their heads in the hay to choose the choice morsels, they imbed a lot of nasty debris in their fleeces. They may also rub off patches of their fleeces when they lean against pipe rails while eating.

The Great Pyrenees is one of the most popular flock-guardian breeds of dog.

from France, the Yugoslavian Sharplaninetz, the Hungarian Komondor and Kuvasz, the Turkish Akbash and Anatolian Shepherd, and the Italian Maremma Sheep Dog. These dogs are raised with and bond with sheep, so they simply remain with their "flockmates" at pasture.

Guardian donkeys live longer than do guardian dogs, they're inexpensive to buy and maintain, and they require no special training. Like llamas, donkeys instinctively dislike the canine clan and protect their herds (be they sheep, goats, poultry, or other donkeys) from stray dogs, coyotes, and even wolves. A donkey in attack mode can gallop and stomp at the same time. Needless to say, a new guardian donkey must be carefully introduced to family dogs—who will quickly learn to give the newcomer a wide berth! Like some dogs, some donkeys will hang around the barn, seeking human contact instead of minding the flock. However, people who like donkeys swear by these long-eared guardians.

Llamas may be the best bet for hobby-farm flock guardians. Llamas bond with their woolly charges extremely well, eat the same feed, and require most of the same vaccinations. Llamas, especially females with *cria* (suckling offspring) and castrated males, instinctively guard their herdmates, even when their "herd" consists of sheep. Guard llamas shriek at, charge, and sometimes stomp marauders. Few small predators (such as dogs and coyotes) willingly tackle the same angry guard llama twice.

- Don't feed hay off the ground. Once one sheep defecates or urinates on it (and one will!) none of the others will eat it. Eating off the same stretch of worm-egg-infested earth encourages worm infestation, too.
- Mount feeders reasonably close to the ground so hay chaff doesn't drift into faces and fleeces while sheep are eating.
- Locate feeders where you can fill them easily with hay or grain without flinging it. Sheep raise their heads when it's least expected; voilà, a face full of hay or grain.

Guarding Your Sheep

Canine flock guardians are not herding dogs or pets; they're specialists bred to guard livestock, and they do it well. Originating in Europe and Asia, guardian breeds include the Great Pyrenees

---

ADVICE FROM THE FARM

SHEEP FEEDING TIPS AND TRICKS

After breeding time, we take our older girls and pen them together in a separate area from our younger ewes. We call it our "geriatric pen." These old ewes get about one-half pound each per day of Equine Senior. It has made an amazing difference in their weight, activity, and the babies they produce. For a treat, they love bread.

—Lyn Brown

My four kids raised the three main species of farm animals for 4-H and Future Farmers of America (FFA) market shows, with numerous grand champions along the way. We always fed the feed that the local feed mill produced for that particular species and our specific location. It worked well for us. The feed contained the trace minerals our area [lacked] and the feed composition percentages the species needed.

—Lynn Wilkins

Most people think that sheep are stupid, but they're wrong. Folks who keep sheep tell of clever ewes teaching lambs to split chestnut hulls with their hooves or a flock banding together to shake apple trees and harvest the succulent fruit. Sheep are good at "sheep things" but sometimes not as adept at performing tasks that their caretakers prefer. For example, a sheep may not happily leave its peers and move quietly from place to place in an orderly fashion. A flock may not thrive indoors in close confinement. These things are foreign to the instincts that make sheep tick.

SHEEP
HANDLING AND SAFETY

A University of Illinois Extension paper titled "An Introduction to Sheep Behavior" ranks sheep "just below the pig and on par with cattle in intelligence." Researchers at the Babraham Institute in Cambridge, England, trained twenty sheep to recognize pictures of other sheep faces. Electrodes measuring their brain activity proved that some remembered at least fifty of the faces for up to two years. "It's a very sophisticated memory system," explains Dr. Keith Kendrick. "They are showing similar abilities in many ways to humans."

Sheep also learn and respond to their names. Club lambs and exhibition sheep lead, stand tied, allow extensive grooming, and pose in the show ring. Pet sheep learn to pull carts; some even do tricks. Sheep, intelligently and quietly handled, are very trainable.

Sheep are not stupid; they are reactive. Their only means of survival is to band together for protection and then to run. Frightened, stressed sheep flee blindly, pack into corners, and get wedged behind gates. Quietly handled sheep generally do not.

Shepherds who have difficulties working with sheep have never learned to think like their charges. These animals operate by an ingrained set of rules that are rarely broken. When we learn to play by sheep's rules, we exist in perfect harmony. Those who understand how sheep communicate and how they perceive and interact with their world find sheep intelligent, ingenious, and easy to handle.

Your first step is to learn what factors have shaped sheep instinct and how sheep typically behave as a result. Next, learn to read your sheep, who communicate through intricate verbal and body signals. This knowledge will help you achieve your goal of handling your sheep effectively.

Why Sheep Do What They Do

Sheep know that they are vulnerable to predators. They don't bite or kick to defend themselves, and they rarely use their horns in self-defense. So they are supremely vigilant and suspicious. When startled, they are wired to flock and run. Historically, Mother Nature culled bold sheep and any others that paused to ponder. Timid, wary sheep bred on.

Although wily coyotes and hungry wolves aren't fixtures on today's hobby farms, sheep remain unconvinced of their relative safety. They know it's easier for a wolf to snag a solitary sheep than to pick one out of a tightly packed, fleeing flock, so they first seek safety in numbers. Although flocking behavior does vary by breed—from supremely gregarious white-faced sheep, such as Merinos and Rambouillets, to self-reliant Cheviots, Leicesters, and most hair-sheep breeds—all sheep flock to some degree.

Flock dynamics apply to mobs of four or more sheep; fewer sheep may not respond as expected. Dr. Ron Kilgour, of New Zealand's Ruakura Animal Research Station, assessed flock dynamics by working sheep in a maze. Single sheep were considerably stressed but eventually mastered the maze quite well. Two sheep went in different directions, and three sheep scattered; both of these groups generally muddled things up. Four or more sheep moved as one and learned faster than single sheep. That's why three sheep are the norm at herding trials; they give herding dogs the acid test.

Flock animals, sheep ascribe to "safety in numbers."

Sheep are gregarious, meaning they crowd together for reassurance and protection. They have a strong inner compulsion to follow a leader. These traits compose their flocking instinct. In most cases, the leader is simply the first sheep that starts moving in a given direction; flock hierarchy rarely enters the picture.

White-faced (wool) sheep are more gregarious than are black-faced (meat) breeds. When stressed, huge

ADVICE FROM THE FARM

RAMS AT A GLANCE

Rams start cracking heads as breeding season arrives—it's their way of showing their dominance.

This summer, one of our young rams took aim on me a couple different times in a week's span. Each time, I responded with a swift kick. One time this fall, he was standing there looking at me from about ten feet away with a funny look in his eye. I kept my eye on him, and he walked up to me, placed his head on my thigh and pushed—he knew not to come at me from a distance and get up a head of steam.

Rams tend to view us as part of the herd, too, and we have to keep our dominance in the flock to remain top dog. Otherwise, we are in trouble.

—Lyn Brown

Never trust them. Your sweet ram that has never hurt you once can turn on a dime and run you down. Be smart; keep your eyes on him.

—Laurie Andreacci

I have a bottle-raised ram that got a little aggressive. What we did was, we bought a set of water blasters. Every time he even suggested he was coming at us, he would get a blast in the face. It didn't take him long to figure out he'd rather stay away. After a while, we didn't need to carry the water blasters any more.

About once a year, he starts acting rammy, and we have to have a refresher course.

—Lyn Brown

Some people have a tendency to think that because rams are relatively small (compared to stallions and bulls), they are just sweet little things, as lambs are portrayed in so many ways. But these are animals with animal instincts, and they will try to dominate you—and often do.

I had a large Suffolk ram keep me in a hay feeder we put out in the middle of the barnyard instead of close to "people getaway places." I was up there for about a half hour before he lost interest.

We had an older North Country Cheviot ram who was the tamest, sweetest ram you could ever imagine. One day, after having not been in that particular pasture for some time, we herded them into the barn, and he rammed me against the fence. Before I could get up and untangled, he got me again. Luckily he couldn't get a good run at me, or he would have broken my hip.

I've noticed that sometimes a good spray of water will stop them long enough to get away. I've even grabbed my husband (now ex) and put him in-between us to get away. It worked, too!

—Connie Wheeler

space behind swinging gates. They pack into them and stay there.

Calm sheep move forward at a walk, and frightened ones stampede or back up. Calm sheep are thinking sheep; sheep in a panic just react. When its flight zone is penetrated by a perceived predator (including a dog or shepherd), a sheep moves away.

Sheep move forward out of darkness, toward light. Their faulty depth perception renders shadows and harsh light/dark situations terribly frightening. They move from confinement toward open spaces, into the wind rather than downwind, and more readily uphill than down. Sheep prefer not to cross water, they dislike passing through narrow openings, and they panic on slippery surfaces, be they natural or manmade (such as wet stall mats and slick concrete). Sheep have long memories. Most of them easily recall bad experiences for a year or more.

flocks of Australian Merinos can pack so tightly that humans swept up in the crush are injured or killed. Weakly gregarious breeds include the Suffolk, Hampshire, Corriedale, Cheviot, Leicester, and Dorset. Because strongly gregarious breeds tend to move as a group instead of scattering, herding them is easier than mustering breeds that are not gregarious, especially when using a herding dog.

While there usually is no flock leader, every flock of sheep does incorporate a pecking order or social hierarchy. This is especially evident at feeding time, when high-ranking members eat first, and those near the bottom eat last, if at all. In large flocks, several hierarchies may exist simultaneously. The ewes maintain a pecking order among themselves, the rams maintain another, and a third power play exists between the sexes. Jostling for position translates into head butting, shoving, and body slamming. Ewes are fairly subtle about it; rams indulge in all-out warfare.

Sheep want companionship, preferably that of other sheep. They'll make do with a goat, a pony, or a donkey companion, but they won't be as happy and might not thrive. They also prefer other sheep of the same breed. In mixed flocks, breed-specific subflocks are the rule. When they can, sheep form family groups within their flock or subflock. A family might include an old ewe and her daughters and granddaughters, along with all three generations' suckling lambs.

Do You Read Me?

Sheep's verbal and body signals vary somewhat depending on the number of sheep in the flock and its members' sexes, ages, and breeds. Skilled shepherds easily "read" sheep, and you can, too. The trick is to spend time among your sheep, simply hanging out and observing. Sheep will teach you a lot that way.

Sheep are more at ease with humans who get down on their level. Consider squatting or upending a bucket and sitting among your sheep to observe them. Don't do this if your ram is aggressive, and don't stare directly into a dominant sheep's eyes. By the same token, if confronted by a testy sheep, try to look big. Stand tall and hold your arms up and out to the sides. If you spy anything to extend

Fleeing

Nervous sheep mob together and then usually flee. If terrified, they flee blindly. If sheep in the front of a flock ram into a barrier, there can be pileups as rear runners tailgate flockmates directly in front of them. Frightened sheep don't "do" sharp corners, including the

DID YOU KNOW?

Sheep on the West Yorkshire moors have learned to roll on their sides, commando-style, across 8-foot cattle grids guarding the town of Huddersfield. Of them, a National Sheep Association spokeswoman said, "Sheep are quite intelligent creatures and have more brainpower than people are willing to give them credit for."

your outward reach—such as a stick, shepherd's crook, or hand tool—use it.

Once in a safe position to read the sheep's signs, look and listen for some of the following:

Ear movement. Sheep depend greatly on their keen hearing. Breeds with upright ears are often more reactive than are floppy-eared sheep, simply because they hear scary things that the others miss. A sheep with erect ears is more easily read than others. If her ears are pricked alertly forward, she's focused on whatever she's looking at, or she may be intently listening. Ears intently upright but swiveled toward the rear mean she's listening to something behind her; she'll probably turn and face that direction. Ears laid back close to the head (pinned) generally signal annoyance or aggression but may indicate fear or excitement. A relaxed sheep's ears face to the sides; she's not focused on anything in particular.

Stamping, nodding, glaring. Annoyed sheep may stamp their hooves, raise or nod their heads, or glare. They usually back up a few paces before charging, although ewes sometimes bash each other from a standstill. Rams or extremely pushy ewes may rub or bump humans with their foreheads; this is a sign of early aggression and should be strongly discouraged. When working among sheep, the wise shepherd does not unnecessarily handle the sheep's foreheads, stomp his feet, or walk backward because these actions can all be misconstrued by sheep as inviting aggression.

Panting. A ram pursuing his heart's desire pants like a freight train. He grumbles at his intended in a low, seductive tone, sniffs her backside, and rolls his upper lip back in the *flehmen* response. This distinctive grimace blocks his normal breathing apparatus and draws air through a pouchlike opening in his upper palate called the Jacobson's organ. By analyzing her scent, he determines if the ewe he's courting is in heat. If she is, she invites his attention by squatting. If not, she strolls off and he evaluates another ewe.

High-pitched baaing. Frantic, high-pitched baas are stress indicators. They're the earmark of sheep in pain, very hungry sheep, or individuals separated from dams, lambs, or flockmates. Medium-pitched baas are exchanged by adult sheep casually seeking nearby lambs, friends, or feed.

Low vocalizations: Ewes utter low, tender, guttural vocalizations to their new lambs. It's a lovely sound, not to be missed.

Handling Sheep

The best way to handle sheep is quietly and gently. Move them no faster than their usual walking speed. Move the leaders, not individuals at the back of the pack. Decide what you want and how you can use the flock's natural tendencies to achieve your goal. Then enter the leaders' flight zones toward their rear until they set off in an orderly fashion; the rest will follow. Don't race about, shout, fling your arms, or otherwise frighten the sheep.

When approaching or driving sheep, don't gaze into their eyes; staring is predatory behavior. Instead, focus on a sheep's nearest shoulder, and it will be far less likely to scuttle away.

In a hobby-farm small-flock situation, it's worlds easier to lead sheep with a bucket of grain than to try to drive them. Treating friendly flock members with apple tidbits or crumbled commercial horse treats, or even scratching their shoulders or chests, will make them like you. Between dewormings, load the drenching gun with liquid molasses or honey and give willing sheep a squirt. They'll be easier to deworm next time. Take a tip from livestock-handling guru Temple Grandin, who trained sheep to enter a scary tilt table and be briefly mobilized for a grain reward. "Fourteen out of sixteen ewes," she writes, "returned for one or more additional passes." It's amazing what sheep will do for food!

Hire shearers who are careful not to gouge or nick your sheep and who treat them kindly. Use a vet who likes sheep and handles them with care. Never snag sheep by their wool or ears; grabbing or hoisting a sheep by its wool is very painful, and doing so can separate layers of skin and leaves nasty bruises. Also avoid catching them by their legs if you can.

A traditional crook-style shepherd's staff carved from a Scottish Blackface horn at Wind Fall Farms in Missouri.

Safety

On occasion, a panic-stricken, cornered sheep will jump high and hard when approached from the front. Be prepared to leap aside. Most adult sheep tip the scale somewhere between 75 and 400 pounds, and some have mighty horns. Frightened or angry sheep can badly damage or kill unwary humans. Never take your safety for granted.

Aggressive individuals generally warn before attacking and usually belong to one of two camps: either protective ewes with new lambs or, more likely, rams. If you breed sheep, rams are a necessity. But if you don't breed, or you have small children or vulnerable adults in your family or neighborhood, please don't keep a ram. The verb "to ram" is derived from the Old English word *ramm,* meaning "an intact male sheep."

Apart from procreating, ramming is what unaltered sheep do best. Being charged by a ram is more than a bump on the backside. It may

Don't isolate a sheep unless you absolutely must. Even a penned ram needs a pal. Keep another sheep within sight to quell an ailing or injured sheep's fear. In their paper "Comfortable Quarters for Sheep in Research Institutions," researchers Viktor and Annie Reinhardt state, "Individuals show a multitude of endocrine, hematological, and biochemical alterations, stereotypic behaviors, and a marked increase in heart rate and respiration when isolated from other sheep." This study's recommendations translate well to small-scale sheep keeping.

It is best to make friends with your sheep so that when you need to do something with an individual, you can walk right up and slip a halter on its head. However, in the real world, that's not always easily done. Previously mishandled or wary sheep may never be hand-tamed; even a gentle sheep, when injured or agitated, may elect not to be caught. Be patient. The old-timer may scoff at these measures, but compassionate handling pays off in contented sheep and fewer hassles.

It's easier to snag a sheep out of a group than to chase down a frightened solitary member of the flock. Having other sheep present helps quell the target sheep's fear, and they'll block her escape route once she's captured.

Quietly herd the group into a corner. Draft an assistant to help, if possible. Because sheep have a blind spot of approximately 70 degrees directly behind them, you should try to approach your target sheep from the rear. Place one hand under her chin and raise her head, cupping the other hand around her rump; you can guide or restrain most sheep in this fashion. Or use a long shepherd's crook to gently snag the sheep around the neck and then move in quickly to restrain her. Leg crooks work but can injure sheep; unless you know your sheep won't struggle, use a standard shepherd's crook instead.

DID YOU KNOW?

It's a fact: once a sheep's feet are off the ground, it will remain stock-still. That's why sheep are "tipped" for shearing and routine tasks such as hoof trimming and treating minor wounds. The chore seems daunting at first, but it's easy when you follow this routine.

Standing on the left side of the sheep, reach across and grasp the right rear flank (don't pull the wool!). Bend its head away from you, against its right shoulder. Lift its flank and pull back toward you. This puts the sheep off balance, and it will roll gently toward you onto the ground.

Hold the sheep on its side and then quickly grasp both front legs and set the sheep up on its rump so it is slightly off center, resting on one hip. If it struggles, put one hand on its chest for support and inch backward until it's more comfy.

One caution: don't tip a sheep right after it has eaten; this position puts considerable strain on a full tummy.

TIPS ON BEHAVIOR

There is a definite pecking order within all species. I have three different breeds of sheep, and they tend to graze within their breed. This could be because they were raised with their breedmates and are familiar with them and have their own pecking order.

I was also told that different colors within breeds will stay together and have found this to be generally true. As with any social species, there are leaders, followers, and airheads that are always in trouble, wandering off or being where they aren't supposed to be.

As for intelligence, we have a few apple and pear trees in our yard. We occasionally let the ewes out to munch on the weeds. One ewe will stand in front of the tree and another will put her front feet on that ewe and start plucking fruit off of the tree. Seems pretty clever to me.

—Lynn Wilkins

When I bring in a new ewe, she's immediately checked out by "the bosses" and most often challenged to see where she will be in the pecking order. It also depends on where I had her for quarantine. I usually put newcomers in one of the horse stalls, away from the flock, to see how she responds to people and whether she's got a disease that I can see, and while she's there, I give her shots. Then I'll move her to a pen inside the barn with the flock around her. They'll come up and see her—at least most of them will; some just don't care—but they can't get to her because she's in that pen. If she's aggressive enough to "butt" through the pen, I'll let her out in a day or two. If she's not aggressive, then I'll sometimes leave her in the pen until they figure she's just one of the flock and won't harass her at the hay feeder. Most of the time, she will just get turned out with the flock after being in the horse stall and establish herself.

I've raised ten to twelve different breeds, and while some breeds are categorized as "aggressive," not all [individuals] of that particular breed may be. It depends on how they were raised, too. A bottle-fed baby may not associate with sheep because she feels she is a person instead of a sheep and just go on her merry way.

—Connie Wheeler

The predominant ewes here seem to take a head-on stance and look directly into the challenger's eyes, and take a couple of steps toward the upstart. If the challenging ewe doesn't back off, the head-butting contest begins.

Our ewes also use foot stomping quite often. Very occasionally with other ewes, but usually when the herding dogs are in the pen. Some seem to trust the guardian dogs more than others. Our Great Pyrenees is very respectful of Mom's wishes. She wants to check out her new wards, but if Mom stomps, she turns and walks away.

And watch those bottle-raised cuties. Males can turn into monsters when they get older because they have no fear of humans. The worst working-over I ever got was from a yearling wether we had bottle-raised. It was a nice September day, and I was picking up in his pen. I bent over, and he took the opportunity to come at me from about 15 feet away. He had me down and hit me three times before I could get to a stick to crack him across the nose.

—Lyn Brown

A hefty sheep is restrained with one hand under its chin, raising its head, and another hand on its rump.

presence of predators (including dogs) also stresses them, as does lengthy confinement. Prolonged stress triggers cortisol production, and elevated cortisol levels significantly reduce immunity. Stressed sheep tend not to thrive. Symptoms include panting, restlessness, teeth grinding, and skittishness. Closely confined sheep, even those penned with their peers, gnaw wood and nosh on wool—their own or that of underling companions.

Sheep maintain personal security zones. Anything scary invading an individual's personal space generates flight. A sheep's flight zone might be 50 yards or nothing at all; breed, gender, tameness, training, and the degree of threat enter each equation.

If something arouses a sheep's suspicion, it will watch closely. Its vigilance will alert others within the flock. If danger stays beyond the most timid observer's flight zone, the flock stays put. When the timid sheep's security zone is breached, it will turn and flee, and its flockmates will then run, too. Alarmed sheep stampede first and ask questions later. Instinct assures them that the sheep that tarries is the sheep most likely to be eaten.

leap and crash into your spine or your chest. It'll bowl you over, and if you can't get back up, it might not quit. Always know where the ram is, even when visiting other flocks. Never, ever take any ram for granted!

Stress and Flight Zone

Because sheep are gregarious, being separated from other sheep stresses them. Harsh handling, loud noises, and the

---

SHEEP ENCOUNTERS

My most exciting sheep encounter involved a handsome two-horned Jacob ram named Waleran. He belonged to my best friend, Barbara, and he took his job seriously: Waleran lived to ram. He rammed fences, he rammed dogs, and, one day, he rammed me.

Waleran and his ewes lived in my friend's spacious yard; they were organic lawnmowers. Because Waleran had a reputation for sending humans flying, I usually watched out for him. But I wasn't paying attention that day—and I was alone.

I'd just passed the horse trailer and was partway across the lawn when a freight train bashed me from behind. Sprawling in the driveway I twisted to find Waleran's face just feet from mine. And he was grinning.

Being a portly person, I'm not one to leap to my feet. By the time I was back up, Waleran knocked me down again. Up, down, up, down, over and over until a light switched on in my brain. As he backed to charge again, I grabbed those magnificent spiral horns, and the adversary hauled me to my feet. As his face expressed surprise and supreme annoyance, I sprinted for the open horse trailer.

It was an improvement—but not much of one. Waleran kept charging up the ramp, and I kept kicking him back. Did I mention I was yelling? But there was no one within miles to hear me.

Finally, he gave up. Just stopped. Went away. He was gone. I cowered in the horse trailer, trembling, for another ten minutes. Then, as I crept down the ramp, some small inner voice whispered, "Look left."

Flattened against the side of the trailer was Waleran, quietly waiting in ambush. A stupid sheep? I think not. He taught me a lot about rams. To this day, I respect them. They can be ornery, and they are *smart*.

If you've been around sheep or shepherds very long, you've no doubt heard the old chestnut, "A sick sheep is a dead sheep," but that's far from true. What is fact is that sheep are exceedingly stoic creatures, and by the time anyone notices they're suffering, they may already have two hooves through death's door. The trick, then, is recognizing and addressing problems in their early stages.

KEEPING
SHEEP
IN GOOD HEALTH

Maintaining a Healthy Flock

The best way to keep a flock healthy is to start with healthy sheep. Almost all illnesses surface after introducing new sheep to the flock. Even seemingly healthy sheep can be carriers of disease. To keep out the serious problems, such as scrapie, ovine progressive pneumonia (OPP), and Johne's (YO-nees) disease, consider buying from flocks certified as free of these illnesses. If your flock eventually becomes certified, you'll have no other option.

Make sure that your sheep are dewormed, shorn, and vaccinated, and that their hooves are kept in the pink. Sanitation is a must. Shelters, paddocks, and pastures should be tidy and free of poisonous plants, junk machinery, and all other debris that could harm sheep.

Feed your sheep quality hay, grain, and minerals, and keep things sheep-friendly. Don't offer grain mixes, minerals (including mineralized salt licks), or milk replacers that are not designed specifically for sheep. Finally, respect those rumens. Sheep can't digest feed until the right set of organisms becomes established in their rumens. Switch feeds (even when going from hay to green grass) over a period of ten to fourteen days.

Know what well and sick sheep look like, and know your sheep. Make time to lean on the fence and watch them every day. Or haul a lawn chair into the paddock and just hang out. You'll enjoy yourself and learn a lot about flock social order. More importantly, when a sheep behaves unusually, you'll know that there's something amiss and can investigate and treat it right away.

A sheep that doesn't eat with the rest or nibbles and then turns away is always suspect. If he spends his day lying down while the others graze, check it out. A depressed sheep, standing hunched over, with head down, is probably ill. Constantly runny or crusty noses and eyes, persistent sneezing or coughing, unexplained weight loss, limping, bizarre movements or head positioning, and intense rubbing on solid objects are definite red flags.

If you suspect there's something wrong that you can't easily fix, call a knowledgeable sheep person for advice. Better yet, phone your vet. Take the sheep's temperature and write it down, along with his other symptoms. Have the sick sheep penned so that examining him doesn't turn into a game of keep-away in the sheep pasture.

Vaccinations

Most shepherds learn to inoculate their own sheep. It's cost-effective, and veterinarians are generally pleased to make fewer farm calls. But you must first consult with your vet to formulate a vaccination program specific to your area and your flock. Needs vary greatly from one locale to the next.

DID YOU KNOW?

Natives called sheep brought by Spaniards to the New World "cotton deer," and they were considered sacred and never eaten. The Tzotzil people of southern Mexico still honor that tradition. Tzotzil shepherdesses constantly pray to St. John the Baptist, the Holy Shepherd, asking for protection and good health for their sheep. Their native Chiapas sheep are named, cared for, and respected as part of the family.

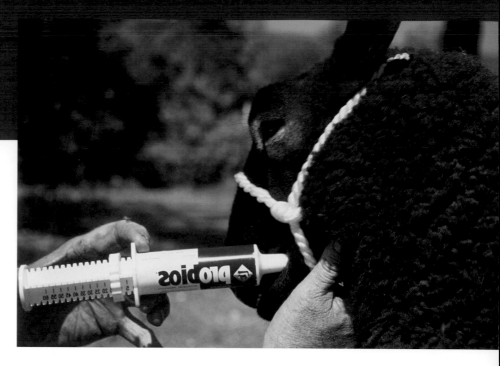

Paste-type medications, such as probiotics and wormers, are administered orally.

You can buy vaccines from your veterinarian, at many feed stores and farm stores, and from online suppliers. It's best to ask your veterinarian or an experienced sheep breeder to show you how and where to give injections, even if you routinely vaccinate other farm animals or horses.

Parasite Control

Like all other warm-blooded creatures (and some creatures that aren't), sheep are plagued by parasites. Some live inside the sheep, and some don't. Following are descriptions of both external and internal parasites.

Flies, Lice, and Ticks

Sheep are annoyed by the same flying insects that annoy humans. Black flies nibble their ears, face flies buzz their noses, and stable flies make them stomp. But some of the worst sheep pests don't have wings at all: lice and the bugs we call *sheep keds*.

Sheep can become infested with sheep-biting lice and several species of blood-sucking lice: the sheep body louse, the African blue louse, and the foot louse. These pests are host-specific, so unless you bring lice home on infested sheep, your flock will remain louse free. Louse bites cause intense irritation, so louse-infested sheep rub their bodies on walls, fences, and anything else solid in an effort to ease the itch. You'll know lousy sheep by their tattered, ragged fleeces; heavy infestations cause anemia and can sap sheep's strength and immune systems. Infestations peak during the winter and early spring months, but they can be problematic all year. Sheep lice are fairly easy to kill by zapping them with a spray, pour-on, dip, or lice-dust product purchased from your vet or feed store.

DID YOU KNOW?

Humans can catch a number of diseases from sheep: sore mouth, ringworm, toxoplasmosis, Q-fever, campylobacteriosis, salmonellosis, cryptosporidiosis, and tetanus. Contracting sore mouth and ringworm from sheep are fairly common occurrences. The others are not as common, but their consequences can be devastating.

Sheep keds are sometimes called sheep ticks, but they're actually blood-sucking wingless flies. They're flat, leathery, and reddish brown, and—at a ¼-inch long—they're much easier to spot than lice. Sheep keds truck around their hosts' bodies and poke their feeding apparatus through the skin to feed on blood and lymph. They cause itching and anemia just like lice; most products that kill lice dispatch sheep keds, too. The easiest time to treat sheep for lice and sheep keds is immediately after shearing. However, heavy infestations should be treated as they occur.

Two-winged pests wreak havoc on sheep: sheep-nose bot flies and blowflies. Female nose bot flies hover around sheep's faces and deposit maggots (larvae) in their nostrils. The maggots crawl up into their hosts' head sinuses and set up camp. When fully developed, they migrate back to the animals' noses and drop to the ground, where they pupate and eventually become adults. Adults are fuzzy, yellowish gray to brown, and a hair bigger than a honeybee. You'll know they're at work when your sheep stampede, stomp, snort, fling their heads, and press their noses to the ground. Sheep infested with maggots sneeze. Their breathing is labored, and blood-flecked snot may ooze from their nostrils. Fortunately, ivermectin effectively kills bot-fly maggots when given as a 0.8-percent liquid oral drench.

Blowflies cause dreaded fly-strike by laying eggs in wounds and on inflamed skin irritated by urine-soaked, filthy wool. Their eggs quickly hatch into maggots. These maggots secrete enzymes that liquefy their host's injured flesh. Fly-strike is incredibly nasty and painful, and it can eventually lead to death from septicemia or toxemia.

To prevent fly-strike, clean wounds as they happen. Clip wool away from an injury and apply antibacterial dressing or another wound treatment of choice and then smear insecticidal cream around the edges. Repeat daily until the wound heals. Eliminate fleece rot by keeping sheep tidy; continual moisture anywhere on a sheep's

Trimming hooves "horse style" can be hard on a person's back.

The meningeal (brain) worm affects all ruminants, wild and domestic; its natural host is white-tailed deer, but its larvae spend time in snails and slugs. When a sheep ingests an infected slug, the larvae travel up the spinal nerves to the spinal cord and the brain. This damages the central nervous system, resulting in a parade of bizarre symptoms and possibly death.

Internal parasitism is a complex topic. Your best approach is to research the topic and then discuss it with your vet. Bring manure samples from several sheep to your vet, who can run fecal egg counts, assess your flock's needs, and devise a custom-tailored deworming program.

Hooves

Horse folks say, "No hoof, no horse." That principle applies to sheep, too: sheep with sore hooves lack vigor and are unhappy.

Trimming Hooves

A sheep with ragged, untrimmed hooves is an unhappy sheep. When hooves get long and roll under, they're uncomfortable and may cause a sheep to limp. When its feet hurt, a sheep won't graze as much or as often as its flockmates. Untrimmed hooves can also contribute to serious sheep hoof disease, such as scald and foot rot.

Affected sheep will lose weight. If an affected ewe is not grazing and thus not providing the nourishment that her in utero lambs require, they will be born smaller than other lambs.

Trimming is probably the most important thing you can do for your sheep's hooves. Soil moisture and type, time of year, and breed influence how fast hooves grow. In general, plan on providing pedicures two or three times a year. Timing them to coincide with other labor-intensive procedures, such as deworming, vaccinating, and shearing, makes good sense. However, avoid trimming hooves during high-stress periods such as extreme weather conditions, late pregnancy, or weaning. Hooves trim more easily when they're moist, whether from dew, rain, or snowmelt.

You'll need the proper tools. Most shepherds recommend hoof shears, as do folks who also keep goats. Shepherds accustomed to trimming horses' hooves often prefer horse-hoof nippers, a hoof knife, and a rasp. Shears or a hoof knife followed by a hoof plane work well, too. It's mostly a matter of taste, experience, and convenience. If you're buying first-time tools, hoof shears are a best bet; they're

body irritates the skin and draws blowflies. Shear dirty crotches and remove manure tags. Wash ewes' bloody backsides if they lamb during fly time. Clip wool away from rams' and wethers' penises. Don't give fly-strike a chance to occur!

Worms

As far internal parasites go, sheep are troubled mainly by roundworms (nematodes), tapeworms, lungworms, liver flukes, and—only occasionally—meningeal (brain) worms. Only roundworms are common; all sheep have them to some degree. The trick is dispatching as many as you can, lest they damage or kill your sheep. Depending on their species, roundworms live in the stomach or small intestine. They are tiny; you're unlikely ever to see a roundworm. They feed on blood and body fluids and interfere with the digestion and absorption of foodstuffs. Badly infested sheep are thin and weak and they usually cough. They're sometimes anemic and often suffer from diarrhea. Because they're run-down, they're wide open to infectious disease. Lambs are more susceptible to roundworms than are adults.

In some parts of the country, lungworms cause problems for sheep that graze lowland and boggy pastures inhabited by snails. Sheep swallow snail larvae while grazing; the larvae then migrate through tissue to the sheep's lungs. Affected sheep cough, and severe infestations cause fluid in the lungs.

Tapeworms, even substantial numbers of them, don't do adult sheep a lot of harm, but they can drastically affect the growth of lambs. Heavy tapeworm infestations, however, occasionally block a sheep's intestinal tract. Such an infestation can result in unnecessary death.

Liver flukes are another denizen of wet, low pastures. Like lungworms, their intermediary hosts are snails. Low to moderate infestations of liver flukes impact lamb growth; heavy infestations can kill adult sheep.

When a sheep kneels, it is experiencing foot pain. Be sure to investigate the cause.

inexpensive and easy to handle, and they make trimming a one-step job.

If your back is good or you have a helper, tip your sheep to trim their feet or ask your helper to support the sheep while you trim. In a solo operation, halter and tie your sheep and trim with the sheep standing up. You can squat beside the sheep, sit on an overturned bucket, or stand and lean over to trim.

Start trimming at the heel and work forward. Trim even with the frog (the soft, central portion of each toe) and then trim the walls level to match. If the frog is especially ragged, touch it up with a knife, taking paper-thin slices until you reach a hint of pink, at which point you stop immediately; the frog is a sensitive structure. When you're finished, the hoof should be flat on the bottom and parallel or nearly parallel to the coronary band.

When trimming a sheep with foot disease, trim the bad hoof or hooves last to avoid spreading disease to healthy hooves. When you're finished, disinfect your tools to avoid infecting other sheep.

Foot Rot

When shepherds spy a limping sheep, they tend to think the worst. But before concluding that your lame sheep has hoof disease, run a diagnostic evaluation. Watch the gimpy sheep from afar. Which foot or feet is it favoring? How badly is it limping? When you catch the sheep, scan for foreign objects lodged in the toes or between them.

Next, examine the scent glands positioned between the toes and to the front of each hoof. Those glands get plugged. Gently squeeze the one on the gimpy foot. If a glob of waxy gunk pops out, that should do the trick. If nothing seems right so far, carefully trim all four hooves. As you do, watch for signs of disease. The ones you're most likely to encounter are foot rot and foot scald.

Foot scald and foot rot are closely related. In fact, they share a causative agent: the bacterium *Fusobacterium necrophorum*. Although foot scald appears in the presence of *F. necrophorum* alone, the addition of *Bacteriodes nodosus* causes full-blown foot rot.

F. necrophorum is a common, hardy bacterium that dwells in soil and manure on virtually every farm on which livestock is kept. It

ADVICE FROM THE FARM

WATCHING OUT FOR THE FLOCK'S HEALTH

When trimming feet, notice the natural shape of the hoof. Trim down to the quick, making sure that each side of the hoof matches the other. Also cut off the point of the hoof to avoid those sharp hooves raking your arms and legs—or lambs, if your ewe paws her lambs to get them up.

—Connie Wheeler

To collect a [urine] sample, pinch the animal's nostrils shut for a minute. Have some sort of container under him so that you can collect the urine. He should pee right after you release his nose. Hang on, because he is going to try to get away.

—Nancy Larsen

I make a splint out of wood, vet wrap, and duct tape. I have never had to deal with a compound fracture, though some had to be set because the bone, at the break, was not meeting. All have survived and are doing well.

—Cathy Bridges

Illnesses can easily pass from sheep to sheep in the close quarters of a livestock auction.

As mentioned, you should not buy sheep at livestock sales. Many producers dump infected sheep at auctions. Even if the one you buy isn't infected, it will have been exposed to infected sheep and held in pens where *B. nodosus* thrives. Foot rot bacteria grow slowly. Because seemingly disease-free sheep can harbor *B. nodosus,* you should consider any sheep from an infected flock verboten.

causes thrush in horses and contributes to foot rot in cattle. It's an anaerobic organism (meaning that it can grow only in the absence of oxygen), so when animals are kept in dry, sanitary conditions, *F. necrophorum* poses no threat. But when hooves are continually immersed in warm mud and muck, bacteria invade the foot, often via a minor scratch. The result is foot scald, a moist, raw infection of the tissue between the sufferer's toes. Foot scald usually affects one but not both front feet. It's nasty and painful, but its worst characteristic is that it often leads to foot rot.

Foot rot occurs when *F. necrophorum* is joined by *B. nodosus,* another anaerobic bacterium that thrives only in the hooves of domestic sheep. It gains access via foot-scald lesions and other injuries. If *F. necrophorum* is present, it sets up house in deeper layers of the skin, where it produces an enzyme that liquefies the tissue around it.

You can't miss foot rot. Affected sheep are very lame. Infected tissue is slimy and smelly. It's an odor you simply can't miss. Infection beneath the wall and sole of the hoof causes the horny walls to partially detach. More than one hoof may be involved.

Foot rot is treatable, but it's a long, costly, and time-intensive process—and, in most flocks, often not an entirely successful one. The key to foot rot control is to keep it away from your flock in the first place. If your sheep don't have it, they can't get it without coming in contact with *B. nodosus* bacteria through an infected sheep.

Caution with New Sheep

If a prospective purchase limps, examine the sheep carefully and think twice before you buy. When buying any sheep, you should play it safe by trimming its feet on arrival and then quarantining the sheep for at least three weeks. Do the same with any sheep returning to your premises, whether they've been in a 4-H show, boarded at a veterinarian's facility, or away from your farm for any other reason.

A Glance at Sheep Afflictions
Abortion

Enzootic abortion of ewes (EAE) is a chlamydial disease transmitted from aborting sheep and fetal tissues to other ewes. Infected ewes abort during the last month of pregnancy or give birth to stillborn or weak lambs that soon die. An effective vaccine is available.

Vibrosis is caused by the bacterium *Campylobacter fetus,* subspecies *intestinalis.* When one or two ewes affected by vibrosis abort, they can trigger an "abortion storm." A vibrosis vaccine is available, often in combination with the EAE vaccine.

Toxoplasmosis, which is caused by the coccidium *Toxoplasma gondii,* is spread when a host cat contaminates feed and water with his droppings. There is no vaccination or treatment for toxoplasmosis.

When a ewe aborts her lamb, you should submit the fetus and tissues to a laboratory for diagnosis; you can't treat the rest of the flock unless you know what you're dealing with. Your vet can tell you where to send the specimens. The material must be fresh, so store it in sturdy plastic bags, pack the bags in a Styrofoam box and surround them with cold packs, and then rush the package to the lab.

Blackleg and Malignant Edema

Blackleg and malignant edema are deadly diseases caused by the bacteria *Clostridium chauvoei* and *Clostridium septicum,* respectively. They are indistinguishable in sheep except by laboratory diagnosis. They occur in sheep of all ages and are caused when soil-borne bacteria contaminate wounds and abrasions. A combination vaccine prevents most occurrences.

Bloat

Bloat is a build-up of frothy gas in the rumen, usually triggered when a sheep fills up on an unaccustomed abundance of grain, rich grass,

Too much clover or other rich forage can cause a sheep to bloat.

or legume hay. Bloated sheep can quickly die of the condition, so if you suspect that your sheep is bloated, call your vet immediately. Signs include a distended, taut abdomen, which may come on suddenly; behaviors indicating pain and discomfort; standing stiffly; excessive panting; difficulty breathing; and difficulty moving/staggering.

Bluetongue

Bluetongue is caused by reoviruses transmitted by *Culicoides variipennis* gnats. It infects sheep, goats, cattle, and wild ruminants such as deer, elk, and moose. The typical form of bluetongue causes fever and swelling of the lips, tongue, and gums. Badly chewed tongues turn purple due to lack of oxygen, hence the name "bluetongue." Afflicted sheep have difficulty breathing and swallowing. Death usually occurs in about seven days, although some sheep survive.

Caseous Lymphadenitis (CLA)

Caseous lymphadenitis (CLA) is a chronic, contagious disease of sheep and goats caused by the bacterium *Corynebacterium pseudotuberculosis*. The bacterium breaches a sheep's body through mucous membranes or via cuts and abrasions. The animal's immune system valiantly tries to localize the infection by surrounding it in one or more cysts. If the ploy is unsuccessful, the sheep will die. CLA presents as lumps near the jaw, in front of the shoulder, and where the ewe's udder attaches to the body. Some sheep develop internal cysts, too.

Contagious Ecthyma

Commonly known as *sore mouth* and also known as *scabby mouth* or *orf*, contagious ecthyma (CE) is a pox-like virus that causes the formation of blisters and pustules on the lips and inside the mouths of young lambs, as well as on the teats of the infected lambs' mothers. The blisters pop, causing scabbing and pain so intense that a lamb will occasionally starve rather than eat. Most lambs recover in one to three weeks without treatment.

Because sore mouth is easily transmissible to humans, you should wear rubber gloves when handling stricken lambs, and keep children away from all infected animals. An effective live vaccine is available, but you mustn't use it unless you already have sore mouth on your property.

Enterotoxemia

There are two types of enterotoxemia in sheep: *Clostridium perfringens* type C and type D. The first is a disease of young lambs caused by *Clostridium perfringens* type C, an anaerobic bacterium found in manure and soil. It enters via newborn lambs' mouths when they suckle dirty wool or manure tags while seeking their mothers' udders. Bacteria produce a toxin that causes rapid death. Treatment is unsuccessful, but lambs from ewes vaccinated for enterotoxemia during late pregnancy develop immunity via their mothers' colostrum.

The second form is caused by *Clostridium perfringens* type D, also present in the soil and manure. It attacks rapidly growing, slightly older lambs that ingest the bacterium while investigating their environment. It, too, causes rapid death and with it tremors, convulsions, and a host of strange neurological behaviors. A vaccine is available alone or in combination with type C or type C and tetanus vaccine.

Johne's Disease

Johne's (YO-nees) is a deadly, contagious, slow-developing, antibiotic-resistant disease affecting the intestinal tracts of domestic and wild ruminants, including sheep. The bacterium that causes Johne's, *Mycobacterium avium* subspecies *paratuberculosis*, is closely related to the bacterium that causes tuberculosis in humans. Infected sheep are dull, depressed, and thin. Also known as paratuberculosis, Johne's disease is incurable.

Mastitis

Mastitis is an inflammation of the udder caused by bacteria or yeast infections. Usually only one side is affected. One type of acute mastitis can lead to gangrene. Anyone who keeps brood stock or milking ewes should learn to recognize mastitis symptoms.

A ewe needs healthy udders to produce milk for her babies or your refrigerator.

and chronic wasting disease (CWD, which affects deer and elk). No human has ever contracted scrapie (or either of the human equivalents, kuru and Creutzfeldt-Jakob disease) from sheep.

As a sheep owner, you must become scrapie-savvy—it's the law. All sheep older than eighteen months and residing in the United States must be identified through the National Scrapie Eradication Program according to federal and state regulations.

Scrapie appears to be caused by an infectious agent, but genetics also play a part. The disease was recognized in Britain and western Europe at least 200 years ago, and it came to the United States in 1947 with British sheep.

Scrapie is a slow, progressive disease that systematically destroys the central nervous system. Goats exposed to infected sheep sometimes get scrapie, too. Symptoms typically appear two to five years after infection and include weight loss, tremors, bizarre locomotion (such as bunny-hopping or high-stepping like a hackney horse), swaying, stumbling, wool pulling, lip smacking, and intense itchiness. Between one and six months after symptoms appear, infected sheep die.

Ovine Progressive Pneumonia

Ovine progressive pneumonia (OPP) is one of the serious progressive diseases that shepherds fear. Studies have shown that close to a quarter of America's sheep is infected. This North American retrovirus is closely related to maedi-visna, a similar disease found in most other parts of the world. Only Iceland, Australia, and New Zealand are free of maedi-visna and OPP.

Transmission is via infected colostrum and milk, but because of the disease's slow progression, infected sheep do not develop symptoms until much later. Symptoms include weight loss, sluggishness, lameness, a fibrous udder condition known as *hardbag*, and all of the usual signs of pneumonia. Most OPP-infected sheep develop secondary bacterial pneumonia, from which they die. There is no vaccine or treatment for OPP.

Pneumonia

Pneumonia is caused when one of a wide variety of opportunistic bacteria and viruses mix with stressed sheep. Typical symptoms include depression, fever, coughing, and labored breathing. Because so many bacteria and viruses may be involved, accurate identification of the infectious agent is an essential part of successful treatment.

Scrapie

Scrapie is a transmissible spongiform encephalopathy (TSE) similar to bovine spongiform encephalopathy (BSE, or mad cow disease)

Urinary Calculi

Urinary calculi are tiny stones or crystals that form in the urinary tracts of sheep and goats. Ewes get stones, but they pass through the larger female urethra (the tube that empties urine from the bladder) without difficulty. A ram or wether with a blocked urethra is in trouble, however; his bladder is apt to rupture, and he'll probably die.

When ram lambs are castrated, penis growth stops, so wethers castrated early are especially troubled by calculi; their much tinier penises and urethras are easily blocked. A workable solution is not to castrate ram lambs younger than four to six weeks old.

A calcium–phosphorus ratio of 2 to 1 in the diet helps prevent calculi formation, as do minute amounts of ammonium chloride added to feed. Male sheep should drink lots of water. Make it more appealing by keeping water sources readily available, full, and sparkling clean.

White Muscle Disease

White muscle disease, also known as "stiff lamb disease," is caused by selenium deficiency. Ewes grazing on selenium-poor land or eating hay raised in depleted conditions require selenium/vitamin D supplementation during the last two months of pregnancy. Otherwise, their affected lambs will have problems rising and walking. Some will even become paralyzed. Prevention is the key to eliminating white muscle disease.

Many sheep owners reach the point at which they wish to breed their sheep (lambs are cute little critters). If you are thinking about breeding sheep, there are several things to consider and prepare for. You must determine what kind of sheep you want. You need to learn as much as you can about the breeding characteristics of your particular breed and sheep and how to facilitate the process. You also need to know how to care for pregnant ewes and newborn lambs.

PRINCIPLES OF PROPER

SHEEP

BREEDING

Choosing Breeding Stock

When buying breeding stock, remember that you get what you pay for. Buy the best breeding-stock sheep you can afford. If you want meaty lambs for the ethnic holiday market, choose sheep that breed out of season; pure or crossbred hair sheep would be a good choice. If handspinners' fleeces are your game, lay down the bucks necessary to own healthy sheep with the type of fleece you prefer. Pet breeders' sheep should be calm and manageable; berserk sheep give birth to spooky lambs. Always begin with sheep of the temperament, type, and quality you want to produce.

Buy breeding stock from shepherds who keep detailed records. Ask him or her what sort of lambs your prospective purchases have already sired or produced. Choose rams and ewes from multiple births that arrived early in the lambing season. Such ewes tend to produce like their prolific dams, and research indicates that a ram's daughter's performance is strongly influenced by his dam's reproductive characteristics.

Consider buying a proven ram; ram lambs can be a gamble. An older, top-quality ram may cost considerably less than a younger, mediocre one. The older guy can shine in a hobby-farm setting, where he's comparatively cosseted and is required to settle fewer ewes. The same can be said for ewes. Large-scale producers routinely cull six- and seven-year-old ewes. With attention to their diets, some ewes are still lambing when they're twelve years old. Older ewes with outstanding production records are especially good buys for newbie shepherds; the old girls can teach them the ropes.

Don't buy an ornery ram with one testicle. Avoid ewes with pendulous udders and balloon teats. Don't buy problems. Choose brood stock from scrapie-, OPP-, and Johne's-free certified flocks if you can. Foot rot, lice, sheep keds, and excessive internal worm loads affect breeding performance. Sick, gimpy sheep are never a good buy.

If you have your heart set on breeding registered sheep, investigate the breed before you buy. Some carry genetic disorders that you should be aware of (and avoid). Check registration papers before you buy. In most cases, if the seller isn't the last recorded owner, he can't sign a transfer form, and you're stuck with expensive grade sheep.

Keep your stock healthy. Work with your vet to devise a tailor-made deworming and vaccination schedule and stick to it. Consult with your county extension agent and develop a sensible feeding plan. Trim those hooves. Shear in a timely fashion. Provide shelter from winter storms and summer sun. Content, healthy sheep deliver!

Breeding

Most sheep are seasonal breeders. As autumn approaches (around August, in most locales), decreasing daylight triggers hormonal changes in ewes. They begin coming into heat every sixteen or seventeen days until they become pregnant or breeding season ends (usually in January, when days start getting longer). While in heat, they "show" to the ram by squatting and urinating, and generally making themselves available; they're receptive for a day or two. Most ewes ovulate twenty-four to forty-eight hours after a cycle begins.

A full-grown fifteen-month-old Keyrrey-Shee ram. Rams as young as seven months old can be used for breeding.

Some sheep breed out of season, meaning they cycle all year or almost all year. Woolly out-of-season breeders include the Dorset, Merino, Polypay, Rambouillet, and Tunis. Most hair-sheep breeds cycle twelve months of the year. With especially good feed and care, these breeds can produce three lamb crops in two years.

Rams become infertile when their body temperatures soar too high. Steamy, sizzling weather, as well as stress, sickness, and overexertion from caroming around after ewes or battling other rams when it's hot outside can do the trick. Shearing their scrotums and keeping rams indoors (sometimes with a fan trained on them) during especially hot days can help. Don't move or work breeding flocks during the heat of the day, and make certain they have access to a ventilated, shady area.

Rams become sexually active between five and eight months of age. One mature ram can service thirty to thirty-five ewes in a typical sixty-day breeding season. Some mature rams in stellar condition can impregnate up to seventy-five ewes per season (it pays to keep breeding rams cool!). Ram semen can be collected and successfully frozen.

Ewes reach puberty between six and eight months of age, but they generally shouldn't be bred (depending on breed and maturity) until they're eight to twelve months old. Most ewes come into heat seasonally from early fall through early winter. Some breeds, such as the Dorper, Dorset, Katahdin, and St. Croix, come into heat year-round.

A ewe's heat lasts twenty-four to thirty-six hours. Unless impregnated, the ewe will cycle every fourteen to nineteen days throughout the breeding season. Ovulation occurs twenty-four to thirty hours after the cycle begins. A ewe may be bred by natural cover (by a ram) or, more rarely, by artificial insemination, using fresh or frozen semen.

Most ewes lamb 145–155 days after conception; most shepherds figure five months from their last known breeding. They can produce from one to as many as seven lambs, depending on age and breed. Twins and triplets are common. Ewes have only two teats, but good milkers can successfully rear triplets without help. Lambs are precocial, meaning that they stand, nurse, and frisk about soon after birth. Commercial ewes are often culled when they reach five or six years of age, but, depending on breed and husbandry, a ewe can often lamb until the age of twelve.

To be certain that breedings occur, many shepherds fit their rams with marking harnesses. A marking harness has a crayon affixed to its chest assembly that marks ewes when they're mounted by the ram. It's important to use only crayons designed for this purpose; others might not scour out of shorn fleeces. Change crayon colors every sixteen days, using the lightest colors first, to know exactly when each ewe settles.

Flushing is a practice endorsed by many shepherds; its purpose is to increase twinning (the USDA claims 18–24 percent more lambs are born to flocks where flushing is the norm). The process consists of feeding grain for seventeen days before exposing the ewes to a ram, then continuing for about thirty days longer. Grain is introduced slowly and tapered off when the flushing period ends. If you flush, don't overdo it. One-half to one pound of grain is plenty, depending on your ewes' weight and size.

Ewes shouldn't be overfed during their early pregnancies, when lamb growth is minimal. However, because 70 percent of the lamb's growth occurs five to six weeks before birth, ewes—particularly ewes carrying twins and triplets (or more)—require supplementary feeding and careful monitoring. During these final weeks of pregnancy, ewes occasionally suffer rectal or vaginal prolapses and develop pregnancy toxemia or milk fever. Knowing how to recognize these conditions and how to treat them just might save a ewe's life.

Bred ewe lambs should be at least twelve to fourteen months old when they lamb. They need some cosseting, too. They should be kept in their own flock, away from bossy older ewes, and given a small amount of grain. It's best to breed ewe lambs to a smaller-breed ram to cut down on problems at lambing time.

Crossbred lambs are often worth a bit less than purebred ones. However, it's better to produce an easily deliverable crossbreed

LAMBING KIT

Lambing is the most rewarding part of the shepherd's year. The process usually goes smoothly, but the shepherd should assemble a lambing kit in case of emergency.

We pack our lambing supplies in two containers: one that we take to the barn and another that keeps our lambing supplies organized inside the house. The bucket for the barn contains the following:

- **Sharp scissors** to trim the umbilical cord to an inch or so in length. We disinfect them after each lambing and then store them in a plastic zipper bag.
- A **hemostat (disinfected and kept in the same zipper bag as the scissors)** to temporarily clamp on the cord if it continues bleeding.
- **7-percent iodine (individually double-bagged)** to prevent infection in the cord after trimming. Keep a **shot glass** in the lambing kit to fill with iodine and dip the navel in while the lamb is standing.
- **Two flashlights,** stored in a plastic zipper bag. We like packing a backup in case the first light malfunctions.
- **Lots of lubricant** for repositioning lambs. We keep two squeeze bottles of **SuperLube (individually double-bagged)** in our kit.
- **Betadine scrub** to swab a ewe's vulva before repositioning lambs.
- **Shoulder-length OB gloves**—sterile and individually packaged. They're harder to find than nonsterile gloves, but worth the search.
- A **sharp pocket knife** for routine cutting chores.
- A **digital thermometer**—the kind that beeps.
- A **bulb syringe**—the type designed for human infants works well for sucking mucus out of tiny nostrils.
- A **length of strong, soft rope** to pull lambs. We store this, the thermometer, the knife, the gloves, and other small items together in a single plastic zipper bag.
- A **scourable paint stick** designed for sheep or some other type of ID marker. You may not need this if you have a small flock and can easily tell which lambs go with which ewes.
- **Nutri-Drench f**or sheep (individually double-bagged) for weak lambs and exhausted ewes, along with a **catheter-tip syringe** with which to give the stuff.
- An **adjustable halter and lead**—it's easier to move most ewes with one than without one. We use flat nylon alpaca halters for our girls.
- A **lamb sling**—quite a back saver!
- **Towels**—soft, old, toweling comes in handy for cleanup.

Our other container holds the following items:

- **Ewe's milk replacer**—we repackage it into plastic zipper bags and store 3–4 pounds in the lamb-supply bucket and the rest in a tightly sealed tin.
- A **plastic lamb bottle** with a **Pritchard teat** and **several spare teats**.
- A **flexible plastic feeding tube**, a **felt-tip marker**, and a **60-cc syringe** for tube-feeding.
- A **16-ounce measuring cup**—it's also great to store the spare Pritchard teats in. It's a big one, so it can double as a milking pail, too.
- A **small whisk** for mixing milk replacer. Place all feeding supplies, including the measuring cup, are stored in a second plastic zipper bag.
- **Syringes and needles** go into another plastic bag, along with an **elastrator and rings**.
- A **prolapse retainer and harness** resides at the bottom of the bucket. We've never used it, but it's there if we need it.

lamb than a dead one—or end up with a dead ewe—because the purebred lamb was too large for its immature mother to deliver. The practice of crossbreeding for this reason is very common in the livestock world. Cattlemen frequently breed heifers of the large exotic breeds to Angus bulls because Angus have small heads, thus producing calves that are more easily delivered.

Here Come the Lambs

As lambing time approaches, you'll want to prepare for the big event. You'll have much to prepare before, during, and immediately after the birth(s). Don't hesitate to ask for assistance, particularly from those who have had some experience with the blessed event.

Preparing for the Lambs

About five or six weeks before lambing begins, administer booster vaccinations and dewormer. Immunized ewes transfer passive immunity to their lambs via colostrum, but those boosters must be current to be effective. The CD/T (*Clostridium perfringes* types C and D and tetanus) vaccine is a must; your vet can tell you which others are necessary in your locale.

A newborn lamb takes a break after its first feeding.

Four to six weeks before lambing, carefully and gently shear your ewes or crutch them. Crutching involves shearing only the wool surrounding a ewe's vulva and udder. With their warm fleeces gone, shorn ewes are more likely to lamb in a shelter, which is where you want them instead of out in the snow, behind a bush. After lambs arrive, shorn ewes seek shelter during foul weather instead of letting their little ones chill. Finally, shearing or crutching off filthy wool tags makes it easier for new lambs to find a teat.

At the same time, begin supplementing the ewes' diets with grain. Start with ¼ pound per ewe per day, gradually increasing rations until they're up to between ½ and 2 pounds apiece, depending on their size and body condition. If fat, bossy ewes eat more than their share, separate the ewes into smaller groups.

If the hay that your ewes eat or the pasture that they graze is selenium deficient, they'll need selenium/vitamin E (Bo-Se) shots four weeks before lambing. Ask your veterinarian about these. Also, watch ewes closely for the last two weeks before lambing. Should they succumb to milk fever or pregnancy disease, it's likely to happen then.

Assemble a lambing kit or update the one you already have. Build or set up jugs (individual post-lambing pens for each ewe and her lambs) in a well-ventilated, draft-free area in a shed or barn. On hobby farm, allow one jug for each five ewes. Build 4 × 4-foot jugs for small breeds, 5 × 5-foot jugs for larger ones. Use wood or welded hog panels to construct them; if drafts might be a problem, opt for solid wooden walls. Equip each jug with a hay feeder and water bucket, and bed it with bright, fresh, long-stemmed wheat straw or other suitable bedding (not sawdust).

The Lambs' Arrival

Now keep an eye on those ewes! As lambing day approaches, their udders enlarge; by lambing day, the teats appear round and firm. Vulvas get floppier and take on a rosy hue. As lambs drop into position, most ewes become slightly swaybacked, and hollows appear in front of their hipbones. Just before lambing, many hie themselves off to a far corner of the pasture or seek a dark, cozy corner in the barn. Don't pen them up unless you must; ewes like privacy and want to choose their own birthing spots. As labor begins, a ewe will paw as though nesting, throw herself down, get up and circle, and then lie down again. Within about four hours, barring complications, the lamb or lambs will arrive.

When the first lamb is born, the mom begins murmuring to the new babe. She'll lick and lick the lamb. Don't wipe the newborn dry! Licking is part of the bonding process. However, as soon as the lamb appears, carefully strip fluid from its nostrils and then get back out of the way. Don't imprint newborn lambs; their mother might decide they're yours and let you raise them! Wait to see if another lamb is coming. If one is, the ewe will dig, circle, and flop down again, leaving her first lamb to deliver the second. Creep forward and retrieve the first lamb. Use sharp scissors to trim the umbilical cord about 1 inch from the tummy and then dip the cord in 7-percent iodine. *Don't omit this step.*

When all of her lambs have arrived, check the ewe's udder. Strong wax plugs in the teats sometimes prevent lambs from nursing. Make certain that milky fluid squirts out of each teat and then watch until the lambs begin suckling. Next, move the little family to a lambing jug. Carry the lambs at ground level where their mom can see them. If she's halter-trained, halter and lead her while someone else carries the lambs.

It doesn't always work like this. While most lambings proceed without incident, sometimes things go terribly awry. Lambs are stuck or malpositioned; ewes die or reject their lambs; lambs are born too weak to stand and suck. You must be prepared for every eventuality.

You must understand what a normal delivery looks like so you can spot problems early. If a ewe needs help—if her lamb or lambs are positioned incorrectly or if for any other reason she can't deliver them—you'll need to call the vet (pronto) or assist her yourself. Buy a book on lambing or print out one of the excellent free lambing

---

WATER SAFETY

Don't use 5-gallon food-service buckets for water; lambs have been known to drown in them.

In some instances, you will have to step in for the ewe to care for the lambs.

way corner of the barn. Keep checkups low-key, but check often. Every few hours is sufficient.

Caring for New Lambs

Lambs are susceptible to diseases and conditions as diverse as constipation and scours (diarrhea), pneumonia, acidosis, enterotoxemia, tetanus, polio, and white muscle disease. Learn all you can about these problems before your lambs arrive; the resources in our Appendix will point the way.

guides available online from various extension services and stow it in your lambing kit.

Be there when lambs are born. Monitor them periodically throughout the day and roll out of bed to check your ewes at intervals throughout the night—or set up temporary camp in an out-of-the-

Lambs are delicate, wee creatures and may need your help to survive.

Feeding Help

Weak lambs may need to be tube-fed until they're strong enough to stand and suckle. Passing a feeding tube is a daunting task to most

---

ADVICE FROM THE FARM

LAMBS AND LAMBING

Lambs are rarely so far in that you can't reach them with your hands, and even better if you can just put your finger in to pop out a stuck leg or something. In the seven years that I've been here, I've only assisted once (twins, each with stuck front legs, and I just popped them out with a finger), and that same ewe had a *very* mild prolapse, which went right back in after lambing. I used a ewe spoon to keep it in while she was carrying.

—Laurie Andreacci

To practice, put a stuffed animal inside a paper bag and, without looking, reach in and figure out what is head and front legs. With the lamb puller, you loop the string over your fingers and reach in, so you'd have to do it all with one hand to wiggle the noose off of your hand and onto the lamb. You probably won't need to do it, ever, but if you practice with the paper bag, you at least have an idea of how it works. Keep in mind that there isn't that much "free space" inside a ewe, and you'd have insides working against you.

—Laurie Andreacci

If they're not slab-sided and aren't sunken in, they're usually fine. Mom most generally won't let them nurse for very long at time unless her udder is really full of milk. If their mouths are warm (a sign that they're healthy; a cold mouth is a sign of hypothermia), feel their tummies. Are they full? If they are, the lambs are eating. I wouldn't bottle-feed them unless they start having problems or it becomes very obvious they're not getting enough milk.

—Kim McCary

One thing I have in my lambing kit is the lamb sling. Since we usually have twins and often triplets, one person can gather up the lambs in slings and hold them in front of the ewe, and she will follow along into the jugging pen with little or no hesitation. If you pick up the lamb, the ewe gets confused; ewes do not understand the concept of "up." Nature did not equip them to look for flying lambs.

—Lyn Brown

first-time shepherds. Ask your vet or a sheep-savvy friend to show you how in advance.

Orphan and rejected lambs as well as weak lambs with a suckle reflex can be bottle- fed. Every shepherd should keep sheep's milk replacer and feeding bottles on hand. Because newborn lambs must ingest their mother's colostrum (the first milk, which is liberally laced with antibodies) for roughly twenty-four hours, hand milking is another skill best learned in advance.

Keeping Warm

New lambs must be kept reasonably warm. Some shepherds use heat lamps above wintertime lambing jugs, but this can be risky. If you use heat lamps, make certain they're fastened very securely (don't hang them by their cords) and far enough from flammable bedding that they won't trigger a fire. Blankets are an option but not usually a good one; blanketing newborns masks their scent, and sniffing is how ewes recognize their lambs. Lambing jugs with solid side panels, set in a draft-free section of the barn and deeply bedded with hay or long-stemmed straw are sufficiently warm in most cases.

Tails

Lambs are born with long tails; tradition demands that we whack them off. But there is method to this madness. Long tails become soaked with and encrusted with manure, and the mess attracts

---

BOTTLE-FEEDING ORPHAN LAMBS

What I usually do to get my new lambs started is sit with my legs crossed and tuck the lamb in the middle in a sitting position (front legs straight and butt on ground). I cup my left hand under the lamb's jaw and open the mouth and insert the nipple with my right hand. Once the nipple is in the mouth, I balance and steady the nipple with the left hand that is still under the jaw.

In other words, I keep the bottle and nose aligned so that the lamb doesn't spit or move the nipple to the side or back of the mouth. I elevate the bottle with the right hand only enough to avoid the lamb sucking air. In this position, you can feel the lamb's throat with the heel of your hand, and you know if it is swallowing.

If you elevate the bottle too much, the milk can pour into the mouth, and if the lamb is not swallowing, the milk could enter the lungs. I try to keep the bottle as level as possible while keeping milk in the bottle cap and nipple. Of course, that means the more the bottle empties, the more tilt there needs to be.

Most people kill their first bottle baby with kindness; they overfeed it because the lamb cries and they think it must be hungry. I know I did. I follow this feeding schedule strictly (no exception). If our lambs cry between feeds, we feed them Pedialyte or Gatorade. That won't hurt them as far as enterotoxemia goes and gives them electrolytes while filling the void for them.

- Days 1–2: 2–3 ounces, 6x/day (colostrum or formula with colostrum replacer powder)
- Days 3–4: 3–5 ounces, 6x/day (gradually changing over to lamb milk replacer)
- Days 5–14: 4–6 ounces, 4x/day
- Days 15–21: 6–8 ounces, 4x/day
- Days 22–35: work up gradually to 16 ounces 3x/day

At about six weeks, I begin slowly decreasing the morning and evening feedings and leave the middle feeding at 16 ounces until I eliminate the morning and evening bottle entirely (remember, they are eating their share of hay or pasture by now). I continue with the one 16-oz bottle for about two weeks and then eliminate the bottle feedings entirely.

By making changes gradually, you can observe changes in the condition of the animal and judge and adjust accordingly. Gradual changes also avoid the complications (some of which can be fatal) of sudden changes in diet. Whatever you do, when you buy milk replacer, use lamb replacer. All-purpose milk replacers and calf replacers do not work well with lambs.

—*Lyn Brown*

  HOBBY FARM ANIMALS

A new lamb with its natural tail before being banded.

use Elastrator bands—thick, strong rubber bands that cut off circulation to the tail, causing it to slough off in a week or so. It's considered the most humane method.

Until recently, tails were docked close to sheep's bodies. However, close-docked ewes often prolapsed before or while giving birth. Nowadays, tails are left longer, often long enough to cover ewes' vulvas. A good way to determine where to apply a band is to lift the lamb's tail and band it where the web of skin beneath the tail meets wool.

blowflies. Blowflies lay eggs in the filthy mass, and hatchling maggots create fly-strike. Not a pretty sight.

Lambs are usually docked (their tails removed) when they're a day or two old. Although some shepherds merely lop them off, most

Banding is probably more than a little uncomfortable; most lambs race around for a little while or so until their tails go numb. But after that, they're fine. Banding may not be fun for the lambs, but it certainly beats fly-strike.

ADVICE FROM THE FARM

BANDING TAILS

I don't band the tail on the first day because when I've done it in the past, the mom smells the alcohol on the band and then she won't lick the lamb's bottom. When lambs first poop, it's not solid but very soft, and if mom doesn't lick it away, it sticks. That's what happened with some lambs last winter. The mom never did lick her lambs' butts, so I was there to wipe away anything that came out (I won't be doing that again!).

—Laurie Andreacci

We use the elastrator band, and we band the first day. It has been our impression (no scientific evidence, just observation) that newborns are equipped with pain-blocking mechanism for the birth process, and if we do the tails that first day, they don't seem to feel it as much as if we wait a day or two.

Either way, it is only painful until the tail goes numb, which in our experience is only a half hour or so. We've had no

complications or infections from the banding process. That's with from forty to a hundred lambs per year.

If you decide that you are going to leave tails on, which is perfectly alright, then you must be observant about sheep developing loose bowel movements. The tail can stick to the body, blocking off the anus and putting the animal in serious medical jeopardy. You even need to watch for this on the young ones, before the tails fall off. Make sure you keep their little bottoms clean.

—Lyn Brown

I agree with Lyn about it being less invasive the first few hours of life. Even on day two or three, there seems to be a difference in the lamb's reaction versus that in the first twenty-four hours. I absolutely will not band a weak lamb, though, and I am sure that Lyn will concur with me on that.

—Cathy Bridges

Unless your sheep are hair sheep, they absolutely must be shorn. Harvesting fleece is just one of the reasons for regular shearing.

YOUR
SHEEP
BUSINESS

Fleece: Shearing, Selling, Spinning

Shearing your sheep is the healthy and humane thing to do for several reasons. For one thing, most unshorn sheep keep growing wool. For another, rain-logged fleece pulls on tender skin, causing skin separation, pain, and sores. Masses of filthy fleece invite fly-strike. Megawoolly sheep are prone to heat stroke. Unshorn pregnant ewes may lie down, roll onto their backs, get stuck, and die.

Whether made from the fine wool of a Merino or the felt from scraps obtained during your first attempt at shearing, you can make a wide variety of garments. You can also pick up some hefty bucks from others, such as handspinners, who wish to use your fleece.

That the wool has to come off is a given, but shearing is a stressful time for sheep (and shepherd, too). Make the process as neat and painless as you possibly can. First, learn which time of the year is best for shearing your sheep. Next, decide whether to hire a professional shearer or to learn to do the job yourself.

Time It Right

Although mild-climate wool producers sometimes opt for winter shearing, sheep are traditionally shorn from March through May. Some breeds are shorn twice a year, in spring and again in early fall. Your shearing schedule will depend on a number of factors, including the type of sheep you have and your locale.

Many shepherds shear their ewes before lambing. Otherwise, new lambs seeking their mothers' teats sometimes find and suck wool tags instead, missing out on vital early meals of colostrum.

Shorn ewes take shelter when it's cold, whereas sheep with warm, heavy wool may stay outdoors where their tender, newborn lambs are apt to freeze. The downside to prelamb shearing is that late-term ewes must be handled more carefully than other sheep. It's wise to have them shorn well before lambing.

Unless you can temporarily place them in a warm barn or blanket them, don't shear sheep until the last winter storms have passed. It takes most sheep six weeks to grow an insulating layer of wool, and they're especially vulnerable for several days post shearing. By the same token, in steamy southern locales, you should shear before warm weather sets in. Heavily fleeced sheep, especially dark-colored individuals, sizzle under a sultry summer sun.

The Professional Shearer

If you can hire a competent shearer, you're wise to do so. The job requires know-how, a strong back, and specialized tools. Shearers are in very short supply, and good ones are scarcer still. If you can hire a shearer, choose one with a sterling reputation. The shearer should handle sheep humanely and remove their fleeces in readily marketable condition. To locate shearers, ask other shepherds for recommendations, scout for business cards and flyers on the bulletin

DID YOU KNOW?

Today's method of shearing sheep—the Bowen Technique—was pioneered by New Zealander Godfrey Bowen in the 1940s. His brother, Ivan Bowen, won the first International Golden Shears title in 1960, along with four more world championships. At eighty-four, Ivan was still competing in Golden Shears veteran competitions, shearing a sheep in about sixty seconds.

Shearing the old-fashioned way, with traditional hand shears.

boards of feed stores and large-animal veterinary practices, call your county agricultural agent or state vet school, and peruse online livestock directories.

Ask for and check shearers' references, and explain your expectations up front. Shearers are paid by the head, so they want to work quickly and be on their way; some aren't willing to slow down to bow to humane demands. If that's the case, look elsewhere or learn to shear your own sheep.

The epidermal layer of sheepskin is thin and tender, so even the best shearers sometimes nick, scrape, and ding a sheep. But multiple superficial wounds, deep cuts, and teat and penis injuries are unacceptable, as is hauling sheep around by their wool. A fleece should be removed in a single piece. Reshearing a given area results in short bits called "second cuts." They drastically reduce market value and are the mark of a careless or inexperienced shearer.

As mentioned, shearers charge by the head. They figure fees based on how far they must travel to your farm, the number of sheep they'll process, and the type of wool they'll be handling. If you have a small flock, be willing to pay premium per capita, mileage, or setup fees to obtain quality service. It's not cost-effective for professional shearers to drive many miles to shear a flock of fifty head or fewer.

Another option is to check with other small flock keepers in your locale and arrange everyone's shearing for the same day or weekend. If you do, ask the shearer to visit each farm; don't gather sheep at a central location. Commingling sheep easily spreads disease from flock to flock.

DID YOU KNOW?

Wild sheep have hair rather than fleece; they grow woolly winter undercoats, which they shed each spring. However, a 6,000-year-old figure of a woolly sheep was unearthed at an archaeological dig in Sarab, Iran, proving that wool sheep had been developed by that early era. By the ninth century BC, the Greek poet Homer praised the whiteness and quality of wool produced in Thessalia, Arcadia, and Ithaca.

In days past, many shearers accepted shorn fleece in lieu of fees. Today's commercial wool prices are too low to make this economically feasible, although some will charge less if they take your wool.

Shearing Prep

Once you've scheduled the services of a recommended shearer, start preparing for his or her arrival. The night before the visit, pen your sheep in a clean, dry, roomy section of a barn or shed. Sheep can't be shorn if they're wet—and that includes dew or frost.

Sheep shouldn't eat for 8–12 hours before shearing. The positions that the shearer places them in are fairly comfortable for sheep as long as their stomachs aren't packed. Uncomfortable sheep are likely to struggle, and struggling sheep often get nicked.

Provide a clean, level shearing surface in place before the shearer arrives. Two possibilities are a clean tarp or two sheets of plywood with the space between them sealed with duct tape. Shake the tarp or sweep the wooden surface between sheep. The shearing floor should be situated in a well-lit, well-ventilated, covered area. Have heavy-duty extension cords on hand if your shearer needs them; flimsy household cords won't do.

Arrange for additional help if you need it. The shearer's basic fee doesn't include snagging reluctant sheep from the flock or hauling them to a pen when finished. Have hot or cold drinks available for your shearer and helpers; if shearing overlaps lunch- or dinnertime, provide something light, such as sandwiches.

Pen your nonworking dogs, and keep herding dogs away from the shearing floor. Dogs worry many sheep, causing undue stress and fidgeting. Stay with your flock until the job is finished. Make certain it's done to your satisfaction.

Keep antiseptic at the ready to treat nicks and dings. The shearer shouldn't have to wait while you fetch it. If you trim hooves and

PRODUCING BETTER FLEECES

Whether or not you jacket your sheep, there are ways to produce better fleeces.

- Clean up your pastures. Weeds, burdock burrs, thistle fluff, and seed heads all work deeply into fleece.
- Avoid hay and straw put up with polypropylene strings. Short bits of twine sucked into the baler end up tangled in fleeces, and it's the hardest contaminant to remove. Keep an eye out for stray plastic feed-sack fibers, too.
- Don't use feeders that allow sheep to bury their heads in hay. Standard big-bale feeders are especially poor choices.
- Don't toss hay at feeders when there are sheep in the path. When feeding concentrates, try to avoid pouring grain onto the heads of sheep jostling for position at the feeder.
- Choose contaminant-free, long-stemmed straw for bedding; reject straw full of weeds, seeds, and chaff. Waste hay will work if it's free of dust, mold, and fleece contaminants. Don't use sawdust or wood shavings.
- If you use a marking harness on your ram or you spray paint numbers on ewes and lambs for identification purposes, choose chalk, crayon, or paint that easily scours out of fleeces.
- Never allow your sheep to be shorn when they're wet. Damp fleece yellows and molds.
- Keep your sheep dewormed and free of sheep keds, lice, and other skin irritants.
- Follow a well-balanced feeding program. Well-fed sheep grow twice as much wool as scrawny, sickly sheep.
- Vaccinate for diseases prevalent in your locale. Sick and otherwise stressed sheep suffer wool break when their fleece growth is temporarily interrupted and then resumes, leaving a weak spot in each wool fiber.
- Choose a breed that produces the type of wool you want to market.
- Consider black sheep. Most handspinners prefer black (which actually ranges from ebony to shades of pale gray) or uniquely colored fleeces such as those grown by Icelandics, Navajo-Churros, and Shetlands. Quality white fleece sells, too, but colors are all the rage.

deworm on shearing day, do it after the sheep are shorn. Don't expect the shearer to hold each sheep while you trim or treat it.

Advise the shearer of special needs before shearing begins. Elderly, chronically lame, and recently ill sheep require gentler handling, and a lone wether in a flock of ewes could have his penis zipped off if a speedy shearer doesn't know it's there. If it's cold and you haven't proper shelter for shorn sheep, ask the shearer to use winter combs that leave a short layer of fleece on your animals. And, when the shearing is done, pay up.

When the Shearer Is You

Amateurs can certainly shear their own sheep, but don't expect a professional-looking job the first few times. Invest in quality tools. Determine which shearing position—sitting or standing—is more comfortable for you. Your back and your sheep will be grateful you did.

When you have only a few sheep to shear, or if money is a concern, traditional hand shears will do the trick. Purchase a high-quality, triple-ground shear with 6½-inch blades, a set of mini shears to make touch-ups a snap, and an inexpensive sharpener to refresh their edges between professional grindings, and you're in business!

Keep your shears sharp. Hold them parallel to the animal's body. While you're learning, work slowly and cautiously, or you're likely to cut your sheep. As you gain experience, you can work considerably faster.

If you're short on time, if you have a lot of sheep, or if cost isn't terribly important, invest in good electric shears. Shears are not horse or dog clippers; they are heavy-duty machines equipped with

heads specifically designed to shear sheep. Oster, Andis/Heiniger, Lister, and Premier build quality shearing machines.

You'll need twice as many cutters because they dull much faster than combs do. Pros shear up to twenty sheep per comb and ten per cutter; novices should allow two cutters and a comb for each pair of sheep. Combs and cutters can be professionally resharpened about ten times; don't try to resharpen them yourself! You'll also need clipper oil; choose a clear product to avoid staining fleece.

While in use, apply clipper oil to the cutters at 3- to 5-minute intervals. Clean and coat the blades with clipper oil within an hour after you finish shearing; store them in plastic clipper-blade boxes. Stow everything in a sturdy case or padded bag. Take care of your shearing gear to make it last.

Professional shearers follow a set shearing protocol, swiftly setting and swiveling sheep into a series of positions whereby they're immobilized, and marketable fleece is zipped off in a single piece. Hobby farmers can learn this technique or shear their sheep while standing. Weekend shearers of limited strength and battered back will likely prefer the latter method.

A shearing stand makes the job a whole lot easier. A goat-milking platform or the sort of stand designed for beautifying show lambs works very well. If you don't have one, not to fret; you can shear sheep standing on the ground.

Secure the sheep's head. Show sheep stands are fitted with a cradle apparatus that does just that; a fence-mounted version is available from some suppliers. If you have neither, use a halter and

Invest in quality equipment if you will do your own shearing.

lead to tie the sheep to a sturdy fence. Use a slip knot in case the sheep gets tangled and needs to be set free fast.

Be patient. Most sheep aren't thrilled with this situation. Start at the head and work carefully to remove desirable fleece in a single piece. When it's off, shear the face, lower legs, and as much belly wool as you can and then set the sheep on its butt to finish the rest. Take long, bold strokes with electric shears, and keep the comb flat against your sheep's skin. If the comb rises, don't make a second sweep—at least not until the fleece has been removed.

Don't expect instant success. Your first shearing will probably ruin the fleece for commercial purposes and embarrass both you and your sheep. But take heart; you can use the wool for home felting projects or quilt batts, and the more you shear your sheep, the better you'll get.

Whether a professional shears your sheep or you do it yourself, you must take proper care of the shorn fleece if you plan to sell it. This includes caring for the fleece while it's on the sheep as well as after it's removed. Some ingenious ideas have been developed for such purposes.

DID YOU KNOW?

Remember the following if you're doing your own shearing:

- Wear old shoes. Sooner or later, a sheep will pee on your foot.
- Don't wear shorts. Scrabbling sheep hooves hurt—a lot.
- If you thing the gentlest sheep will be the easiest to shear, you are mistaken.
- Newbies don't peel off nice, neat one-piece fleeces— ever.
- After the sheep are shorn, touch up your motley job with horse clippers. But clean them between each sheep!
- Mistakes grow out, and sheep forgive you—no matter how funny they look.

Sheep Chic

In 2001, Australian Wool Innovation, Ltd., commissioned a study to determine the economics of jacketing sheep. Twelve hundred Australian sheep in six flocks (including a control group) took part in the study. Its findings: covers significantly improved wool yield, reduced fleece contamination, and drastically reduced fleece rot and fly-strike. They didn't affect fleece weight, micron count, or staple length and strength. The paper recommends that sheep coats be constructed of tightly woven ripstop nylon treated for UV resistance and have expandable fronts, rump coverings, and roomy leg straps. The experiment was a rousing success.

Most American small-scale producers of handspinners' fleeces jacket their sheep. Besides keeping fleeces clean, covers keep sheep warmer after shearing and cooler in the hot summer months (white jackets on colored sheep are especially useful). Their failings: shepherds spend considerable time and effort making, fitting, laundering, repairing, and changing sheep coats. Some felting of fleece may occur in some breeds, especially along the spine and around necklines. Considering that coated fleece routinely sells for considerably more than the price of unjacketed fleeces, most shepherds consider it a pretty fair trade-off.

It's fairly easy to sew your own sheep coats, or you can buy them. You'll need several sizes per sheep to allow for fleece growth. However, the size that one sheep outgrows may neatly fit a flockmate—at least for a while.

You have some choice in materials. If coats aren't crafted of breathable fabric, the wool they cover can mildew or mold. Uncoated nylon and woven plastic fabrics work well, but canvas is usually a poor choice, especially in damp climates.

With most commercial models, you slip the jacket over the sheep's head and then thread its legs through the leg straps. It's fairly easy to jacket sheep that are accustomed to wearing covers.

Take good care of the fleece after it's been shorn from the sheep.

Things can get terribly exciting, however, the first time your flock gets dressed.

Selling the Fleece

Nowadays, it usually costs more to have sheep shorn than their wool is worth. Yet a growing legion of small-scale sheep keepers—hobby-farm shepherds, if you will—show a tidy profit (or at least pay for their flocks' annual upkeeps) by producing quality fleece for handspinners. There are other markets and uses for the fleece as well.

The Lingo

To deal in wool, you must learn the lingo. First, familiarize yourself with terms that refer to the fiber itself. The value of handspinners' fleece is determined by its *grade* (the fineness of the fiber), *class* (its staple length, meaning the length of the fiber), and *quality* (its freedom from contaminants and the character of the fiber).

Fiber diameter is measured in *microns* (a micron is 1/1,000 of a millimeter, and about 1/25,000 of an inch; this is called the Micron System) or *numerical count* (also called the English or Bradford Spinning Count System). Numerical count refers to the number of 560-yard hanks of wool that could be spun from 1 pound of clean wool. For example, Cheviots grade 27–33 microns and 46s–56s (46–56 hanks of yarn).

The American Blood Count System is a largely obsolete third way to measure fiber thickness. This method compares wool is compared to Merino fleece, which is considered fine wool (2.5-inch staple with lots of teensy crimps per inch). The seven accepted grades are: fine, 1/2 blood, 3/8 blood, 1/4 blood, low 1/4 blood, common, and braid.

Staple length is the length of a lock (staple) of greasy (unwashed) wool, measured in inches in the United States and in millimeters abroad. *Staple strength* is the force needed to break a staple measured in newtons per kilotex. *Crimp* is measured in crimps per inch in the United States and per centimeter elsewhere. Fine wool measures more crimps per inch than do coarser wools. Yarn spun from closely crimped fiber is more elastic than yarn spun from uncrimped wool.

Luster (*lustre* in Great Britain) equals sheen. Coarser wools from breeds such as the Leicester Longwool, Teeswater, and Wensleydale are more lustrous than wool from fine-wool and down breeds. Sheep coats keep junk such as dirt, burrs, grain, hay, and bedding out of the best parts of a fleece. Although coats don't completely cover a sheep, the exposed wool is on parts *skirted* (cut) off of handspinners' fleeces anyway. *Yield* is how much fleece is left after skirting. Wise producers skirt generously, removing all low-grade and contaminated wool. Skirted material isn't waste; it can be packaged separately and sold at a lower price to felt makers. *Grease wool* is unwashed fleece, just as it comes off the sheep. Some breeds are greasier than others; Merinos and Rambouillet produce heavy-grease wool. After *scouring* (washing), a fleece weighs considerably less than when it was "*in the grease*" (unwashed).

Handspinning

Handspinning as a craft has taken the country by storm. If you're interested in producing handspinners' fleeces, there's a ready market for a quality product. The best way to get a feel for handspinning is to try it yourself. Start with simple drop-spindle spinning and if you like it, take some classes and buy a spinning wheel. Due to the burgeoning interest in handspinning, classes are

DID YOU KNOW?

When stroked through fingertips from base to tip, the shaft of a single wool fiber feels smooth; from tip to base, rough. This scaly outer covering is called the *cuticle*; it causes fibers to interlock with one another. Inside each fiber, a cablelike cortex imparts strength. A wool fiber is so springy and elastic that it can be bent more than 30,000 times without harm. Stretched, considerable crimp allows it to spring back to shape.

SHEEP COATS AND WOOL SELECTION

I've seen sheep coats made of canvas and have used them, as well as coats made out of lawn-furniture fabric, but I like nylon ones best—they're light, so the wool doesn't felt underneath; they're puffy, so the wool doesn't press; and they shed water, so the wool doesn't rot.

I've never seen my coated sheep pant any more than noncoated sheep, and if I stick my hand under their coats, the sheep aren't sweaty.

—Laurie Andreacci

I did a lot of looking via the Internet and discovered that there is an up-and-coming market for fiber animals to meet the needs of the resurgence of spinning and weaving. There is a demand for both the animals and the fiber.

—Lynn Wilkins

increasingly easy to find. Consider reading Paula Simmons's *Turning Wool into a Cottage Industry*.

Mutton or Milk?

Humankind has dined on lamb and mutton since time began. Artifacts from China's New Stone Age (4,000–3,000 BC) indicate that mutton had replaced venison as China's primary meat source, and archaeologists recovered the bones of at least 500,000 sheep from a 2,600-year-old ritual feasting area on England's Salisbury Plain.

Early humans milked sheep, too. A Mesopotamian clay tablet (circa 2,000 BC) documents a farmer's production of sheep's milk, butter, and cheese, proving sheep dairying was established by then.

When most folks think of milk, they rarely think of sheep—but they should. More than 100 million ewes are milked worldwide, particularly in France, Italy, Greece, Turkey, Israel, and Eastern Europe. France alone has more than 1 million ewes in dairy production. Lamb, mutton, and sheep's milk products remain dietary staples throughout today's world. Shrewd hobby farmers can turn a tidy profit producing them for America's burgeoning lamb and sheep-cheese markets.

Lamb and Mutton Production

Many new shepherds expect to turn a profit selling lamb. Being able to do so means finding the right market and the right breed. Compared with citizens of other nations, the average American eats very little lamb—less than 1 pound per year, in fact—and almost no mutton (mutton is the stronger-flavored flesh of more mature sheep). Lamb is more popular in certain ethnic communities in the United States, however, so a lucrative way to earn hobby farm income with sheep is to niche-market grass-fed, natural, or organic lamb.

Ethnic Communities

A report from the USDA's *Trends in US Sheep Industry* says, "Lamb as a staple seems more typical of Middle Eastern, African, Latin American, and Caribbean consumers. Consumption has remained constant within these groups. The typical lamb consumer is an older, relatively well-established ethnic minority who lives in a metropolitan area such as New York, Boston, or Los Angeles on the West Coast, and prefers to eat only certain lamb cuts."

Producing lamb for ethnic consumers is a best-bet proposition for small-scale sheep producers. An especially lucrative market exists prior to holidays, including Western (Roman) and Eastern (Greek Orthodox) Easter, Christmas, Passover, and the Muslim observances of Ramadan, Eid-al-Fitr, and Eid-al-Adha. However, producers must familiarize themselves with the sort of lamb that each ethnic group requires. The heavy, fatty American market lamb often isn't right.

Specialty Lamb

Health-conscious buyers are willing to pay a considerable premium for specialty meat. Organic lamb generally fetches the highest prices, but it's notoriously difficult to produce.

To call your lamb "organic," you must be enrolled in the USDA's Certified Organic Program, and qualification is tough. Every nibble of feed and grass your that sheep eat must be certified organically grown. Synthetic dewormers are verboten; antibiotics and synthetic

DROP-SPINDLE SPINNING

As many as 10,000 years ago, even before they domesticated sheep, New Stone Age humans gathered bits of shed wool and spun them into yarn. Archaeologists working Neolithic and Bronze Age digs throughout the world uncover stone whorls from ancient drop spindles. All wool was processed with drop spindles until the spinning wheel was invented during the Middle Ages. But the drop spindle never fell out of vogue. It's inexpensive, portable, and easy to use. Drop-spindle spinning is still the mode of choice in many developing countries, and it's enjoying a revival right here at home, too.

Barbados Blackbelly crossbreds, shown here, are a good choice for hobby farmers interested in producing meat.

- Contact TV and radio stations catering to your target group. They're always seeking newsworthy stories, and they might use yours.
- A month or so before a religious holiday, phone mainstream newspapers and TV stations. Your human interest story may be precisely what they're seeking.

medication (other than vaccinations) are taboo. To qualify for organic certification, you must keep detailed flock records and conform to specific handling practices. If a sheep becomes ill, you are required to provide adequate treatment, but if synthetics are used, the sheep must be removed from your certified flock upon recovery. During processing, certified organic lamb can't come in contact with non-organically produced meat, nor may synthetic coloring, preservatives, flavoring, texturizers, or emulsifiers be added to certified organic meat.

A less exacting product is "natural" lamb. Natural lamb is generally raised without the use of antibiotics, growth hormones, or stimulants. It is usually grass fed, although some producers feed animals by-product-free grain for a few weeks or months prior to slaughter.

Grass-fed lamb is, well, grass fed. That is, lambs remain on pasture (and sometimes on stored forage such as hay or silage) from birth to market. Like the others, they're raised without growth hormones. Grass-fed lamb is considerably leaner than commercial supermarket lamb, and it's consistently tested higher in heart-healthy CLA (conjugated linoleic acid).

Selling Methods and Restrictions

Lambs can be sold at auction prior to religious observances, marketed through an intermediary such as a broker, or—as in many cases—direct marketed straight from farm to buyer. If you market lamb from your farm, you have to sell it on the hoof because federal law prohibits direct-from-the-farm sales of home-processed meat. In most cases, buyers take their live lambs with them to slaughter themselves or take to a butcher trained in their religion's beliefs.

To help buyers find you:
- Place ads in metro area newspapers as well as in minority publications serving your target group within that locale.
- Post flyers and business cards in ethnic neighborhoods and at farmer's markets.

The Right Breed

If you're eager to enter the ethnic or specialty lamb meat market but haven't settled on a breed, think hair sheep. In North America, this means Barbados Blackbellies, Dorpers, Katahdins, St. Croix, and Wiltshire Horns. Some have hair coats, and others shed their wool, but none of these sheep have to be shorn.

To various degrees, all hair-sheep breeds are resistant to internal parasites and foot rot, tolerant of heat, and uncommonly productive. They are peerless foragers and browse more than wooled sheep. All breed out of season; with proper care, they can lamb three times in two years. They're attentive mothers, give oodles of milk, and generally produce triplets or twins. According to university studies in Georgia, California, and Mississippi, hair-sheep flesh is milder, thus tastier, than that of wooled-sheep lamb and mutton. Some ethnic minorities greatly prefer hair-sheep lambs and will pay premium prices for them even when auctioned.

Milk: The Nutritious Stuff

At 5.6 percent, sheep's milk is higher in protein than cow's or goat's milk (3.4 percent and 2.9 percent, respectively), and it packs up to twice as much calcium, phosphorus, and zinc. It also boasts a hefty 6 percent fat content and contains considerably more solids than other dairy milks, making it an ideal medium for crafting luscious, full-bodied cheeses.

Because sheep cheese breaks down into smaller molecules than other types, it's more easily digested. Lactose-intolerant diners often enjoy sheep's-milk cheese. Its hearty flavor quickly satisfies

DID YOU KNOW?

In addition to holiday consumption, the birth of Muslim babies often warrants a lamb feast, as do weddings, birthdays, and wakes.

Beautiful Icelandic sheep are in demand for both milk and fleece.

the palate, so less is eaten and thus less fat is consumed per sitting.

For these reasons, and because it just tastes so good, the United States imports more than 75 million pounds of sheep's-milk cheese per year. You've probably eaten it. Sheep's-milk cheeses include Le Berger de Rocastin, Ossau-Iraty-Brebis, and Roquefort (sheep's milk brie) from France; Brin D'Amour from Corsica; Iberico, Manchego, Roncal, and Zamarano from Spain; Greece's Feta, Kasseri, and Manouri; Pepato and Ricotta Salata from Sicily; and Italy's Pecorino, including word-famous Romano.

Now American artisan and small-scale cheese makers are turning to sheep cheese in droves. To supply their demand for raw product, more sheep dairies are flocking to the fold each year. The need will only increase.

The Advantages of Producing Sheep's Milk

There are many advantages to producing sheep's milk. It costs less to equip a brand-new sheep dairy than a comparable cattle operation, and existing cattle facilities can be converted to sheep dairying at relatively low cost. Quality dairy ewes cost less to buy and maintain than cows of the same caliber. They are also easier to handle and less likely to damage costly equipment.

Where a market exists, sheep's milk generally sells for about four times the price of cow's milk. Sheep's milk freezes well for up to six months; it can be stockpiled until the supply warrants shipment.

Ewe's milk registers 18.3 percent solid content, as opposed to cow's milk's 12.1 percent. It takes 10 pounds of cow milk—and only 6 of sheep's milk—to craft a 1-pound block of cheese.

But there are disadvantages, too. Production of sheep's-milk-cheese is still a developing market in North America, and processors are few and far between. According to Penn State's Milking Sheep Production, most processors require a steady supply of milk from at least 750 ewes. Producers may need to form sheep dairy co-ops to meet demands or produce value-added products, such as cheese or yogurt, and market it themselves. Due to stringent North American livestock importation regulations, purebred European dairy sheep are relatively expensive—and scarce. Where a market exists—or if you're willing to create one—sheep dairying can be a viable small-farm moneymaker. But it's not for everyone.

Talk to people in the business—lots of them. Join sheep dairying organizations and attend their functions. Read everything you can lay your hands on, including online resources. Do the homework carefully before you commit.

Household Dairy Sheep

Sheep make fine kitchen dairy animals, too. Depending on their breed and individual productivity, two to five ewes can keep a typical family in luscious milk, cheese, and yogurt year-round. But before you buy any dairy animal, be absolutely certain that you want one. Milking ties you down. Someone has to milk the ewes twice a day, every day, at the same time—no exceptions—for their entire 3- to 7-month lactation period. The milk has to be strained, cooled, and possibly bagged and frozen daily. If you get sick or hurt or are called away on

HEAVY MILKING NONDAIRY-SPECIFIC EWES*

(from Penn State's bulletin *Milking Sheep Production*)

- Coopworth
- Dorset
- Finnsheep
- Katahdin
- Polypay

- Rambouillet
- Romanov
- Shropshire
- St. Croix
- Targhee

Averaging 120–150 pounds of milk per lactation

SHEEP'S MILK VERSUS GOAT AND COW MILK

		Sheep	Goat	Cow
Whole Milk %	Total Solids	18.3	11.2	12.1
	Fat	6.7	3.9	3.5
	Protein	5.6	2.9	3.4
	Lactose	4.8	4.1	4.5
	Calories/100g	102.0	77.0	73.0
Vitamins mg/l	Riboflavin B2	4.3	1.4	2.2
	Thiamine	1.2	0.5	0.5
	Niacin B1	5.4	2.5	1.0
	Pantothenic acid	5.3	3.6	3.4
	B6	0.7	0.6	0.6
	Folic acid ug/l	0.5	0.06	0.5
	B12	0.09	0.007	0.03
	Biotin	5.0	4.0	1.7
Minerals mg/100g	Calcium (Ca)	162–259	102–203	100–120
	Phosphorous (P)	82–183	86–118	90–90
	Sodium (Na)	41–132	35–65	56–60
	Magnesium (Mg)	14–19	12–20	10–12
	Zinc (Zn)	0.5–1.5	0.18–0.5	0.2–0.4
	Iron (Fe)	0.02–0.01	0.01–0.1	0.02–0.06

urgent business, you'll have to find someone to cover for you. If that doesn't sound like your cup of tea, buy your milk at the supermarket.

To produce milk, your ewes must lamb every year. Are you willing to keep a ram or find and pay for stud service for your girls? Hauling them to another farm could expose them to disease, as could bringing a ram to them. Will you keep the lambs? Sell them? Eat them? Will you milk with a machine or by hand?

Hand-milking a gentle ewe or two can be the most relaxing, meditative time of your day, and it's easier than you think. If you've never milked before, start with trained dairy ewes or other docile sheep that are accustomed to being handled. Battling flighty, rambunctious, "don't-touch-me!" ewes twice a day will burn you out fast. If production isn't an issue, any easygoing ewe you already own can be milked if she's trained or easily trainable. She may not give a lot, but it's bound to be mighty good!

If you'd like to hand-milk a ewe or two or three, start your milking string with newborn lambs. Choose ewe lambs from productive, "milky" mamas and then bottle-raise them so they grow up tame. Halter your future milkers; teach them to lead and stand patiently tied. Pet them a lot and massage their udders so they accept your

touch. When they're old enough to lamb, they'll be ready-made milkers. It's the easiest way to train milk sheep.

Dairy Sheep Breeds

Whether you milk two sheep or two hundred, choosing ewes from proven dairy stock is time- and cost-effective. Buy from a reliable producer. Many quality dairy ewes are crossbreds, so registration papers won't matter unless you breed registered lambs. Detailed health and production records, however, are essential.

East Friesian, Lacaune, and British Milk Sheep ewes are the pros of the dairy-sheep world; they give 400–1,000 pounds (or more) of milk per 120- to 240-day lactation. Heavy-milking nondairy breeds produce 100–200 pounds of milk in 90–150 days. Crossbreds strike a happy medium at 250–650 pounds per lactation, and they cost far less than their purebred sisters do.

Another promising dairy sheep is the purebred Icelandic. In addition to her milk-producing capabilities, she's a beautiful animal. Icelandic fleece is coveted by and readily marketed to handspinners, and registered Icelandic breeding-stock lambs are in high demand.

INDEX

Note: Page number in **Bold** typeface indicate a photograph.

They Won't Forget (movie), 209
"This Steer's Life" (Pollan), 23
ticks, 42, 348
torticollis, 300
toxic mold, 133
toxic plants. *See* poisonous plants
toxoplasmosis, 249
transmissible gastroenteritis (TGE), 249
transmissible spongiform
   encephalopathy (TSE), 195
Trends in US Sheep Industry (USDA), 368
trough feeders, 80, 176, 229, **230**
tuberculosis, beef cattle, 39, 45
Turbyfill, Dawn, 146, 155, 161
Turner, Lana, 209
twisted wing, ducks, 145
Tyzzer's disease, 302

U

undulant fever, 40
urinary calculi, 173, 195, 353
urine burn, 297–298
USDA
   airline animal shipping standards, 223
   Certified Organic Program, 207, 368
   health certificates, 293
   meat cut standards, 263
   National Agricultural Statistics
      Service, 205–207
   National Organic Program standards,
      132, 159–160, 263–264
   National Poultry Improvement Plan, 67

National Scrapie Eradication Program, 331
on pork production, 214
quality grades, 55–56
scrapie eradication program, 195, 331, 353
small-scale building plans, 228
Trends in US Sheep Industry, 368

V

vaccinations
   beef cattle, 39–41, **40**
   ducks, 142
   goats, 187–188, 191
   pigs, 243
   sheep, 347–348, **348**
vent disease, 302
vermicompost, 319
Victoria, Queen, 209
viral enteritis, ducks, 147
viral hepatitis, ducks, 147

W

Walters, Robin L., 192, 202
warthog, **214**
warts, beef cattle, 45
water requirements
   beef cattle, 15, **15**, 28–29, **28**
   chickens, 80–81
   ducks, 138
   goats, 181
   pigs, 228–229, 258
   rabbits, 280

waterers
   chickens, 78, 81, **81**, 103
   ducks, 125
   pigs, 229
waterfowl traits, 135–137, **136**
Weaver, Sue, 59, 163, 321
weepy eye, **300**, 301
Wendland, Josh, 215, 220, 226, 229, 239, 253,
   254, 256, 264
wet dewlap, 300
wheat, 232
Wheeler, Connie, 335, 340, 344, 350
white muscle disease, 195, 353
Wilfong, Michelle, 167, 180
Wilkins, Lynn, 329, 337, 344, 368
Williams, Bud, 33
wire cages, 277–278, **277**
wood railing fences, 225
wool block, 299–300, **299**
wool categories, 324–325
worm control. *See* deworming
wryneck symptom, 300

Y

Yorkshire pig breed, 219

Z

Zemny, Chris, 312
zoonotic diseases, 39–40, 139, 246

PHOTO CREDITS

Front cover photos:
top, Sandra.Matic/Shutterstock; center, left, symbiot/Shutterstock; center, right, grafvision/Shutterstock; bottom, left, Lim ChewHow/Shutterstock; bottom, right, smart.art/Shutterstock

Back cover photos:
left, Jose Manuel Revuelta Luna/Shutterstock; center, AGF-stockphoto/Shutterstock; right, l i g h t p o e t/Shutterstock

Shutterstock:
2009fotofriends, 282; A_Lesik, 354; ADA_photo, 110; Helder Almeida, 293; Florian Andronache, 117 (third from top); Anneka, 98 (all three); Matt Antonino, 230; antos777, 83 (bottom); Aprilphoto, 291; Ariene Studio, 87, 152; Artist1704, 278; Aumsama, 259; Tim Belyk, 118, 210-211, 215; BENSZA, 30; beornbjorn, 102; bikeriderlondon, 284; Bildagentur Zoonar GmbH, 335; BMJ, 249; bofotolux, 136; Bohbeh, 21; Simon Bratt, 217; Joy Brown, 219; c12, 236; Carbercal, 182; John Carnemolla, 324; CHAIWATPHOTOS, 101; Simeon Chatzilidis, 147; Eugene Chernetsov, 35, 290; chinahbzyg, 240; wim claes, 286; constructer, 314; CoolR, 270; Steve Cordory, 119; Erika Cross, 273 (top); Linn Currie, 317; Jack Dagley Photography, 12, Pablo Debat, 49; francesco de marco, 367; dollythedog, 265; Mr Doomits, 117 (top); eastern light photography, 11; egschiller, 131; Elenarts, 159; Gergana Encheva, 274; Ewa Studio, 6, 90; Alicia Fabregas, 144; Stephen Farhall, 224; Eric Fleming Photography, 76; Fotokostic , 83 (top); Four Oaks, 342; JORGE FRANCO, 15; frantab, 280; fullempty, 50; Tymonko Galyna, 292; Gam1983, 28; Linda George, 351; Goldika, 120; Mariia Golovianko, 137; Dmitri Gomon, 86; Ralf Gosch, 52; Ana Gram, 192; Arina P Habich 77; Jancz Habjanic, 29; Denis Habler, 114; Kassia Halteman, 288; Margo Harrison, 329; Ross Gordon Henry, 45; Steve Herrmann, 111; Patricia Hofmeester, 212; Horse Crazy, 106; Anna Hoychuk, 67, 81; Iam_9November, 340 ; iBrave, 51; Imageman, 92; imanolqs, 10; In Green, 266-267; InavanHateren, 257; INSAGO, 186; Eric Isselee, 298; janecat, 242; Jarous, 104; Rosa Jay, 308; joyfuldesigns, 74; Jumnong, 158; jurgenfr, 79; KAppleyard, 153; Wendy Kaveney Photography, 233; Klaus Rainer Krieger, 325; Martin Kucera, 100; Keattikorn, 316; Jesus Keller, 53; Nancy Kennedy, 216; Rita Kochmarjova, 299; Kokhanchikov, 319; Gerard Koudenburg, 33, 34, 37, 42; KPG_Payless, 162-163; krisgillam, 353; Yuriy Kulik, 122; Heather LaVelle, 281; L F File, 166; LeonP, 22; Lex-art, 294; Poly Liss, 234; Torsten Lorenz,

366; macgyverhh, 322; Susan Mackenzie, 241; maoyunping, 268; MAR Photography, 168; martinkay, 36; Marykit. 93; marylooo, 273 (bottom); Mediagram, 221; Nataliia Melnychuk, 164; Dudarev Mikhail, 271; Erika J Mitchell, 220; Monkey Business Images, 44; Ulrich Mueller, 222; Christian Musat, 238; MVPhoto, 132; My Good Images, 276; NAS CRETIVES, 130; Jodie Nash, 94; Steve Oehlenschlager, 48; Zhukov Oleg, 23; Tyler Olson, 297; oneo, 255; originalpunkt, 139; Jakkrit Orrasri, 25; Nataliya_Ostapenko, 196; Jens Ottoson, 125; jakkrit panalee, 277; hal pand, 232, 246; Panya7, 174; patboon, 334; pavla, 250; Nataliya Peregudova, 272; Catalin Petolea, 96; PHOTO BY LOLA, 214; Cornelia Pithart, 359; Zeljko Radojko, 38; RadVila, 65; Michael Ransburg, 370; Alexander Raths, 260; rCarner, 9; riganmc, 331; Jamie Rogers, 126; Mirko Rosenau, 117 (second from top); rtem, 248; RTimages, 244; Guy J. Sagi, 143; sanddebeautheil, 16; Julija Sapic, 313; sarkao, 117 (bottom); schankz. 142, 160, 239; science photo, 4-5; Eugene Sergeev, 287; Seqoya, 338; Becky Sheridan, 80, 344; Dmitriy Shironosov, 303; Bryan Sikora, 24; Starcea Gheorghe Silviu, 19; Dr Ajay Kumar Singh, 208; Emily Skeels, 108; Yury Smelov, 223; smereka, 14, 320-321; Smileus, 8; Carolina K. Smith MD, 78; NigelSpiers, 326; spiphotoone, 279; Terry Straehley, 352; Igor Stramyk, 258; stoonn, 27; Nick Stubbs, 145; symbiot, 32; Ferenc Szelepcsenyi, 310; Diana Taliun, 264; Daniel Tay, 289; Robert Taylor, 140; TFoxFoto, 227; The Len, 68, 304; TonyV3112, 18, 20; Nancy Tripp Photography, 112-113; Tsomka, 58-59; TTphoto, 46; Tupungato, 109; Ulga, 129; Vishnevskiy Vasily, 309; Volt Collection, 218; voylodyon, 300; vvoe, 124; Chuck Wagner, 134; Wasu Watcharadachaphong, 40; Piotr Wawrzyniuk, 156; Ivonne Wierink, 30, 88, 341; WilleeCole Photography, 307; Paul Wishart, 228; xstockerx, 148; yaibuabann, 151; yevgeniy11, 245; Dora Zett, 84; Zurijeta, 318

Additional photographs by:
Rachael Brugger, 62, 64, 95
Jean Madden-Hennessey, 60
The Original Turtle/Flickr, 311
Gage O'Sheen, 82
John and Sue Weaver, 169, 170, 172, 176, 177, 179, 180, 181, 184, 185, 188, 189, 190, 191, 193, 194, 195, 198, 199, 200, 202, 204, 206, 327, 328, 330, 332, 336, 337, 343, 345, 346, 348, 349, 350, 356, 358, 361, 362, 364, 369

Illustrations:
Tom Kimball, 63

ABOUT THE AUTHORS

Beef Cattle

Ann Larkin Hansen manages the Hansen hobby farm in west-central Wisconsin. Ann has served as a section editor and reporter for the *The Country Today* weekly newspaper and has contributed to publications such as *The Organic Broadcaster* and *Mother Earth News*. She also has authored books in the *The Farm*, the *Farm Animals*, and the *Popular Pet Care* series, published by Abdo and Daughters of Minneapolis.

Pigs

Arie B. McFarlen, PhD, is a co-owner of Maveric Heritage Ranch Co. in South Dakota, a ranch dedicated to saving and promoting endangered livestock breeds. She is the author of several articles on endangered breeds and their care featured in *Hobby Farms* magazine, *Rare Breeds Journal*, and *Small Farm Journal*.

Chickens, Goats, and Sheep

Sue Weaver has written hundreds of articles and numerous books about livestock and poultry. She is a contributing editor to *Hobby Farms* magazine and writes the Poultry Profiles column for *Chickens* magazine. Sue lives on a small farm in Mammoth Spring, Arkansas, which she shares with her husband, a flock of Classic Cheviot sheep and a mixed herd of goats, horses large and small, a donkey who thinks she's a horse, two llamas, a riding steer, a water buffalo, a pet razorback pig, guinea fowl, and Buckeye chickens.

Rabbits

Chris McLaughlin is a hobby farmer and freelance writer and author living in Northern California. She's raised rabbits for about twenty years, and many of those years were spent breeding and showing American Fuzzy Lops as well as leading rabbit projects for 4-H clubs. Chris also founded a rabbit rescue in the Sierra Foothills and worked at Sierra Wildlife Rescue in Placerville, California, where her specialty was rehabilitating and releasing wild rabbits.

Ducks

Cherie Langlois is a freelance writer and photographer who specializes in farm, pet, and travel topics. She's had more than 150 articles published and is a contributing editor to *Hobby Farms* magazine. A former zookeeper and veterinary assistant with a BA in zoology, Cherie has worked with a variety of waterfowl and has raised Muscovy ducks on her 5-acre farm in Washington state for fifteen years. Her much-loved menagerie, managed with the help of husband Brett and daughter Kelsey, also includes horses, sheep, goats, chickens, cats, a dog, a house rabbit, and a cockatiel.